Also by Michael Hiltzik

A Death in Kenya

Dealers of Lightning:
Xerox PARC and the Dawn of the Computer Age

The Plot Against Social Security:
How the Bush Plan Is Endangering Our Financial Future

Colossus:
The Turbulent, Thrilling Saga of the Building of Hoover Dam

The New Deal:
A Modern History

BIG SCIENCE

Ernest Lawrence and the Invention That Launched the Military-Industrial Complex

MICHAEL HILTZIK

Simon & Schuster
New York London Toronto Sydney New Delhi

Simon & Schuster
1230 Avenue of the Americas
New York, NY 10020

First Simon & Schuster hardcover edition July 2015

Illustration credits are on page 513.

Interior design by Robert Ettlin

Manufactured in the United States of America

10 9 8 7 6 5 4 3 2 1

Library of Congress Cataloging-in-Publication Data

Hiltzik, Michael A.
Big science : Ernest Lawrence and the invention that launched the military-industrial complex / by Michael Hiltzik.
pages cm
1. Cyclotrons. 2. Physicists—United States—Biography. 3. Lawrence, Ernest Orlando, 1901–1958. I. Title.
QC787.C8H55 2015
530.092—dc23
[B] 2014017463

ISBN 978-1-4516-7575-7
ISBN 978-1-4516-7603-7 (ebook)

To Deborah, Andrew, and David

Contents

BIG
SCIENCE

Introduction

Creation and Destruction

On July 4, 2012, a pair of international scientific teams announced that they had discovered an elementary particle known as the Higgs boson with the help of one of the most complex research machines on earth: the Large Hadron Collider. The Higgs boson had been the target of an intensive search by physicists for nearly a half century, or since its existence had been posited in 1964 as the carrier of a field that gives mass to matter in the universe. But it took the collider to find it.

The scheduled announcement, at the Geneva, Switzerland, headquarters of the European Organization for Nuclear Research (CERN), the collider's builder and owner, drew spectators from around the world and the highest echelons of physics. Present was Peter Higgs, eighty-three. The British physicist who had predicted the existence of the particle that bore his name stared, like every other guest, at a screen at the front of a CERN lecture hall. On it were displayed PowerPoint slides of data produced from the almost unimaginably violent collisions between beams of energized protons that the LHC experimenters had aimed at one another point-blank, hoping to coax the Higgs boson into showing itself for an infinitesimal moment within the resulting maelstrom of energy. The numbers told them, to within a convincing range of probability, that the experimenters had found the Higgs boson. When the presentation ended, there was a standing ovation for the research teams and expressions of awe for the incredible apparatus that brought them their victory.

Everything about the Large Hadron Collider is big. Its construction, from conception to the generation of its first proton beam, took twenty-five years and cost $10 billion. Buried three hundred feet beneath the pastoral landscape on the border of France and Switzerland, the machine occupies a concrete tunnel seventeen miles in circumference. Inside the tunnel, 9,600 magnets chilled cryogenically to nearly minus 300 degrees Celsius guided the protons toward their head-on collisions at velocities approaching 99.99 percent of the speed of light.

The collider, and the discovery announced that summer day in 2012, stood then as the ultimate expressions of Big Science: the model of industrial-scale research that has driven the great scientific projects of our time—the atomic bomb, the race to put a man on the moon, the dispatch of robotic probes beyond the confines of the solar system, investigations of the workings of nature at the microcosmic scale of subatomic particles. To this day, Big Science guides research in academia, industry, and government. It addresses gigantic questions, and therefore requires gigantic resources, including equipment operated by hundreds or thousands of professional scientists and technical experts. Its projects often cost more than what a single university can afford, or even a single country; CERN's collider draws its financial and technical support not only from the organization's twenty-one member states but also from more than sixty other countries and international institutions. Those are the dimensions of Big Science today. As physicist Robert R. Wilson has written, research on this scale cannot be achieved by solitary efforts: "It is almost as hard to reach the nucleus by oneself as it is to get to the moon by oneself."

Yet the creation of Big Science was itself a solitary effort. The birth of this new way of probing nature's secrets can be traced to the day nearly nine decades ago in Berkeley, California, when a charming and resourceful young scientist with a talent for physics and perhaps an even greater talent for promotion pondered a new invention and declared, "I'm going to be famous!"

His name was Ernest Orlando Lawrence. His invention would revolutionize nuclear physics, but that was only the beginning of its impact. It would transform everything about how science was conducted, in ways that still matter today. It would remake our understanding of the basic building blocks of nature. It would help win World War II. Lawrence called it the cyclotron.

The Large Hadron Collider is a direct offspring of Lawrence's invention, though few today would recognize the family resemblance. The first cyclotron fit in the palm of Lawrence's hand and cost less than one hundred dollars. The LHC comprises several advanced cyclotrons as well as synchrocyclotrons and other advanced accelerators designed to propel subatomic particles to unnatural velocities, all descending from the original design. Lawrence's Radiation Laboratory in Berkeley employed about sixty scientists and a couple of dozen technicians at its peak. That seemed like a veritable army to Lawrence's professional forebears, such as Sir Ernest Rutherford of Cambridge University's legendary Cavendish Laboratory, who made earthshaking discoveries with two assistants, employing handmade tools—some of which could fit comfortably on his workbench—in the first decades of the twentieth century. But it would look like a paltry brigade to the two teams that announced the Higgs discovery, which numbered three thousand members each.

Lawrence's role as the creator of Big Science was widely acknowledged by his peers but is largely overlooked today. Yet it is worth reexamining for several reasons. One is that the instincts and ambitions that drove him in his research, along with his personal management style, gave Big Science its lasting character. But there is more: his is a compelling story of a scientific quest that spanned a period of unprecedented discovery in physics and placed him at the crossroads of science, politics, and international affairs.

From the late 1930s on, there was scarcely a question of national scientific policy on which the views of Ernest Lawrence were not sought. As the inventor of the world's most powerful atom smasher and leader of the nation's greatest research laboratory, his influence expanded with the onset of World War II. By placing his personal commitment behind the Allied effort to build the atomic bomb, he saved the program from nearly certain cancellation at a crucial moment in its history. Then, after the war, it was his prestige and influence that helped launch the program to build the hydrogen bomb. The world we live in today, poised uneasily under a thermonuclear sword of Damocles, surely stands as Ernest Lawrence's bequest, albeit an equivocal one, to modern civilization.

Lawrence knew on the day of his brainstorm in 1929 that he had happened upon an astoundingly effective new way of accelerating subatomic particles. His goal was to use them as probes to discover the structure of the nucleus, the charged kernel of protons and neutrons that accounts for most of the atom's mass, as someone might wield a screwdriver to probe a desktop radio's electronic innards. His cyclotron was a conceptually simple solution to the riddle of how to pump up the energies of subatomic particles—specifically protons, the nuclei of hydrogen atoms—so they could penetrate the protective electric field of the nucleus. Scientists and engineers all over the world were working on this problem. Lawrence solved it.

Physics then was undergoing a difficult transition. The geniuses of small science, like Rutherford and Irène and Frederic Joliot-Curie, the daughter and son-in-law of Marie Curie, had worked to the limit the humble tools nature had given them. With his handmade apparatuses, Rutherford had discovered the nucleus and intuited the existence of the neutron, which later would be found by his deputy, James Chadwick, in another feat of small-scale experimentation. The Joliot-Curies, working in their own modest lab, continued Marie Curie's investigations into the mysteries of radioactivity, learning to transmute one element into another by bathing the first in radioactive emissions. Both labs relied on naturally radioactive substances such as radium and polonium to produce their invisible subatomic probes.

Their achievements were brilliant, but they could not escape the realization that further investigations of the structure of the nucleus would require bullets that were speedier, more powerful, and more precisely aimed than the rays pulsing haphazardly from blocks of radioactive minerals. What physicists needed, in other words, were man-made projectiles. Mustering high-energy beams and training them on their targets required not equipment that could fit on a laboratory bench but machines that could barely be contained inside buildings. Rutherford and the Joliot-Curies knew that they were the last magnificent leaders of an era of hands-on science, and soon they would have to yield to a new generation.

These physicists of the old school would contemplate the changes Lawrence wrought in their science with awe. As Maurice Goldhaber, whose eminent career spanned the heydays of small science and Big Science, recalled the transition: "The first to disintegrate a nucleus was Ernest Rutherford, and there is a picture of him holding the apparatus in his lap. I then always remember the later picture when one of the famous cyclotrons was built at Berkeley, and all of the people were sitting in the lap of the cyclotron. Roughly speaking, that gives you an idea of the change."

Goldhaber was not exaggerating. The cyclotron to which he referred was a behemoth housed in a building of its own erected in 1938. The machine's enormous electromagnet weighed 220 tons and stood eleven feet high. The photograph Goldhaber mentioned did indeed depict the entire staff of Lawrence's laboratory—twenty-seven grown men—standing or seated under its horseshoe-shaped iron span.

Ernest Lawrence's character was a perfect match for the new era he brought into being. He was a scientific impresario of a type that had seldom been seen in the staid world of academic research, a man adept at prying patronage from millionaires, philanthropic foundations, and government agencies. His amiable Midwestern personality was as much a key to his success as his scientific genius, which married an intuitive talent for engineering to an instinctive grasp of physics. He was exceptionally good-natured, rarely given to outbursts of temper and never to expressions of profanity. ("*Oh, sugar!*" was his harshest expletive.) Raising large sums of money often depended on positive publicity, which journalists were always happy to deliver, provided that their stories could feature fascinating personalities and intriguing scientific quests. Ernest fulfilled both requirements. By his mid-thirties, he reigned as America's most famous native-born scientist, his celebrity validated in November 1937 by his appearance on the cover of *Time* over the cover line, "He creates and destroys." Not long after that, in 1939, would come the supreme encomium for a living scientist: the Nobel Prize.

Lawrence upended the stereotype of the man of science as a wild-eyed mystic buried obsessively in his lonely work, isolated in a remote laboratory (typically of Gothic architecture), his creations always on the verge of blowing their maker to bits. The defining characteristic of the scientist in popular culture was unworldliness: *Time* had portrayed Albert Einstein as an oddball genius laboring alone in an attic behind a clanging iron door, "haggard, nervous, irritable . . . Mathematician Einstein cannot keep his bank account correctly."

Lawrence, by contrast, bristled with intellectual energy and physical vigor. His success eventually brought him a laboratory that was no dark Gothic castle but a modern shrine to science on a hillside above the bustling Berkeley campus of the University of California, blessed with a stunning view of San Francisco across the bay. Far from solitary, he presided over a team of energetic young scientists and graduate students—physicists, chemists, medical doctors, and engineers, all toiling and cogitating in interdisciplinary harmony—and managed millions of dollars with the assurance of a corporate executive. He embodied the muscular brashness of the New World, with its ambition, verve, ingenuity, and wealth. The progressive journalist Bruce Bliven, who normally plied his trade among cynical politicians and world-weary pundits, was disarmed by the renowned Professor Lawrence, begetter of scientific miracles, upon finding him "easy to talk to and as completely American as you could imagine."

• • •

The term *Big Science* was coined by the physicist Alvin Weinberg in 1961, three years after Ernest Lawrence's death. Weinberg surveyed the previous decades of scientific research from his vantage point as director of Oak Ridge National Laboratory (which had been built to Lawrence's specifications to produce enriched uranium for the atomic bomb) and defined the period as one that celebrated science with monuments of iron, steel, and electrical cable—towering rockets, high-energy accelerators, nuclear reactors—just as earlier civilizations had paid obeisance to their celestial gods and temporal kings with spired stone cathedrals and great pyramids.

Only a bureaucratic style of management could keep these monuments to science functioning. In Lawrence's Radiation Laboratory, the central apparatus, the cyclotron, was so technologically complex and operationally willful that it required full-time engineering attention. "The logistics of keeping the place going—whether this means the scientific machinery or the elaborate organization that tends the machinery—[became] an essential ingredient of the activity," Weinberg recalled. That these grand commitments were dictated by the daunting complexity of the questions that science confronted became an article of faith among those who tended the machinery: "We simply do not know how to obtain information on the most minute structure of matter or on the grandest scale of the universe . . . without large efforts and large tools," observed Wolfgang K. H. "Pief" Panofsky, a former physicist in Lawrence's lab.

The drive toward bigger and better created its own logic. Every discovery made with a cyclotron opened new vistas for physicists to explore; solving every new riddle demanded machines of even greater power. Every new discovery brought new prestige to the institution that claimed it, creating both the motivation and the opportunity for more construction, more scientists, more renown—and more fund-raising.

What ultimately validated Big Science as a model for scientific inquiry were the two great technical achievements of the Second World War:

radar and the atomic bomb. It is probable that neither could have been developed—and certain that they could not have been developed in time to affect the war's outcome—without the interdisciplinary collaboration and virtually limitless resources that already were the hallmarks of the new paradigm. The atomic pile in which the first nuclear chain reaction was observed—a reaction crucial for the development of the plutonium bomb later dropped on Nagasaki—is commonly credited to Enrico Fermi, who conceived it and supervised its construction. But realizing Fermi's conception called for an army of "physicists, mathematicians, chemists, instrument experts, metallurgists, biologists, and the various engineers who could translate these scientists' findings into practice," observed Weinberg. "The chain reactor was much more than one nuclear physicist's experiment."

The changes that Lawrence's style of research wrought in science inspired not only awe but also disquiet, as they still do today.

Even early in Lawrence's career, when Big Science was still in its formative stage, scientists, university presidents, and other experts were beginning to worry about its effect on the quest for knowledge and its dissemination. In 1941 Karl Compton, president of the Massachusetts Institute of Technology and himself a physicist with a cyclotron at his disposal, lamented the "abnormal competitive element" that the scramble for money and renown had introduced into academia. As he uneasily confided to a friend, "To maintain an active program and a well-rounded staff has required more aggressive salesmanship than the scientific profession relishes." Some scientists found the hypercompetitive, factory-floor style of research hopelessly uncongenial, and fled Big Science institutions like Berkeley for universities where Old World manners and procedures still prevailed. Others, like Panofsky, accepted that Big Science was necessary to address the big questions of physics; they trained themselves in the new system at Berkeley, and then left to spread the Big Science gospel far and wide. (Panofsky brought it to Stanford University.)

Concerns about how Big Science might permanently alter the way scientists worked were shelved during the war, when the scientific and technical communities focused themselves on the drive to victory. With the advent of peace, however, scientists would again ponder the changes Big Science would bring. Some wondered if there would be any place left for the kind of individual inspiration that had yielded the breakthroughs of the past: "Could the theory of relativity or the Schrödinger equation have been discovered by an interdisciplinary team?" asked the Hungarian physicist Eugene Wigner. He was concerned, as were many others, that the burgeoning demands of management would take the most talented scientists out of the laboratory. The researcher who in the era of small science devoted himself purely to investigating his subject and teaching it to his students now had to juggle many other duties. He had to manage large inflows of donated capital, write grant applications, serve on committees, haunt the corridors of Congress and executive agencies in Washington to pry appropriations loose. Research leaders had to be not only scientists but also ringleaders, cheerleaders, salespersons.

Money was abundant, but it came with strings. As the size of the grants grew, the strings tautened. During the war, the patronage of the US government naturally had been aimed toward military research and development. But even after the surrenders of Germany and Japan in 1945, the government maintained its rank as the largest single donor to American scientific institutions, and its military goals continued to dictate the efforts of academic scientists, especially in physics. World War II was followed by the Korean War, and then by the endless period of existential tension known as the Cold War. The armed services, moreover, had now become yoked to a powerful partner: industry. In the postwar period, Big Science and the "military-industrial complex" that would so unnerve President Dwight Eisenhower grew up together. The deepening incursion of industry into the academic laboratory brought pressure on scientists to be mindful of the commercial possibilities of their work. Instead of performing basic research, physicists began "spending their time searching for ways to pursue patentable ideas for economic rather than scientific reasons," observed the historian of science Peter Galison. As a pioneer of Big Science, Ernest Lawrence would confront these pressures sooner than most of his peers, but battles over patents—not merely what was patentable but who on a Big Science team should share in the spoils—would soon become common in academia. So too would those passions that government and industry shared: for secrecy, for regimentation, for big investments to yield even bigger returns.

It was Lawrence who had helped plant the seed of industry's involvement in research by feeding the ambitions of his patrons with visions of how the cyclotron would serve their favored goals. For biological research institutions, he played up its capacity to produce large quantities of the artificial radioisotopes needed to comprehend the complexities of photosynthesis and to attack cancer cells. He plied industrialists with visions of the atomic nucleus as a generator of electricity that would be unimaginably cheap and almost infinitely abundant. As for those philanthropic foundations still devoted to basic research, he offered the prestige of association with projects aimed at unlocking the secrets of the natural world as its own reward. Raymond B. Fosdick, president of the Rockefeller Foundation, delivered perhaps the most concise distillation of this aspect of Big Science. "The new cyclotron is more than an instrument of research," he stated in 1940. "It is a mighty symbol, a token of man's hunger for knowledge, an emblem of the undiscourageable search for truth which is the noblest expression of the human spirit." That year, the nonprofit foundation's board had voted to grant Lawrence more than $1 million to build the most powerful cyclotron on earth.

There was nothing cynical about Lawrence's appeals to the interests of his financial backers. His most assiduous fund-raising efforts would have come to naught had he not been able to back up his promises with a record of genuine achievement. Berkeley's Radiation Laboratory pioneered the new science of nuclear medicine to fight disease. Its cyclotrons often ran overtime to produce radioisotopes for researchers all over the world. Lawrence's conviction that energy from the atom might someday heat and illuminate millions of homes and factories and send seagoing vessels around the globe was visionary, but no less heartfelt for that—and, of course, it turned out to be true.

The successes of Big Science brought great public esteem to scientists, who became honored and admired as the men and women who had helped win the war and who served as living repositories of mankind's impulse to learn nature's secrets. That degree of lionization could never last, for science's knowledge is imperfect and the public always primed for disillusionment. Scientists began to totter on their pedestals just as the projects of Big Science, growing ever bigger, threatened to consume an outsized share of the public resources needed to address more urgent social problems.

Toward the end of the twentieth century, Big Science's grip on the confidence of society started to ebb. Many of its achievements seemed, in retrospect, equivocal: yes, the atomic bomb won the war, but at the price of a permanent nuclear cloud hovering over the human race. The peaceful atom brought electricity, but at a much higher price than had been forecasted by its promoters—and it also brought us the disasters at Three Mile Island, Chernobyl, and Fukushima, raising the question of whether the technology of nuclear power ever could be reliably tamed by mankind. Men walked on the moon, but after that spectacular moment, public interest in space exploration drained swiftly away. All that expense—for what?

In the same 1961 essay in which he coined the term *Big Science*, Alvin Weinberg outlined the emerging doubts about its impact on research, the university, and society. He asked, quite properly, if massive expenditures to erect monuments to Big Science would suck up scarce resources and distract scientists from inquiries more relevant to the human condition: "I suspect that most Americans would prefer to belong to the society which first gave the world a cure for cancer," he wrote, "than to the society which put the first astronaut on Mars."

In the United States, such doubts energized the debate in the 1980s and early 1990s over the Superconducting Super Collider, an accelerator to be located near Waxahachie, Texas, which would have been as much as three times as powerful as CERN's Large Hadron Collider. The project eventually foundered on the shoals of regional and budget politics, but it had already been mortally wounded by public skepticism about its purpose. In 1993 the SSC was killed by Congress.

The Large Hadron Collider is so vast, complex, and costly that some scientists wonder whether it might mark the end of Big Science on the international level. Its discoveries raise questions about the natural world that can be answered only by bigger, more powerful colliders, in the same way that each of Lawrence's cyclotrons established the need for the next bigger one. Like the Large Hadron Collider, the next machine, if it is to be built, will require a consortium of nations. Getting them to collaborate on a quest that to the layman seems hopelessly abstract will not be easy.

Ernest Lawrence never expressed such misgivings. His goal was to address "the problem of studying nature," as Robert Oppenheimer put it, and his career achieved that end. That it is left to us to deal with its implications does not diminish his achievement. But it does compel us to examine how it came about. The story begins with the towering figures of the small-science world.

Part One

THE MACHINE

Chapter One

A Heroic Time

Ernest Rutherford was one of science's Great Men, a towering figure who drove developments in his era rather than riding in the wakes of others. To an acquaintance who observed, "You're always at the crest of the wave," he was said to have replied: "Well, after all, I made the wave, didn't I?" He was loud, with a boisterous laugh and a hearty appreciation of what was known in his time as "smoking-room humor." C. P. Snow, a youthful associate of Rutherford's who would win literary fame with novels set in the corridors of academia and government, remembered Lord Rutherford as "a big, rather clumsy man, with a substantial bay window that started in the middle of the chest" and "large staring blue eyes and a damp and pendulous lower lip."

Born in 1871 to a handyman and his wife in New Zealand when it was a remote outpost of the British Empire, Rutherford became an intuitive theorist and the preeminent experimental physicist of his age. No one could question his talent for divining the significance of the results produced by his elegant handmade equipment. "Rutherford was an artist," commented his former student A. S. Russell. "All his experiments had style."

Rutherford was twenty-four when he first came to Cambridge University's storied Cavendish Laboratory on a graduate scholarship. It was 1895, a fortuitous moment when physicists were pondering a host of strange new physical forces manifested in their apparatuses. Only a month before Rutherford's arrival, the German physicist Wilhelm Roentgen had reported that a certain electrical discharge generated radiation so penetrating it could produce an image of the bones of a human hand on a photographic plate. Roentgen called his discovery X-rays.

Roentgen's report prompted the Parisian physicist Henri Becquerel to look for other signs of X-rays. His technique was to expose a variety of chemical compounds to energizing sunlight. He would seal a photographic plate in black paper, cover the paper with a layer of the candidate compound, place the arrangement under the sun, and check back later to see if a shadow appeared on the sealed plate. During a stretch of overcast Paris weather in February 1896, he shut away in a drawer his latest preparation: a uranium salt sprinkled over the wrapped plate, awaiting the sun's reemergence from behind the clouds. When he developed the plate, he discovered it had been spontaneously exposed by the uranium in the darkened drawer.

Marie Curie and her husband, Pierre, soon established in their own Paris laboratory that Becquerel's rays were produced naturally by certain elements, including two that they had discovered and named polonium, in honor of Marie Curie's native Poland, and radium. They called the phenomenon "radioactivity." (Becquerel and the Curies would share the 1903 Nobel Prize for their work on what was originally called "Becquerel radiation.")

Other scientists launched parallel inquiries to unravel the mysteries lurking within the atom's interior. Cavendish director Joseph John "J. J." Thomson, Ernest Rutherford's mentor, discovered the electron in 1897, thereby establishing that atoms were divisible into even smaller particles—"corpuscles," he called them. Thomson proposed a structural model for the atom in which his negatively charged electrons were suspended within an undifferentiated positively charged mass, like bits of fruit within a soft custard. Irresistibly, this became known as the "plum pudding" model. It would prevail for fourteen years, until Rutherford laid it to rest.

Rutherford, meanwhile, had busied himself examining "uranium radiation," his term for the emanations discovered by Becquerel. In 1899 he determined that it comprised two distinct types of emissions, which he categorized by their penetrative power: alpha radiation was easily blocked by sheets of aluminum, tin, or brass; beta rays, the more potent, passed easily through copper, aluminum, other light metals, and glass. Rutherford had relocated to Montreal and a professorship at McGill University, which featured a lavishly equipped laboratory funded by a Canadian businessman, in an early example of scientific patronage by industry. Working with a gifted assistant named Frederick Soddy, who would coin the term *isotope* for structurally distinct but chemically identical forms of the same element, Rutherford determined that the radioactivity of heavy elements such as uranium, thorium, and radium was produced by decay, a natural transmutation that changed them by steps—in some cases, after minutes; in others, centuries, years, or millennia—into radioactively inert lead. Eventually alpha rays were identified as helium atoms stripped of their electrons—that is, helium nuclei—and beta rays as energetic electrons. The work earned Rutherford the 1908 Nobel Prize in chemistry. By then, he had already returned to Britain to take up a professorship at the University of Manchester.

There he would make an even greater mark on science by taking on the core question of atomic structure. "I was brought up to look at the atom as a nice hard fellow, red or grey in color, according to taste," he remarked years later of the plum pudding model. But although he speculated that the atom was mostly empty space rather than a homogenous mass speckled with charged nuggets, he had not yet conceived an alternative model. With two Manchester graduate assistants, Hans Geiger and Ernest Marsden, he set about finding one, using alpha particles as his tools. As he knew, these were deflected somewhat by magnetic fields but, curiously, even more on their passage through solid matter—even through a thin film such as mica. This suggested that the atomic interior was an electromagnetic maelstrom buffeting the particle on his journey, not a serene, solid pudding.

Rutherford experimented by bombarding gold foils with alpha particles emanating from a glass vial of purified radium. Geiger and Marsden recorded the particles' scattering by observing the flash, or scintillation, produced whenever one struck a glass plate coated with zinc sulfide. This apparatus displayed Rutherford's hallmark simplicity and style, but the procedure was unspeakably onerous. The observer first had to sit in the unlighted laboratory for up to an hour to adjust his eyes to the dark, and then could observe only for a minute at a time because the strain of peering at the screen through a microscope tended to produce imagined scintillations mixed with the real ones. (Geiger eventually invented his namesake particle counter to relieve experimenters of the tedium.)

The experiment showed that most of the alpha particles passed through the foil with very slight deflection or none at all. But a tiny number—about one in eight thousand—bounced back at a sharp angle, some even ricocheting directly back at the source.

Rutherford was astonished by the results. "It was almost as incredible as if you fired a fifteen-inch shell at a piece of tissue paper, and it came back and hit you," he would relate years later, creating one of the most cherished images in the history of nuclear physics. It was not hard for him to understand what had happened, for the phenomenon could be explained only if the atom was mostly empty space, with almost all of its mass concentrated within a single minuscule, charged kernel. The deflections occurred only when the alpha particle happened to strike this kernel directly or come close enough to be deflected by its electric charge. The kernel, Rutherford concluded, was the atomic nucleus.

Rutherford's discovery revolutionized physicists' model of the atom. But it was by no means his ultimate achievement. That came in 1919, he reported an even more startling phenomenon than the tissue-paper ricochets of 1911.

Rutherford had again relocated, this time to Cambridge, where he assumed the directorship of the Cavendish. The laboratory had opened in 1874 under the directorship of James Clerk Maxwell, who was a relative unknown at the time of his appointment; within a few short years, however, he had published the work on electricity and magnetism that made his worldwide reputation and established the Cavendish by association as one of Europe's leading scientific centers. Maxwell's conceptualization of electricity and magnetism as aspects of the same phenomenon, electromagnetism, would stand as the bridge between the classical physics of Sir Isaac Newton and the relativistic world of Albert Einstein, and his Cavendish would reign as the living repository of the British experimental tradition in physics.

In Rutherford's time, the Cavendish reveled in its tatty grandeur, the epitome of small science in an institutional setting. The building was shaped like an L around a small courtyard: three stories on the long side, the top floor, with its gabled windows, crammed under a steeply raked roof. Inside the building were a single large laboratory and a smaller lab for the "professor," a room for experimental equipment, and a lecture theater. There Rutherford held forth three times a week to an audience of about forty students, occasionally consulting a few loose pages of notes drawn from the inside pocket of his coat. Physicist Mark Oliphant, arriving at the Cavendish from Australia in the mid-1920s, remarked on its "uncarpeted floor boards, dingy varnished pine doors and stained plaster walls, indifferently lit by a skylight with dirty glass." As for the director, he described Rutherford as "a large, rather florid man, with thinning fair hair and a large moustache, who reminded me forcibly of the keeper of the general store and post office." The lab adhered strictly to the European "gentlemen's tradition" of closing its doors for the night at six o'clock regardless of whether any experiments were in progress, with an elderly timekeeper assigned to glower at the lab bench of any scientist still working, rattling the lab keys to remind him of the time. Working late was considered "bad taste, bad form, bad science."

The Cavendish treasured its history of having made great strides with scanty resources. Its entire annual budget was about £2,000, worth about $80,000 in twenty-first-century US currency and meager even in the old days for the magnitude of its work. What took up the slack was the shrewdness and craft of Rutherford's associates, their ability to extract the maximum results from experimental apparatus of marked simplicity and elegance. The 1919 experiments would exemplify the Rutherford style.

Working with James Chadwick, whose experimental skills matched his own, Rutherford trained his alpha particles on a series of gaseous targets: oxygen, carbon dioxide, even ordinary air. With their apparatus, a refinement of the Marsden-Geiger box of 1911, they found that ordinary air produced especially frequent scintillations resembling those of hydrogen nuclei, or protons. Rutherford surmised correctly that the phenomenon was related to the 80 percent concentration of nitrogen in the air.

"We must conclude," he wrote, "that the nitrogen atom is disintegrated . . . in a close collision with a swift alpha particle, and that the hydrogen atom which is liberated formed a constituent part of the nitrogen nucleus." These circumspect words produced a scientific earthquake, for what Rutherford described was the first artificial splitting of the atom. It would eventually be recognized that the reaction entailed the absorption of the alpha's two protons and two neutrons by the nitrogen nucleus—seven protons and seven neutrons—followed by the ejection of a single proton, thereby transmuting nitrogen-14 into the isotope oxygen-17. But what really set the world of science on a new path was the vision that Rutherford set forth at the close of his paper. "The results as a whole," he wrote, "suggest that if alpha particles—or similar projectiles—of still greater energy were available for experiment, we might expect to break down the nucleus structure of many of the lighter atoms."

In other words, alpha particles produced naturally by radium and polonium had exhausted their usefulness as probes of the nucleus. They simply weren't powerful enough. Some way had to be found to impart greater energies to the projectiles: man's cunning had to augment nature's gifts to create a new kind of nuclear probe. Rutherford had drawn a road map for the future of nuclear physics. Off in the distant horizon lay the reality that the task of reaching the necessary energies would overmatch the elegant bench science of Rutherford's generation.

Rutherford's discoveries launched a surge of ingenuity in physics. J. Robert Oppenheimer would later describe this as "a heroic time," not merely because of the intellectual energy focused on the challenge Rutherford posed but because the work took place in an atmosphere of intellectual crisis. Physicists were forced to confront astonishing paradoxes roiling their conception of the natural world. Through much of the 1920s, they were wracked with doubt that they would be able to resolve them at all.

The words of eminent physicists of the era bristle with intellectual despair. The German physicist Max Born, one of the earliest disciples of the new theory of quantum mechanics, wrote in 1923 that its multiplying contradictions could mean only that "the whole system of concepts of physics must be reconstructed from the ground up." The Viennese theorist Wolfgang Pauli, who combined rigorous intellectual integrity with an acerbic tongue—his famous critique of a sloppily argued paper was that it was "not even wrong"—lamented in 1925 that physics had become so "decidedly confused" that "I wish I . . . had never heard of it." Even the level-headed James Chadwick recalled experiments at the Cavendish "so desperate, so far-fetched as to belong to the days of alchemy."

Despite the complexity of their quest—or perhaps because of it—their work enthralled the public. For laypersons in the twenties, physics was invested with an aura of drama, even romance. The postwar decade had begun with Sir Arthur Eddington's spectacular confirmation of Einstein's theory of relativity at a joint meeting of the Royal Society and Royal Astronomical Society in November 1919. "Revolution in Science / New Theory of the Universe / Newtonian Ideas Overthrown" declared the *Times* of London in a historic headline. Eddington's painstaking publicity campaign launched the theory of relativity into popular culture and its father, Albert Einstein, into a life of international renown. But that only whetted the public's appetite for news about the search for the fundamental truths of nature, while fostering the image of modern physicists as intrepid individuals given to collect their data by trekking to the ends of the earth—as Eddington had journeyed to the far-off African island of Príncipe to witness a relativity-confirming eclipse.

Newspaper editors evinced a voracious appetite for news of the latest breakthroughs. Scientists became celebrities. In 1921 a six-week tour of the United States by Marie Curie and her two daughters, Eve and Irène, inspired outbursts of public admiration. The visit was the brainchild of Mrs. Marie Mattingly Meloney, a New York socialite and magazine entrepreneur who had been shocked to learn that Madame Curie's research was hobbled by a meager supply of radium. Meloney conceived the idea of raising $100,000 to acquire a gram of the precious mineral—about as much as would fit in a thimble—and bringing Curie to America by steamship to accept the gift. "Mme. Curie Plans to End All Cancers," declared the front page of the *New York Times* on the morning after her arrival (a bald assertion that the newspaper quietly retracted the following day). The climax of Madame Curie's visit was a glittering reception at the White House attended by Meloney and the cream of Washington society, including Theodore Roosevelt's socialite daughter Alice Roosevelt Longworth. There Marie Curie received the beribboned vial of radium directly from the hands of President Warren Harding, after which she expressed her gratitude (the *New York Times* reported) "in broken English." Such were the demands of fund-raising even in the era of small science.

The public came to imagine that physics held the key to all phenomena of the natural world, including the chemical and the biological. Wrote Rutherford's biographer, Arthur S. Eve, physicists were "endeavoring, with some initial success, to explain all physical and chemical processes in terms of positive electrons, negative electrons, and of the effects produced by these in the ether." If they were right, he observed, "such phenomena as heredity and memory and intelligence, and our ideas of morality and religion . . . are explainable in terms of positive and negative electrons and ether."

Not all the physicists were quite so confident. As the decade wore on and they delved more deeply into the intricacies of atomic structure, their picture of the natural world grew only murkier. Their perplexity stemmed from two related and equally perplexing phenomena. One was the so-called wave-particle duality of nature at the infinitesimal scale: experiments sometimes showed light and electrons behaving like particles, and other times as waves.

Einstein's earlier pathbreaking work on the photoelectric effect suggested strongly that light was composed of a stream of "light quanta," or particles. But he acknowledged that manifestations such as diffraction, interference, and scattering were inescapably wavelike. Instead of reconciling these contradictory observations, he had laid the issue before his colleagues. "It is my opinion," he declared at a scientific convocation in Salzburg, Germany, in 1909, "that the next phase of the development of theoretical physics will bring us a theory of light that can be interpreted as a kind of fusion of the wave and mission [that is, particle] theories."

Physicists grappled with the mysteries of subatomic behavior into the mid-1920s, hoping that the steady accretion of observed results would lead them to the truth. But the opposite was the case: the more data they acquired, the less they seemed to know for certain. "The very strange situation was that by coming nearer and nearer to the solution," reflected the promising young German theoretical physicist Werner Heisenberg, "the paradoxes became worse and worse." The only answer seemed to be the one proposed as a joke by the British physicist Sir William Bragg: "God runs electromagnetics on Monday, Wednesday, and Friday by the wave theory; and the devil runs them by quantum theory on Tuesday, Thursday, and Saturday."

It would be Heisenberg and his mentor, the soft-spoken but rigorously logical Dane Niels Bohr, who finally divined the solution, in a process

Heisenberg likened to watching an object emerge from a thick fog. Their conclusion was that anything one could know about an event taking place at a quantum scale was limited to what one could observe—and this knowledge depended on the means of observation. In other words, if one used equipment designed to examine electrons as particles, they would appear to behave as particles; if one used equipment best suited for detecting waves, they appeared as waves. Electrons as particles and electrons as waves were equally valid manifestations of the same thing; there was no contradiction, but rather, in Bohr's term, "complementarity."

Theoretical breakthrough that it was, complementarity did nothing to resolve the paradoxical results emanating from the atomic nucleus. Its structure was the second great mystery vexing physicists in the twenties.

Ernest Rutherford depicted the atom as a miniature solar system, with negatively charged electrons surrounding a tiny yet massive nucleus consisting of positively charged protons and negatively charged electrons. The alluring simplicity of this model helped it become received truth, especially after Niels Bohr augmented it in 1913 with the premise that the electrons could orbit only at certain distances from the nucleus associated with specific energy levels; this seemed to reconcile the classical mechanics governing orbital motion with quantum mechanics, which dictated the energy levels and thus the orbital "shells" that electrons could occupy. The atom as a whole carried a neutral charge: the negative charges of its orbital electrons balanced the positive charge of the nucleus, the latter created by an excess of protons over electrons. By Rutherford's reckoning, therefore, the helium atom had two orbital electrons and a nucleus comprising four protons and two electrons; radium had 138 electrons and a nucleus of 226 protons and 88 electrons.

It soon became obvious that this model created more problems than it solved—and the heavier the atom, the greater the problems. By 1923, the tenth anniversary of Bohr's atomic model, physicists were questioning its general applicability. Bohr's model corresponded to experimental observation only for the very simplest atom, hydrogen, which had only one proton and one electron. At the next-heaviest atom, helium, the model began to break down, creating the anomalies that drew from Max Born his expression of despair.

The troublemakers were those nuclear electrons. No one could explain how such massive particles could fit in the nucleus; or how, once wedged there, the devilishly energetic particles could be made to stay put. Bohr himself was driven to concede that his treasured quantum mechanics might not apply to the nucleus after all, or that some even more novel and confounding mechanics might have to be developed to explain the proliferating experimental anomalies.

The ready solution came from Rutherford. The grand old man of the Cavendish had been mulling over the riddle since the beginning of the decade, when he theorized that electrons became "much deformed" under the intense forces within the nucleus, so that they took on a very different character from orbital electrons. He was thinking that under such circumstances an electron might combine with a proton to form an uncharged, hitherto undetected compound particle he dubbed the neutron.

Rutherford dragooned the ever-faithful Chadwick into the search for the elusive neutron. "He expounded to me at length . . . on the difficulty in seeing how complex nuclei could possibly build up if the only elementary particles available were the proton and the electron, and the need therefore to invoke the aid of the neutron," Chadwick related years later. "He freely admitted that much of this was pure speculation . . . and seldom mentioned these matters except in private discussion." But "he had completely converted me."

As the search proceeded, it became more obvious that solving the mystery of nuclear structure required probes of higher energies than nature provides. Rutherford was not reluctant to state the implications of this rule publicly. Radium emitted alpha particles at a meager 7.6 million electron volts and beta rays—that is, electrons—at only 3 million volts. "What we require," Rutherford declared, "is an apparatus to give us a potential of the order of 10 million volts which can be safely accommodated in a reasonably sized room and operated by a few kilowatts of power . . . I recommend this interesting problem to the attention of my technical friends."

But generating the voltage that Rutherford specified was only part of the problem, and the easiest part at that: nature could meet that specification, as the voltage of a single lightning bolt ran to hundreds of millions of volts. These enormous, fleeting voltages made for pretty spectacles but not much else, however. The problem was harnessing the power, sustaining it, and manipulating it for an assault on the nucleus. "There appears to be no obvious limit to the voltages obtainable" by arrangements common in the power industry such as transformers connected in series, Rutherford declared, adding that a power plant could produce "a torrent of sparks several yards in length and resembling a rapid succession of lightning flashes on a small scale." But this technology was still not capable of "approaching, much less surpassing, the success of the radioactive elements, in providing us with high-speed electrons and high-speed atoms."

Scientists who tried managing energies of the necessary magnitude often ended up with equipment blown to smithereens and laboratories littered with glass shards. Some chose to brave nature's fury: three men from the University of Berlin strung a pair of seven-hundred-yard steel cables between two Alpine peaks and waited for a thunderstorm. When it arrived, they measured the electrical potential at 15 million volts—but an errant lightning strike blasted one of them off the mountain to his death.

American universities, already beginning to enjoy the fruits of collaboration with big business, duly put this relationship to work. The California Institute of Technology received from Southern California Edison Co. the gift of a million-volt transformer so Caltech might develop high-voltage technologies Edison could exploit to transmit electricity to Los Angeles from a proposed dam on the Colorado River, three hundred miles away. (This would be Hoover Dam.) Caltech physicists used the machine to generate X-rays, but it was nowhere near as compact as Rutherford had specified; rather than fitting into "a reasonably sized room," it filled a three-story, nine-thousand-square-foot building, and had to be anchored in an excavated pit to fit under the roof. And still it was not a serviceable producer of high-energy particle beams. In the end, the unit became best known for the spectacular displays it put on during Caltech's annual community "exhibit day," when it could be made to produce a "long sinuous snarling arc" of electricity accompanied by a thunderous report.

One of the foremost figures in the quest was physicist Merle Tuve, who was determined to put a million volts in a vacuum tube at a time when that much energy would blow existing vacuum tubes to bits. "All of us youngsters were, I believe, extremists," he explained later. "We always wanted to go to extreme of temperature, extreme of pressure, extreme of voltage, extreme of vacuum, extreme of something or other." His chosen instrument was the Tesla coil, a high-voltage transformer invented by the visionary physicist Nikola Tesla in the 1890s. Tuve's version was made from copper and wire wound about a hollow three-foot glass tube submerged in a pressurized vat of oil to suppress sparking. He and his colleagues at the Carnegie Institution of Washington managed to produce 1.5 million volts and even to demonstrate the production of beta rays and the occasional accelerated proton, but the device was quirky, erratic, and uncontrollable, and before long, Tuve abandoned it as unsuitable for nuclear research and cursed it as an "albatross."

Tuve moved on to an electrostatic generator invented by a Princeton University engineer named Robert Van de Graaff. This apparatus consisted of a large hollow metal sphere situated atop a tower through which a continuous belt turned, picking up an electrical charge at the bottom and spraying it out at the top, so that the sphere eventually acquired a suitable voltage. The Van De Graaff produced copious volts and sparks, which would turn it into a staple of many a Hollywood mad-scientist set, but did no better than any of the previous efforts at producing Rutherford's "copious supply" of high-energy bullets. Tuve and Van de Graaff struggled to make it work with the vacuum tubes and other apparatus necessary to produce a focused beam of charged, energetic particles. Eventually they succeeded, but by the time they did, Van de Graaff's technology had been outrun by something entirely new.

Its developer was Ernest Lawrence, who had shared his boyhood fascination with electric gadgetry with Merle Tuve, his schoolmate and friend from across the street in a compact South Dakota town named Canton. It was Ernest's destiny to begin his career at a moment when physics had hit a brick wall in its understanding of the atomic nucleus. The obstacle was galling; physicists could peer over the wall at a murky landscape on the far side, shrouded seductively in mist. Lawrence would breach that wall and clear away the mist, marking at that same moment the transition to Big Science from small science. He did so by inventing a serviceable method for artificially driving subatomic particles into the nucleus with enough energy to give physicists a clear picture of what it was made of. To their colleagues, Rutherford and Lawrence would be known as "the two Ernests," and their work would bookend an epochal quest for knowledge of the natural world.

Chapter Two

South Dakota Boy

Turn of the century Canton, South Dakota, was a thriving agricultural town of two thousand residents located near the southwestern wedge of the state, where it met Iowa and Nebraska at the confluence of the Missouri and the Big Sioux Rivers. There Ernie Lawrence and Merle Tuve grew up together in tidy houses facing each other across the street, Merle the elder by a scant six weeks. Both their families occupied prominent social positions in the town: the Tuves by virtue of Dr. Anthony Tuve's post as president of Augustana Academy, the local prep school, and the Lawrences from Carl Lawrence's position as the county superintendent of schools.

As the children of families that valued scholarly pursuits, Ernest and Merle both were inculcated with the virtues of study and knowledge from an early age. This was by no means an eccentric upbringing in the American Upper Midwest of the era. A tradition of academic learning had been brought to the region by its Northern European and Scandinavian immigrants; years later, Ernest's own expanding laboratory would be staffed with accomplished young researchers who had been introduced to the natural sciences in the rural school systems of Minnesota, Montana, or the Dakotas, and who continued their training at the ambitious land grant colleges of the same states. Their education complemented their uniquely American facility with machines and technology, for they had spent their boyhoods surrounded by mechanical gadgetry: farm machines, radios, and cars. "Most of us were radio hams and had taken apart Model T Fords," recalled Stanley Livingston, another Midwesterner (Wisconsin) who would play a crucial role in helping Ernest Lawrence to launch his career. So it was less of a coincidence than it might seem that two of the nation's most eminent physicists would emerge from the same little prairie town or that they would spend the rest of their lives as friends, colleagues, rivals, and adversaries in parallel quests to divine the laws of nature.

The Lawrence family was pious in the manner of northern Midwestern Lutherans, religion serving as much as the warp and woof of the community fabric as a source of personal succor or theological speculation. Carl taught a Bible class in the Sunday school, but throughout their married life, it would be the role of his wife, Gunda, to maintain a socially conservative atmosphere in the household and Carl's role to flout it cheerfully, if carefully. Ernest's younger brother, John, would recall his father's partaking of the occasional cigar and even the occasional scotch, and observing with a grin, "If a man doesn't have some bad habits—at least one or two bad habits—there's something wrong with him."

Ernest was, as Gunda recalled, a handful. "He was born grown up" was her pet description of her willful and self-reliant elder son. A family yarn told of the time that he demanded to visit a cousin in Sioux City, Iowa, seventy miles from home, on his own. As he was only eight, Gunda rejected the idea out of hand. ("You're just a little bit of a boy," she told him.) Carl reasoned tolerantly that Ernest could ride the train to Sioux City to see his cousin and return the same way. "Mother, let him try his wings," he told his wife, and his view prevailed. Ernest's even temperament, his disinclination to brood over setbacks or overdramatize failure, his composure in the face of challenges that might undo a more emotional soul—whether they involved money or technology—surely was rooted in the serenity of his Midwestern upbringing. The one crack in his placid bearing was a persistent stutter. Family opinion held that this reflected not emotional frustration, as post-Freudians might have it today, but a mind that worked so quickly that his thoughts raced ahead of his ability to express them in speech. The stutter was largely cured by a speech therapist during his early teens, although it would resurface occasionally in adulthood, typically at moments of intellectual excitement.

Photographs from Ernest's adolescence portray a handsome young boy, with his mother's full lips and his father's expressive blue eyes. His slightly bucktoothed grin projected an easy self-confidence; the round spectacles he wore from his schooldays gave him a professorial air even then. He was a tall youth whose clothes hung loosely on his bony frame, as though waiting patiently for him to grow into them. Ernest would inherit his father's vigor but perhaps not all his athletic finesse. His sole year on his school's football team yielded a bump on the forehead visible even in adulthood. Later in life, his enthusiastic enjoyment of weekend skiing and hiking often would be evidenced by his limping into the lab on crutches come Monday. He would terrify passengers with his breakneck piloting of a succession of roadsters and speedboats, and his notable success on the tennis court would owe more to his tireless and aggressive style than to refined technique.

Ernie and Merle became steadily more enthralled with electronics, especially the emerging technology of radio transmission. Characteristically, Ernest's interest manifested itself through hands-on tinkering—one day he took the opportunity of his mother's absence from home to mount a telegraph key in holes drilled through her dining room table—and Merle's through diligent perusal of hobbyist magazines such as *Modern Electrics* and *The Electrical Experimenter*. It was a difference in approaches that would follow them throughout their careers.

Radio was in its infancy. Transmission was dependent on spark-gap generators that operated at low power, produced copious interference, and failed utterly in damp conditions. With the technology yet incapable of carrying information as complex as voice—that would not become practical until the widespread introduction of vacuum tubes after World War I—communication was by telegraph key and Morse code. The system was dependent on the size and power of the antenna, which had to be grounded; Ernest and Merle dug a thirteen-foot hole for their antenna's ground outside the Tuve house, despite Dr. Tuve's uneasiness about the electrical project.

The boys spent their free time huddled in the Tuve attic, surrounded by cast-off batteries, glass tubes, and coils of wire. In 1917 they were tempted to imagine they were capturing messages destined for warships or submarines at sea. That was especially so when they tuned in radio station POZ, a station in Nauen, Germany, which had established a famous milestone in 1913 by transmitting a readable signal 1,550 miles—almost enough to blanket the European continent. But the fun ended abruptly late one night shortly after the United States entered the war, when Dr. Tuve discovered Merle copying down Morse code in the attic. President Woodrow Wilson had decreed that private wireless transmissions must cease for the duration, and despite Merle's protestations that he had only been receiving, not transmitting, his infuriated father informed him, "When the president says something, you obey." He demanded on the spot that Merle dismantle the equipment and pack it all away in its original boxes, which he proceeded to seal with his own hands.

Soon after that, college beckoned. Having acceded to her husband's indulgent discipline throughout Ernest's childhood and adolescence, Gunda now put her foot down, insisting that he attend St. Olaf College, a cloistered Lutheran institution in frigid Northfield, Minnesota. For a boy just turning seventeen, she maintained, the important thing was to keep him sequestered from "the wickedness of the state university."

Predictably, Ernest found St. Olaf a stultifying place. He bristled at the hours wasted in Bible study, chapel, and military drill. Other than a chemistry class that kept him marginally interested, his grades revealed the hopeless tedium of his life: a C in religion and a D in electricity and magnetism—fields, as family members would wryly observe years later, in which he was destined to become a world authority.

That was the end of Ernest's St. Olaf experience. For the summer of 1919, he found himself a laborer's job in Vermillion, South Dakota, the seat of the state university. Vermillion was a mere fifty miles from the family home but a world apart; stealthily, he enrolled at the University of South Dakota for his sophomore year and presented the transfer to his parents as a fait accompli.

At USD, Lawrence encountered his first truly inspiring teacher. He was Lewis Ellsworth Akeley, an implausibly cosmopolitan figure on the rural campus. Akeley hailed from a prominent and adventurous family in upstate New York: his brother was the noted explorer and conservationist Carl Akeley. In Vermillion, Lewis Akeley taught a broad-minded curriculum of chemistry, physics, Latin, and physiology. Ernest had sought him out to pitch the establishment of a campus wireless station. So taken with Lawrence's enthusiasm that he burbled the new student's praises to his wife the evening after their first encounter, Akeley invited him back for another meeting and artfully steered him away from premedical study and into physics. He would hold up Ernest as an example to his other students—sometimes to an embarrassing degree. "There's a fellow here, and I want you to all take a look at him," Akeley told his class on one occasion, so vivid in the memory of one student that he could recount it to John Lawrence four decades later. "That's Ernest Lawrence. He's going to be famous someday."

After receiving his bachelor's degree, Ernest moved on to graduate study at the University of Minnesota, where Merle Tuve was already enrolled. As if by an act of providence, Ernest promptly came within the orbit of a genuinely free-thinking pioneer of advanced physics. His name was William Francis Gray Swann.

At thirty-eight, Swann ranked as one of the nation's leading experts in the theory of relativity, a subject on which he had maintained a lengthy personal correspondence with Einstein himself. An imposing figure with an aquiline nose and penetrating eyes glaring from beneath heavy, dark eyebrows, Swann was a highly cultured individual, sufficiently accomplished on the cello that it had been a close call whether he would make his professional career in science or music.

Born in the West Midlands of England, Swann had crossed the Atlantic in 1913 to join the Department of Terrestrial Management of the Carnegie Institution of Washington, an independent center for scientific research founded in 1902 by the steel baron and philanthropist Andrew Carnegie. Swann had moved on to the University of Minnesota in 1918, but would not stay long. Starting the year after Lawrence met him, his renown as a theorist would land him appointments first at the University of Chicago and then at Yale University's Sloane Physics Laboratory as director. Lawrence would tag along at each step.

Swann's pedagogical approach was conspicuously nonconformist. Disdaining what he called the "cult for the glorification of facts," he viewed the proper goal of physics education to be the inculcation of a certain "attitude of mind": a spirit of mental inquiry that would guide the student to an understanding of the natural phenomena underlying abstract principles. Facts, he maintained, could always be retrieved from reference books or recalled through simple laboratory procedures; it was creativity that must be stimulated in the student, because that is what opens the door to the realm of ideas. Years later, Swann would write: "I like to think that if I should go to bed tonight and wake up in the morning to find that I had forgotten everything that I had ever learned, but had succeeded in retaining such experience as I have in thinking, I should not have suffered very much by the loss."

One can see Ernest Lawrence's career as a recapitulation of Swann's precepts in action. Lawrence would never allow established facts to stand in the way of an idea if the latter conformed better to his perception of the natural world; as a result, he would remake the world of ideas again and again. Numerous times he would be informed by colleagues with more eminent credentials than his own that the facts as they were commonly understood militated against his own intuition; in almost every case, he would follow his hunch. Often enough, and sometimes spectacularly, he would succeed in showing that the "facts" were wrong.

Swann also may have bequeathed Lawrence another lesson, this one negative. For Swann's frequent moves from institution to institution hinted at an aspect of his personality that offset his blazing intellect: he had a tendency to rub his peers the wrong way. "Swann was unhappy anywhere he could not be the prima donna," recalled the Berkeley physicist Leonard B. Loeb, who was a faculty colleague of Swann's at Chicago. To be a Swann loyalist, as Lawrence was, meant being prepared to pack one's academic bags on short notice. It was an uncongenial existence for someone who had been raised in a stable, close-knit family environment. Unlike his mentor, Lawrence would be an academic homebody all his life, preferring to remain in place rather than move on. For months, he would resist Berkeley's blandishments to leave Yale, despite the obvious benefits associated with writing his own ticket. Once he reached Berkeley, new offers would come almost every year; Lawrence would give some of these careful consideration while making sure to let Berkeley know that he would rather stay, if the university would only give him a little bit more freedom and resources. Whether he seriously entertained the offers that came his way from institutions as august as Harvard University is hard to say; but in the end, he turned them all down.

Swann helped Lawrence apply his raw tinker's skills to the fashioning of effective experimental equipment. The process began with Swann's interest in a particular phenomenon of electromagnetism. This was the unipolar effect, which concerned the electromagnetic field associated with a cylindrical magnet spinning along its lengthwise axis—that is, rotating like a top, with one pole remaining in contact with the table and the other facing up into the air. Swann assigned Lawrence to design an experiment to investigate the effect, acknowledging cheerfully that previous efforts all had yielded inconclusive results. "Every two years, someone comes up with an experiment in which he rotates a magnet or something and expects to get the effect he wanted, but he doesn't and can't understand why," he told Lawrence. "We'll shed some light in the darkness."

Lawrence built an elegant mechanism out of brass and steel. His speed in designing and building the apparatus was more impressive than the results he obtained, which matched Swann's expectations and therefore advanced science by a tiny increment, if at all. Nevertheless, he did get results, which was encouraging enough; the more so in that his report constituted his master's thesis. Swann delivered the thesis to the *Philosophical Magazine,* which published it in 1924 as Ernest Orlando Lawrence's very first scientific paper.

Yale, to which Lawrence followed Swann in 1924, was a significant step up from Minnesota in intellectual atmosphere. The peerless laboratory facilities and distinguished faculty of New Haven, Connecticut, attracted an especially promising crop of young researchers. Among them was Donald Cooksey, an urbane Californian whose brother, Charlton, was a member of the Yale physics faculty. The younger Cooksey was a competent physicist then investigating the elusive qualities of X-rays, but he knew he was no match for the newcomer.

"He was different," Cooksey would recall. "He was aggressive, interested, and knowing." Lawrence was also, in Cooksey's eyes, still very much a hayseed. "He had an engaging naiveté about the East. He had never seen a skyscraper." The bonds forged then between the two young scientists would last all their lives. Cooksey educated Lawrence in the ways of cosmopolitan society, but his real calling would be as Lawrence's lifelong acolyte and deputy—codesigner of some of Lawrence's most sophisticated accelerators, manager of Lawrence's laboratory, guardian of Lawrence's legacy after his death. From Cooksey's standpoint, it would be the perfect relationship. "I knew that I would never be a great physicist," he acknowledged many years later. "But I thought I might be of some help to one who, it seemed obvious to me, would become one."

At Yale, Lawrence's closest research collaborator was Jesse Beams, another rural Midwesterner—his grandparents had migrated from West Virginia to Kansas in a covered wagon. Like Ernest, who received his doctorate from Yale in 1925, Beams was a newly minted PhD. Their joint project was exceptionally ambitious and, for reasons that scientists would not fully comprehend for another year or two, doomed to failure. Their goal was to investigate the structure of light by measuring the interval from the moment a quantum particle of light, a photon, strikes a target to the emission of an electron from its surface. They also were attempting to "chop" photons into pieces by passing light beams through a rapidly rotating mirror. The project neatly combined work each had been doing previously—Lawrence's research under Swann into the photoelectric effect in gas vapor and Beams's investigation of short-lived physical phenomena. "We decided that if you could take some of the apparatus I'd been using and some of the apparatus he'd been using and we put these two together, we could get some idea of how long the quanta was," Beams related.

The results they obtained were perplexing. Sometimes the light quanta seemed to measure out at three or four meters long; sometimes they appeared to be "just a little bundle of energy," Beams recalled. In fact, they had encountered the same wave-particle duality of light that was confounding their elders in physics.

The paper they eventually published described their confusion in neutral technical language, but still their disappointment leaked through. "There is no definite information on the length of time elapsing during the process of absorption of a quantum of energy photo-electrically by an electron," they reported, "and the so-called length of a light quantum—if such a concept has meaning—is equally unknown experimentally." For all that, the pair did succeed in measuring the time lag of the photoelectric effect at about two billionths of a second. As this was much shorter than the figure proposed by such eminent theoreticians as Niels Bohr, the results represented a gutsy challenge from two young scientists with the ink barely dry on their doctorates. Their figures held up, however, and would be confirmed decades later by more sophisticated measuring equipment than they had at hand in 1926.

Jesse Beams's yearlong collaboration with Ernest Lawrence left him with a lasting impression of Lawrence's inexhaustible vigor. "He worked me to death, practically," Beams recalled. Their experiments sometimes required equipment running under close supervision round the clock for days, yet Ernest seemed to have no trouble finding time and energy for dates or games of tennis or squash, or a Sunday ride on a spirited horse from the Yale stables. In this period, he also began to pay serious attentions to the sixteen-year-old daughter of George Blumer, Yale's medical school dean. Mary Blumer—or Molly, as she was known—was an attractive teenager and a brilliant student destined to lead her class at Vassar College and obtain a place at Harvard Medical School. At first she resisted the attentions of the tall, impossibly thin and rather unsophisticated Ernest Lawrence, whose distracted air set her sister Elsie to giggling; he was the sort of boy who could fall into a reverie in the midst of tying his shoelaces. But Molly soon gave in to his vitality, his intelligence, and the tenacity of his courtship.

Lawrence's precocious talent for experimentation and his engaging character were already attracting attention from recruiters in academia and industry. He published prolifically; his areas of interest, which included the propagation of electron beams and their usefulness as ionization agents, were among the most intriguing in physics; and his resourceful laboratory technique was widely praised. At General Electric's laboratory in Schenectady, New York, where he and Beams had been invited to spend two successive summers by Albert W. Hull, the company's top research executive, he showed such a facile grasp of even the most recondite technical problems that he was given a brief as "a sort of roving ambassador" to peer over the shoulders of older researchers and make suggestions where needed.

In the spring of 1926, a scout dispatched to the Washington conference of the American Physical Society, the leading professional organization, reported home: "I felt out one of the most brilliant experimental young men in the East—a lad whose name is on everyone's lips on account of his recent papers on Ionizing Potentials . . . [and] personally one of the most charming men I have met—a first class mathematician and thoroughly alive." The scout was Leonard B. Loeb, professor of physics at the University of California. Berkeley's quest to land Ernest Lawrence was about to begin.

• • •

In the mid-1920s, the University of California was at a crossroads. It had plenty of money and superb facilities, yet it was struggling to build a reputation for scholarship commensurate with its riches. The money flowed from private donors and from government grants, the latter partially the result of contributions by Berkeley chemists and engineers to the American war effort. The country's success in the Great War inspired a new appreciation across the land for the value of scientific research as a national endeavor and as a foundation for industrial growth. The National Research Council, which had been founded in 1916 as a conduit of government funds to academic institutions but had been hobbled by political infighting and academic mistrust, became revivified in the postwar years. Universities that had shown their value during the war, such as Berkeley, stood at the head of the line for grants for faculty research and laboratory construction. Academia was just beginning to sense the United States government's potential heft as a patron of scientific research.

The man given the task of boosting Berkeley's academic stature was William Wallace Campbell, who became its president in 1923. One of the nation's leading astronomers, Campbell had been director of the university's Lick Observatory for twenty-three years. He had been a major figure in one of the great scientific dramas of the 1920s, the quest to confirm Einstein's theory of relativity by astronomical observation. The theory predicted that light reaching the Earth from distant stars would be bent by the gravitational field of the sun to a much greater degree than was predicted by classical Newtonian physics. The phenomenon would appear as an apparent shift in the location of the stars, if they could be observed in close proximity to the sun during a solar eclipse, when the darkening of the sun's disk would make them visible. This was solidly within the expertise of Campbell, who, since 1898, had made a half dozen eclipse expeditions to remote locations in India, the Ukraine, and the South Pacific island of Kiribati, sponsored by the Lick and backed financially by the California railroad magnate Charles Frederick Crocker.

Bad weather confounded the relativity observations during a Lick expedition to Brazil in 1918, forcing Campbell to concede the honor of confirming relativity to the British astronomer Arthur Eddington, based on his observations from the African island of Príncipe. But Campbell achieved the final confirmation, made in September 1922 from the northwestern coast of Australia.

Campbell could not disagree with Europeans' condescending view of American science as a backwater rich in money and manpower but poor in theoretical understanding. Every year, he dispatched his most promising graduate astronomers to the great European centers of scientific learning, such as Cambridge, Manchester, Paris, or Göttingen, Germany. These were the obligatory so-called *Studienreisen*, or study tours, their goal being the students' acquisition of the necessary grounding in "the modern developments in Physics" that was simply unattainable in the United States. Campbell felt the deficiency all the more sharply between 1914 and 1918, when the Great War cut off all travel to Britain and the continent.

In the postwar years, when an influx of demobilized students threatened to burst the seams of Berkeley's outdated scientific facilities, California's legislature and the university's private patrons responded with unprecedented generosity. One spur surely was the 1921 transformation of Throop Polytechnic Institute, a modest private technical school in Pasadena, into the California Institute of Technology. The upgrade was the brainchild of the eminent astronomer George Ellery Hale. The money came from Hale's cadre of loyal philanthropic backers, and the institution's sudden leap into the first rank of universities worldwide was certified by its recruitment of Nobel laureate Robert A. Millikan, then the head of the University of Chicago, to become its president.

Not to be outdone, Berkeley unveiled the centerpiece of its surge of spending in 1924. On the day it opened, LeConte Hall, named after John LeConte, a physicist and the very first faculty member appointed upon the founding of the university in 1869, was one of the largest physics buildings in the world. But filling its spacious offices and laboratories with appropriately distinguished professors was not as easy as getting the structure built. In vain, the university put out feelers to Niels Bohr, American Nobel laureate Arthur Holly Compton (who accepted an offer to become physics chairman but reneged after receiving a counteroffer from the University of Chicago)—even Swann.

The department placed its quandary in the hands of two junior professors, Loeb and Raymond T. Birge, who devised a new strategy of snaring scientific prodigies on their way up, before they had a chance to cement themselves into comfortable sinecures elsewhere. As source material, they mined the roster of National Research Fellows. The fellowships had been established by the National Research Council to support candidates from among the top 5 percent of PhDs in the country, making it an invaluable guide to a prospective candidate's scientific ability. They compiled a wish list of a half dozen candidates, among them the young Yale instructor widely considered back East to be the best of the current crop of fellows: Ernest O. Lawrence.

After Loeb made the initial contact with Lawrence in Washington, Birge took over. He peppered Lawrence with pointed gossip about rival institutions that might have designs on him. "I got the following apparently authentic information about Cornell, which may interest you," he informed Ernest in one letter. "Richtmeyer is very much at outs with the rest of the Dept. and is very unhappy there . . . I should advise you to stay at Yale. (For the present, at least.)" Birge's misspelled reference presumably was to one of two eminent physicists, Floyd K. Richtmyer, a long-term professor at Cornell, or his son Robert, who was just coming to the end of his graduate education at the university but soon moved on to MIT.

That summer, in 1927, Lawrence joined Beams on a visit to Europe, where Swann was spending a sabbatical. It was Lawrence's first visit abroad. The young scientists' expectation was that Swann would escort them around Europe, introducing them to his friends and acquaintances among the eminent figures of European science in a sort of belated and truncated *Studienreise*.

Promptly after greeting them in Paris, however, Swann deserted them in favor of a master class with the cellist Pablo Casals. So they spent the summer making their own frugal way among the scientific landmarks of Europe. They stopped at Niels Bohr's Institute for Theoretical Physics at the University of Copenhagen. But the great man was not in attendance, and they contented themselves with walking its halls almost as tourists. In Berlin and Göttingen, they had cordial meetings with leading physicists; in Paris, they met Swann's friend Marie Curie at her Radium Institute, where despite support from the French government and philanthropists in Europe and the United States, she was working with almost shockingly simple, even unsophisticated equipment—certainly nothing approaching the quality of furnishings they worked with at Yale. The same impression struck them in Britain, where the most groundbreaking advances were being achieved in Manchester and Cambridge with the most modest hand-made gear.

Small science was in its glory. Ernest did not at that moment absorb the lesson that high-quality research and solid theoretical reasoning could yield wonders, even with poor equipment, while even the best-furnished lab would produce little of value if the equipment was not deployed intelligently; he would have to learn the hard way. Yet he came away feeling that the Europeans had less to teach him and Beams than he had expected. The United States is "not behind except in fame," he groused to Beams. It would not be long, he predicted, before Europeans would be coming to the United States for *their Studienreisen*.

Upon returning home, Ernest found a letter from Birge portraying Berkeley as an academic paradise on earth. "Now I have an idea that you will like California and California will like you," he wrote, making sure to emphasize that "the teaching schedules are as light here as any place in the country . . . far lighter than those of the average state University, and apparently far lighter than at Yale." Birge's goal was to wean Lawrence from Yale's aura of prestige, which could seem overwhelming compared with that of a distant public university. Nor was Yale allowing the bidding for Lawrence to go unanswered. Just before his departure for Europe, the university had granted him an assistant professorship at a salary of $3,000. Beams, for one, was convinced that Yale would never let Lawrence go.

Not so Birge, who kept up the pressure on a young man he perceived to be in a hurry. "The . . . chance to get somewhere in reasonable time is everything," he wrote Lawrence to reinforce a firm offer from physics chairman Elmer Hall of an associate professor's post paying $3,300, not counting a faculty-wide raise of $500 due in the coming year. At Berkeley, Birge assured him, "the younger men now are being appointed and advanced on an entirely different plane from that of the older men." And that was for *ordinary* talents. "You are not at all concerned with the *average* scale at this or other universities or with the average rate of promotion . . . I think *you* are concerned only in the way the exceptional men are treated. I doubt if any man has ever been offered the permanent position of Asso. Prof. at this University with as short a period of teaching and research experience as in your case. That proves . . . how highly we regard you."

Two factors finally prompted Ernest to cut the cord with Yale. One was Swann's decision to accept the directorship of the small, poorly funded Bartol Research Institute of Philadelphia. The second was Yale's complacent refusal to accommodate its sought-after assistant professor's request for a promotion to associate professor. Such rapid advancement cut against the grain in New Haven. Ernest's academic superiors were unable to see past his youth, much less rid themselves of the conviction that faculty members simply did not leave Yale to accept positions at state universities on the distant West Coast. Beams, who was one of the first members of Lawrence's circle to learn that he was accepting Berkeley's offer, marched into physics chairman John Zeleny's office to announce that Yale was about to commit "the biggest mistake they ever made." Zeleny replied funereally that he agreed but could not convince the dean. On March 12, 1928, Lawrence accepted the Berkeley offer by telegram.

That summer, just before motoring across country to take up his new post, he stopped in Washington to see Merle Tuve. Merle had received his doctorate from Johns Hopkins University and been ensconced for two years at the Department of Terrestrial Magnetism of the Carnegie Institution of Washington, Swann's former haunt. There he was immersed in an effort to produce high-energy protons to probe the atomic nucleus, using his balky Tesla coils and Van de Graaff generators. Inside his lab, amid the stink of petroleum fumes and the clatter of motors, Tuve asked Lawrence what research he intended to do at Berkeley, and got what he thought was a woolly answer.

"He responded, rather vaguely, with some small notions about high-speed rotating mirrors, chopping the tails off quanta and other single-shot ideas," Tuve would recall. It was the old Ernie Lawrence, distracted by gadgetry and apparatus instead of burying his nose in scientific journals.

Tuve upbraided Lawrence with all the authority of his six weeks' advantage in age and with all the frank liberty afforded to one childhood friend speaking to another. "I said it was high time for him to quit selecting research problems like choosing cookies at a party; it was time for him to pick a field of research that was full of fresh questions to be answered." The right choice was inescapable. "Any undergraduate could see that nuclear physics using artificial beams of high-energy protons and helium ions was such a field, and . . . he should stake out a territory there to work and grow in."

Lawrence listened soberly, his eyes wandering over the magnets, Tesla coils, and vacuum tubes littering the lab. "He was vaguely searching for an identifiable field full of specific problems," Tuve concluded. Tuve could not have known it at that moment, but he had set his friend on the path to a career.

Chapter Three

"I'm Going to Be Famous"

For such a seminal moment in a field that demands rigorous procedural records, the full circumstances of the cyclotron's birth remain frustratingly indistinct. We know that Ernest Lawrence acquired the basic idea from an article in an obscure German technical journal, and we know the journal, the article, and its author. We know that the moment occurred in the spring of 1929, though the exact date is murky.

We know less about what prompted Lawrence to pick up the journal and page through it. Had he been pondering the question of how to accelerate particles for a long time, or did the concept that was to guide his life's work occur to him on the spot? His contemporaries are divided on the question, and his own notes are contradictory. Did he come upon the *Archiv für Elektrotechnik* issue for December 19, 1928, in the Berkeley science stacks utterly by chance, or was the journal there because he himself had ordered the subscription? There is evidence for both possibilities. Did he happen on the article by the Norwegian physicist Rolf Wideröe while "glancing over current periodicals in the University library," as he related in his Nobel Prize acceptance speech? Or did he stumble upon it while trying to stave off the boredom of an interminable faculty meeting, as he told Wideröe at their only face-to-face meeting many years later?

What is agreed upon is that Lawrence understood almost immediately the import of what he had seen, even though Wideröe's report was written in dense technical German, a language that was largely beyond Ernest's ken. "I merely looked at the diagrams and photographs," he said in his Nobel speech. From these, he gleaned that Wideröe's "general approach" was to accelerate ions on a straight line by delivering to them repeated kicks from a series of charged electrical gaps. For Lawrence, the stepwise acceleration of particles by multiple small impulses "immediately impressed me as the real answer which I had been looking for to the technical problem of accelerating positive ions." More precisely, it appeared to solve the problem of how to produce high-energy particles without applying high voltages. Laying aside the article, he sketched out a linear accelerator to drive protons to a million volts, the first major step toward satisfying Ernest Rutherford's demand for 10 million volts. Yet he was not quite at the goal: simple math told him that reaching that energy would require a linear tube many meters in length—"rather awkwardly long for laboratory purposes."

Then came the real brainstorm. What if he could force ions into a circular path, thereby passing them repeatedly across a single electrical gap? This conception of a compact and electrically efficient accelerator synthesized several established principles into a novel whole. The first principle was that charged particles moving in a perpendicular direction through a magnetic field follow a curved path. The challenge then becomes timing the electrical impulses so they energize the gap at the very moment the particles cross, which means generating them from an oscillator working at a fixed frequency.

The second principle—and this was so crucial it became known as the "cyclotron principle"—was that as the particles gain speed, their paths spiral wider. This lengthens the distance they must travel to return to the starting point; but the increase in speed and the lengthening of the path work together, so that as they move faster on the longer path, they will still reach the gap at the same interval. This principle resembles that by which a point on the rim of a bicycle remains in sync with a point on the hub, even though with each rotation of the wheel the point on the rim travels several feet and the point on the hub only a few inches.

Put the principles together, and they indicate that an electrical field at a constant frequency can impart repeated kicks to a stream of spiraling protons without being constantly retuned. The phenomenon by which protons keep time with the oscillator even as they accelerate came to be known as "resonance." To accelerator designers, it is a principle as fundamental as Einstein's $E=mc^2$.

The cyclotron principle was not exactly unknown to physics. Earlier in 1929, the Hungarian physicist Leo Szilard, whose fertile mind produced a stream of fanciful devices in the twenties and thirties, had attempted to obtain a patent in Germany for an apparatus based on the idea. But his design was too vague and perhaps too revolutionary to convince the examiners. For all his creativity, Szilard lacked the tenacity of an Ernest Lawrence, as he acknowledged years later. "The merit," he told a friend, "lies in the carrying out and not the thinking out of the experiment." To Szilard, his device was one of many ideas that might or might not work; to Ernest Lawrence, it was an idea upon which to build a career.

Ernest sprinted back to his bachelor quarters at Berkeley's Faculty Club to share his theory with his fellow residents, most of them unmarried junior faculty members like himself. As a rule, Lawrence was not a boisterous sort, but they had all at one time or another witnessed his ebullient outbursts when he was seized with a new idea. Many of them would retain lifelong memories of the first time they heard Ernest Lawrence describe the cyclotron. Tom Johnson would recall hearing Lawrence describe it in the Berkeley library that first night, the Wideröe paper clutched in his hands. At the Faculty Club, the first person Lawrence encountered was Donald Shane, a mathematician who obligingly double-checked Lawrence's scribbled calculations and agreed that the math appeared to be sound.

"But what are you going to do with it?" Shane asked.

"I'm going to break up atoms!" came the reply.

Lawrence was still on the boil the next day. One faculty wife would always remember his exclamation when she encountered him on a wooded walkway as Berkeley was waking up to a chilly spring morning: he shouted at her, "I'm going to be famous!" Jim Brady, one of his graduate students, was tinkering at his lab table on the second floor of LeConte when Ernest bounded in, dragged Brady to the blackboard, and sketched out the equations. Tuve's electrostatic accelerator would produce high voltages, all right, Lawrence told Brady. "But what can they do with them when they get them?" he asked. If you put a million volts into a vacuum tube, you could be sure only of blasting the tube to bits. But if you put a few *thousand* volts on the tube and built up to a million volts on the particles, the glassware would survive, and the particles would accelerate.

Generating an electrical charge of a few thousand volts was practical. Confining ions in a magnetic field for about one hundred orbits was practical. In Lawrence's expansive vision, Rutherford's goal of 10 million electron volts was already within reach. All he needed to achieve it was an oscillating electrical charge of sufficient power, an electromagnet of sufficient size, and a source of ions to be propelled around what he was already picturing as a "proton merry-go-round." As Lawrence told Shane that first night, he could see nothing wrong with it.

He was right about the concept, right about its importance, and right about one more thing: it was going to make him famous.

Yet transforming blackboard scribbles into a working device was not so simple a matter, as Szilard had understood. Lawrence did not yet command a laboratory staff—only three graduate students, each already working on projects he had approved. Nor were all his friends and colleagues as taken with the proton merry-go-round as his neighbors at the Berkeley Faculty Club on that first bleary-eyed evening. Merle Tuve, who was still struggling to produce high-energy particles with his obstreperous Tesla coil, saw little practical application for Lawrence's curved variation on a linear accelerator. Everyone seemed to have a different reason why resonance could not work, and none a reason why it should. The ion particles would fall out of phase, they said, or would crash into the walls of the accelerating chamber, or would collide with stray air molecules in their way.

Lawrence replied that by accelerating his particles in a vacuum tank and holding them on their path with a magnet, he could eliminate collisions, but this idea only introduced the technical question of how to maintain an airtight vacuum within a maelstrom of powerful electromagnetic forces. As the new term opened at Berkeley, Ernest kept the spiral accelerator on the shelf, as there were numerous other projects to keep him distracted. One of these, an X-ray tube he developed with an exceptionally single-minded grad student named David Sloan, seemed to promise electrons of a million volts—not suitable as a nuclear projectile but useful for a wide range of other experiments.

Then, after Christmas, a couple of jolts prompted him to take the accelerator off the shelf.

The first was a conversation with the distinguished German quantum physicist Otto Stern, a future Nobel laureate who was stopping at Berkeley during the holiday. Stern's expertise was the action of magnetic fields on atoms and subatomic particles. When Ernest described his idea to him at a faculty dinner, he became uncommonly interested. For the first time, Lawrence heard a physicist with a solid grasp of subatomic behavior express support for his idea rather than smothering it in quibbles.

"Why don't you get on with it?" Stern barked impatiently, or as John Lawrence, a witness to the encounter, recalled Stern's words in his own fractured German: "*Sie mussen Zurich gehen!*" His meaning was clear: "Get back to the lab!"

A few days later, the second jolt arrived in the form of the January 1 issue of *Physical Review*. The journal carried a report by Merle Tuve claiming that his Tesla coil could drive alpha particles to energies as high as 10 million electron volts without "any serious difficulty." At that energy, Tuve calculated, the particles delivered power equal to that of 2,600 grams of radium. The significance of Tuve's claim was not lost on any physicist who recalled Marie Curie's 1921 voyage to the United States for her single gram of radium, valued at $100,000. Ernest remained skeptical that Tuve's coil could function as a practical accelerator, for the tendency of the high voltage to break down with an enormous spark placed a hard limit on how much sustained power it could impart to a particle beam. But Tuve's announcement impelled him to show that he had a superior idea.

The opportunity to get on with it arrived when Lawrence's graduate student Niels Edlefsen completed his doctoral work early in the new year. Edlefsen was a late bloomer, six years older than Ernest yet still only a teaching assistant. But he was devoted to Lawrence, who uniquely among the physics professors was not above showing up at LeConte Hall in the middle of the night to lend his students a hand, announcing his presence with the words "Mind if I work with you awhile?"

Lawrence now asked *him* for help. "I've got a crazy idea," he told Edlefsen. "It's so simple, I can't understand why someone hasn't tried it . . . Why don't you line up what we need?" It was the first appearance of Lawrence's method for building the world's first great Big Science laboratory: the remorseless exploitation of cheap graduate-student labor, a resource he would soon have in surfeit.

By mid-January, Edlefsen was hard at work assembling Lawrence's apparatus. Over the winter months and into the spring, he fashioned a series of protocyclotrons out of metal and glass containers, which were flattened, split laterally, fitted with filaments and electrical wires, and slathered over with thick gobs of sealing wax to hold a vacuum.

The first models, small enough to fit in the palm of the hand, betray little family resemblance to their ever more painstakingly engineered progeny, much less the miles-long behemoth humming beneath rural villages in Switzerland in the twenty-first century. Edlefsen's devices resemble whisky flasks run over by a truck. At Ernest's direction, Edlefsen would position these units between the four-inch poles of the physics lab's electromagnet, pump them free of air and fill them with hydrogen gas, and then ionize the gas with a charged tungsten filament.

The results were equivocal. According to Edlefsen's rudimentary ion detectors, something was going on inside the flask, but whether it was protons accelerating in resonance with an oscillating electric field was by no means clear. Whatever doubts Edlefsen harbored, however, were overmatched by Lawrence's optimism. Ernest was willing to assume that he had achieved a proof of concept; verifiable results required only superior detection equipment. As he reported in one of his frequent, dutiful letters to his parents in South Dakota, "If the work should pan out the way I hope, it will be by all odds the most important thing I will have done."

That summer, Edlefsen left Berkeley for a postdoctorate job at another university. Ernest, who had spent the vacation months in "laxity" playing tennis, wrote up a talk on the project for a meeting of the National Academy of Sciences in Berkeley on September 19, 1930, and a report for the journal *Science* to be published a few weeks later. These were carefully phrased to suggest, without saying so outright, that he and Edlefsen had achieved the resonance they had been seeking. The article avoided making any reference to actual readings, for they had none to report. But it bristled with Lawrence's self-confidence and his manner of intuitively grasping possibilities well ahead of realities—in this case, the production of sustained resonant proton beams with a million volts of kinetic energy. "Preliminary experiments," he wrote, "indicate that there are probably no serious difficulties in the way of obtaining protons having high enough speeds to be useful for studies of atomic nuclei." By echoing Merle Tuve's words, he had quite deliberately planted his flag next to that of his boyhood friend. But now the physics world expected him to make good on his boast. Fortunately, the next in a long line of talented collaborators was already on the scene.

Milton Stanley Livingston was a burly farm boy who had been raised in the foothills of California's San Gabriel Mountains, where his father owned an orange grove. Stan had received his physics education at Dartmouth College, which was not known for its science programs; Livingston acknowledged that his master's degree from Dartmouth was perhaps the equivalent of a bachelor's degree at a more rigorous research institution. But he had learned enough to land a teaching fellowship at Berkeley. ("There was not nearly as much competition then," he would recall.)

Unlike others making their way to Berkeley in the fall of 1930, Livingston had never heard Lawrence's name; the first he knew that a Professor Lawrence even existed was when he registered for Lawrence's undergraduate magnetism course during his first few weeks of breakneck studies designed to fill in the holes in his physics education. But he soon got swept up into Ernest's orbit. Having polled the faculty for a suitable topic for his doctoral thesis, he was most intrigued by Lawrence's suggestion that he study "the resonance of hydrogen ions with a radio frequency field in a magnetic field." In short: the cyclotron effect. Leonard Loeb, notwithstanding his role in recruiting Lawrence to Berkeley, warned Livingston superciliously that working on Lawrence's project would be a waste of time. "He didn't think [the cyclotron] would work," Livingston would recall. Livingston carried Loeb's doubts back to Lawrence, who extinguished them with a typically powerful display of self-confidence. Livingston signed up.

It was a fortuitous partnership, for Livingston's skills neatly complemented Lawrence's. The latter contributed the vision and an inspired experimental blueprint; Livingston's farm upbringing trained him to be comfortable with machines and adept at hands-on maintenance and the repair of complicated gear that could not be sent out for servicing. His role would be to transform Lawrence's ideas and Edlefsen's jerry-built apparatus into an accelerator that worked.

Livingston's first task was to scrub the hyperbole from the claims that Lawrence and Edlefsen had made. Whatever Edlefsen had seen inside his little chamber, Livingston concluded, it was not resonance. There was no evidence that hydrogen ions had charged around Edlefsen's "rather sketchy" apparatus at all, much less at any appreciable fraction of 1 million electron volts. Livingston figured instead that Edlefsen had merely created heavy ions from the atmospheric nitrogen and oxygen left inside the device by poor evacuation technique. These may have traced a curved path for a short distance, but those that reached the detector had probably done so after only a single acceleration. Most had probably expired in collisions with the walls of the fist-sized tank. The cyclotron principle was still waiting to be observed and experimentally confirmed. It was Livingston's job "to do just that."

He started from scratch, replacing Edlefsen's lumps of metal and sealing wax with a vacuum chamber fashioned from brass in the shape of a flat cylinder four inches in diameter. One half of the chamber's interior was occupied by a hollow semicircular electrode shaped like the letter D and known forever more as a "dee." The other half was empty except for a copper strip serving as a target. Accelerated particles would strike this strip at the end of their spiral journeys, with their final energies to be measured by an electrometer wired to it. Subsequently a second dee would be nestled next to the first, strengthening the effect of the electrical charge that accelerated the particles every time they passed from one dee to the other—producing, that is, two accelerating kicks for every complete circuit by the particle. Livingston would describe this first formalized device as one in which "all the basic features of the modern cyclotron were present *in utero*."

Indeed, he had overcome the practical obstacles to turning Lawrence's fancy into a real-world accelerator. That was in September. But it would be December 1 before he witnessed anything resembling true resonant acceleration and could write in his workbook, with unabashed relief, "At last we seem to be getting the correct effect." By Livingston's calculations, he had accelerated ionized hydrogen molecules, H_2, twenty times over the course of ten complete circuits. Shortly after Christmas, he managed to cadge on temporary loan from a laboratory colleague a magnet twice as powerful as his paltry four-inch model. It was a timely acquisition, for Ernest had succumbed to a rare spasm of self-doubt. "We are having a bit of trouble with our high speed protons," he wrote to Swann. "We can make them spin around alright but we have not been able to determine how many times and therefore what speeds we have been able to produce."

Livingston's results with the bigger magnet dispelled the clouds. His measurements indicated that he had accelerated ions to 80,000 volts by spiraling them around the vacuum chamber forty-one times, thereby delivering eighty-two resonant jolts. "Lawrence was really excited," Livingston recalled. "You see, we'd proven the point. He was off to the races."

More precisely, Ernest was racing for money to build an even more powerful accelerator. Ernest's habit of thinking ahead to the next step before the last one had been fully played out now materialized, spurred in part by the need to validate his optimistic forecasts of what the next step would bring. Even before Livingston's results had been validated properly, Ernest was mapping out a campaign to acquire a more powerful electromagnet to drive protons beyond the million-volt threshold. The apparatus would cost $700, according to the estimate he submitted to the university's board of research. The board duly passed it on to Robert G. Sproul, who had been inaugurated as the university's president only three months earlier. It was his first bill for the cost of turning Berkeley into the nation's pioneering institution of Big Science.

Sproul had become fully aware of Lawrence's value to Berkeley that fall, when Northwestern University approached the physicist with the startling offer of a professorship at $6,500, more than half again the salary of a full professor at Berkeley—not to say a Berkeley associate professor aged twenty-nine. The prospect of losing Lawrence provoked panic on the science faculty. Observing that Lawrence was "the best experimental physicist among men his age in the country"—the loss of whom would present "a very serious handicap" to the university—a committee of senior science professors unanimously recommended that he be offered a full professorship and a raise to $5,000. Queried at long distance by Sproul, the loyal Swann declared that his protégé was destined to become one of the world's ten leading physicists before the end of the decade. A more direct observation came from Gilbert Lewis, the eminent dean of the chemistry faculty, who informed Sproul that the question at hand was not whether to appoint the youngest professor in Berkeley history "but whether or not we are going to have a physics department."

The debate over Lawrence's future at Berkeley climaxed at an acrimonious faculty conclave pitting the hard sciences, represented by Birge and Lewis, against the humanities and social sciences departments. A promotion of such magnitude for a junior professor, leapfrogging faculty members with greater seniority and longer records of accomplishment, was unprecedented, the dissenters said; to show such favoritism to one department could only demoralize the others. The thirty-nine-year-old Sproul made a show of weighing the pros and cons, but his decision was preordained. "If there is one chance in ten that he'll be one of the top physicists in the country, I'll take that chance," he declared in approving Lawrence's promotion. As it happened, Lawrence would meet Swann's ten-year timeline with room to spare, bringing home the Nobel Prize only eight years later.

In truth, Lawrence had not found Northwestern's offer all that compelling. His goal was not to obtain a large raise for himself but to obtain explicit recognition of his lab's importance to the university under its new administration. In this, he succeeded. The funding request for the new cyclotron was his first chance to exploit his new standing. He was not disappointed. Having made a special case of Ernest Lawrence in one of his first official acts as Berkeley's president, Sproul was not about to skimp on his new star's research budget. He signed off on the $700 requisition with virtually no discussion.

With an additional $500 cadged from the National Research Council, Lawrence ordered a custom magnet with nine-inch pole faces from the Federal Telegraph Company. He placed the responsibility for fashioning a suitable new vacuum chamber in Livingston's hands, and then vanished from the laboratory to go "angling for funds and other support," Livingston recalled. "I didn't know what he was doing exactly, but I could see that he was spending a lot of energy searching for support to make the next step."

When he surfaced, it was not always to ease the way forward. Reappearing one day in March, Lawrence announced, "Stan, you've got to stop now and write up your thesis." With a start, Livingston realized that he had only two weeks to complete the work necessary to receive a doctorate that June. Much was at stake for both him and Lawrence: the doctorate was required before Livingston could be given an instructorship, which in turn was necessary if he were to stay at Berkeley into the fall term. And Lawrence considered that necessary, in its turn, because he needed Livingston to complete the construction of the new accelerator.

Livingston dropped everything to write a hasty thesis based on his work with the four-inch accelerator. Then he sat for an oral exam on radioactivity conducted by Birge and three other professors. It was evident from the very first question that his involvement with the accelerator had left him utterly unprepared for the inquisition. Birge "asked me the outright question had I ever studied Rutherford, Chadwick, and Ellis, the most famous book in the field? And I had to admit that I had not. I hadn't had time." Staggering from the examination room, Livingston sought out Lawrence to warn that he almost certainly would be refused his doctorate, due to his failure to demonstrate a grasp of basic nuclear physics. Lawrence absorbed the news with surprising equanimity, for reasons that soon became obvious. The committee awarded Livingston his doctorate without cavil. Plainly Lawrence had placed his thumb on the scale. "I imagine," Livingston reflected years later, "that he was persuasive."

Livingston promptly returned to his labors on the accelerator. He was ready with a freshly designed vacuum chamber, eleven inches in diameter, when the new magnet arrived on July 3. The chamber was fitted between the pole faces and immediately fired up. Two weeks later, Lawrence was spreading the news that the apparatus had accelerated protons up to 900,000 volts. Only a year after launching the effort, he was knocking on the million-volt door.

• • •

"I am hastening to let you know that the experiments on the production of high speed protons have been successful beyond our expectations," Lawrence wrote to Dr. Frederick Gardner Cottrell on July 17. "The work has advanced to an exceedingly important stage and the greatest difficulty now facing us is no longer of an experimental nature, but one of finance."

This letter signifies the start of a crucial new phase in Lawrence's transformation from hands-on physicist to fund-raising impresario. No longer were the university's resources, parceled out at a few hundred dollars at a time, sufficient for his purposes. The next accelerator would require thousands of dollars, and for that, he needed a new patron. That is where Frederick Cottrell came in.

Cottrell had been a popular professor of physical chemistry at Berkeley in 1908, when he received a patent for a process to precipitate impurities out of smokestack emissions by passing them through an electrically charged grid. He was thirty-one and had accepted a commission from E. I. du Pont de Nemours & Company, the giant chemical manufacturer, which hoped to recover waste sulphuric acid from its smokestacks, because his father's death had left the family deeply in debt. His invention, it turned out, could also cleanse noxious vapors and particulates from smelter effluent, coal particles from mine air, and much more. With three partners, Cottrell established a company to commercialize the process. But the idea of a university professor making money from what were essentially the fruits of academic research left him uneasy. At heart, he was a professor, not an industrialist. This was not an unusual mind-set for the period, when the model of basic research still harked back to the work of eminent scientists devoted to the disinterested pursuit of knowledge—men such as Louis Pasteur, who was famous for having declared, "I could never work for money, but I would always work for science."

The question of what should be done with the discoveries pouring out of university laboratories was widely debated in academia. Wrote Abraham Flexner, a leading authority on the responsibilities of the American university, "[T]he moment that research is utilized as a source of profit, its spirit is debased." Yet the genteel detachment of Pasteur and his contemporaries seemed quaint, even foolish, given the profits to be made in the industrial world. Professors could only envy the millions pocketed by radio entrepreneurs from the discoveries of those pioneers of electromagnetism Michael Faraday and James Clerk Maxwell, who had claimed no patents and earned nothing.

University conferences and professional committees worried over the conflicts and compromises presented by academic profit seeking. Should public universities license discoveries funded with taxpayer dollars so that private corporations could sell them for profit to those same taxpayers? Who really deserved the patent for a discovery that might be based on the work of innumerable scientists over years or decades? What might happen to the unfettered give-and-take among scientific colleagues if they were turned into rivals when the search for truth was overtaken by the race for commercial advantage? How would beckoning wealth affect the scientific method, in which one might learn much from (unprofitable) failure in the disinterested search for truth?

Were scientists to be researchers or businesspersons? Alan Gregg, the medical director of the Rockefeller Foundation, relayed the complaint of "the dean of one of our larger medical schools" that one of his faculty members was "so busy controlling the product made under a patent held by the university that there is no time left for research or teaching." And the distinguished British scientific administrator Walter Fletcher warned an American audience that commercial prospects were certain to exercise "a vicious influence" over academic standards. "The university will be more inclined to reward by pay or promotion him who makes some addition to knowledge of an immediately profitable kind rather than him who works for knowledge itself," he said. "Nothing could be more disastrous than this, as we know, to the advancement of knowledge itself."

Then there was the reality that funding for basic research was always in danger of cutbacks, especially in periods of economic strain. "Science is dependent on wealth for its material support," observed *Harper's Monthly* in 1936. "Laboratories cost money to equip and maintain, and lately it has become increasingly difficult to obtain this necessary wherewithal. Endowments have shrunken . . . the appropriations of the governments for pure science have been curtailed . . . Science creates wealth; why then should it not turn its talents to a program of self-support?"

Many of these issues remain unresolved to this day; one can only imagine the ferocity of the debate that raged when they were new.

It was clear that a way had to be found to capture the profits of laboratory discoveries without undermining academic standards or constraining scientific inquiry. A pioneering model emerged at the University of Wisconsin, where Professor Harry Steenbock had invented a way to fortify foods with vitamin D. Rather than keep the patent rights to himself, he cofounded the Wisconsin Alumni Research Foundation, or WARF, which took possession of his patent in 1925 and promptly licensed it to the Quaker Oats Company. By 1930, the patent was generating $1,000 a day, all of it reinvested in University of Wisconsin research. Other major research universities adopted the WARF model, though not the University of California, which required any faculty member with a patentable invention to present its particulars to the university president, who would then appoint a board to advise him "as to the action, if any, which should be taken by the Board of Regents with respect to the invention." Berkeley's ability to exploit its increasingly inventive faculty's work stagnated under this lumbering regime until 1931, when Sproul met Frederick Cottrell.

Within a year of receiving his own patent, Cottrell was pondering how to use his profits as feedstock for a nationwide "endowment for scientific work"—a foundation that would assemble a broad portfolio of patents and use its profits to support promising scientists all over the country. The main challenge lay in finding an entity to endow. The University of California was a logical beneficiary of his own patent rights, since Cottrell's work had been done in its laboratories. But its administration was utterly unable to reconcile its responsibilities as a disinterested servant of the public interest and those of an owner of a commercially valuable patent. Instead, Cottrell established an independent philanthropy, the Research Corporation, in 1912. As its birthright, the corporation received Cottrell's patent and his directive that it be exploited for profit to fund grants for scientific research. The corporation's directors were empowered to fatten its portfolio with other promising inventions and use them the same way.

Over the following two decades, however, Cottrell's dream faltered. The problem was that out of a surfeit of integrity, he had refused to accept any formal management position with the corporation, instead handing control to a board of industrialists over whom he exercised no authority or even much influence. These hidebound trustees were determined to run the Research Corporation as conservatively as they ran their own businesses—that is, accumulating a large capital cushion before starting to give it away. By 1930, the value of the Research Corporation's patent rights exceeded $1 million. But its grant portfolio was a measly $23,000.

Fortunately for Ernest Lawrence, the old mind-set was already changing. The process had begun in 1927 with the appointment of a new president of the Research Corporation. He was Howard Poillon, a hardheaded businessman who was determined to make the most of the corporation's growing patent portfolio but also more willing to defer to Cottrell's wishes and judgment on philanthropic matters. The two men would work together for fifteen years, during which they created a model for the philanthropic support of science that would spread to the Rockefeller and Ford Foundations and other leading institutional patrons of research.

Under Cottrell's influence, Sproul rescinded Berkeley's patent policy. Henceforth, faculty members with patentable inventions would have "full freedom of action" to exploit them but were quietly encouraged to refer them to the Research Corporation, which would make private arrangements with the inventor for royalties. The university would claim no rights, but it was understood that in recognition of Berkeley's role in productive patents, the corporation might make financial contributions "from time to time at its discretion" to support research on the campus. It was a healthy symbiotic relationship, for the Research Corporation would function as Berkeley's patent agent, and the university would become one of the corporation's most important beneficiaries.

With that relationship now established, and in his role as scientific advisor to Poillon, Frederick Cottrell began to educate himself about what was going on in Ernest Lawrence's lab. It took very little time for him to appreciate its significance. Writing Poillon on July 7, 1931, Cottrell flagged Lawrence's work as something that "may prove to be very big indeed." He described Ernest as "a man we should keep close track of. He is young enough and with a sufficiently good early start to go far. He not only does good work himself, but I have been particularly impressed with how much he manages to get out of his graduate students on the research problems of which he keeps a surprising number going full blast." Of these projects, he identified two—Sloan's X-ray tube and the accelerator, not yet known as the cyclotron—as "rather spectacular and important in their early developments."

Lawrence, who was already angling assiduously for the Research Corporation's generosity, had kept Cottrell apprised of progress on the eleven-inch accelerator, culminating in his July 17 letter announcing success "beyond our expectations" and identifying "finance" as the only limiting factor. Lawrence was now focused beyond the million-volt threshold. Twenty million volts was the next step. In search of the crucial piece of a new, larger accelerator, his eye had settled on an eighty-ton magnet that had been forged for a canceled project in China and now stood derelict in a vacant lot on the premises of its manufacturer, Federal Telegraph Company, in a San Francisco suburb. Since the power production of the cyclotron was directly proportional to the size of its magnet, this would be a major leap forward. But as Lawrence informed Cottrell, the magnet would require its own building and a brand-new array of high-powered oscillators and other accessories. The bill, he estimated, would approach $10,000. In an overt hint that he saw his work as the proper subject of funding from research-oriented charities of all kinds, he mentioned that colleagues had advised him to solicit the money from the Carnegie Foundation, which had "special funds for special research projects." Then he added, reassuringly: "I, of course, immediately thought of you."

The Research Corporation would take the plunge, drawing in as a partner the Chemical Foundation, a more affluent organization that had been formed by the US government in 1918 to manage German chemical patents seized as war booty. The two foundations contributed jointly $7,500 and the university the balance of a price tag that swelled to $12,000 by the time Ernest's new accelerator was ready to start operating in late 1931—even despite his having prevailed on Federal Telegraph to donate the magnet for free, in an early manifestation of his ability to extract more than mere cash from his well-placed patrons. As part of the university's contribution, Sproul turned over to Lawrence a two-story wood structure on campus, just across a narrow alley from Gilbert Lewis's chemistry building. There Ernest installed the new magnet, its pole faces specially milled to accommodate a new vacuum tank twenty-seven inches in diameter. He would christen the building, which had been slated for demolition but featured sturdy concrete underpinnings, as the Radiation Laboratory. (The name has "the advantage of brevity," he advised Sproul.) With that step, Ernest Lawrence became more than just another physics professor with a clutch of graduate students in his charge. He had staked out a personal fiefdom on campus. This was the beginning of what would be known, with even greater brevity, as the Rad Lab, and the launch of a new paradigm in scientific research.

Ernest's burgeoning relationship with the Research Corporation forced him to deal with the novel (to him) concept of patenting. At the outset, he shared the university scientist's traditional antipathy to the patent process, so redolent of commercialism and so distinctly unacademic. This provoked a sharp reaction from Poillon, who urged Cottrell to keep his protégé mindful of the financial value of his work: "If he is one of the men that we are going to make awards to from time to time, it seems that we should develop his protective instincts."

Lawrence reluctantly agreed to meet with the corporation's Los Angeles patent attorney, Arthur Knight. But the last of his resistance was dispelled by a flash of unwelcome news one day in September. Raytheon Company, a maker of radio tubes in Cambridge, Massachusetts, had applied for a patent on a machine that sounded suspiciously like his spiral accelerator. The word came from John Slater, MIT's physics chairman, who was passing on thirdhand information from a National Research Fellow in his laboratory, who heard it from a friend at Harvard working at Raytheon part-time. Slater wrote Lawrence that the application concerned "your proton merry-go-round . . . It never would have occurred to me to patent such a thing but I thought at any rate you would want to know." Lawrence replied that "it never occurred to me to patent the work we are doing either, and I am doing so only at the urgent request of the Research Corporation and the Chemical Foundation." Now he was alive to the risk that a private corporation might steal away with the rights to his invention if he did not promptly establish his prior claim. At Knight's direction, he solicited a statement from his old friend Tom Johnson (now working with Swann at the Bartol Institute), confirming that Johnson had witnessed Lawrence's examination of the Wideröe paper in the university library in April 1929 with his own eyes, and had heard him describe on the spot "his method for spiraling protons around in a magnetic field and increasing their energy at each half revolution." From Otto Stern, Ernest obtained a note attesting that "during my stay in Berkeley [at the] beginning [of] 1930 you had often talked to me about your experiments with the production of very fast light ions in the form in which you have published them now."

Lawrence's relationship with the patent bureaucracy would never be particularly comfortable or, for that matter, lucrative. The idea of patenting the cyclotron cut against the grain of an inventor whose interest lay more in the machine's proliferation across academia than in its commercial licensing, especially since its industrial utility was still hard to divine. A further irritant was the inability of patent office examiners to grasp Lawrence's work, which resulted in skeptical questioning of his patent claims. "It is apparent that the person who reviewed our patent application has not much of any idea of what it is all about," Lawrence grumbled to Knight at the midpoint of the two-year effort to obtain the cyclotron patent, which was finally issued in February 1932. Lawrence eventually accepted the patenting of scientific inventions as a necessary evil. "It is entirely proper, indeed almost a duty, for the research worker to bear in mind the commercial possibilities of his work, to the end that some of the fruits of commercial development will return in the support of his work," he acknowledged to Poillon in 1935. But he found the process "distinctly unpleasant"—especially following a prolonged and ultimately fruitless wrangle with the patent office a few years later over his methods for producing radioactive isotopes. After the cyclotron, Lawrence would not receive another patent until wartime, when federal officials demanded that his inventive process for uranium separation be protected legally—and the patent rights be assigned permanently to the government.

While the twenty-seven-inch accelerator was still on the drawing board, Lawrence and Livingston worked intently to improve the performance of the eleven-inch chamber that Livingston had built for the old magnet. It was a brass box, its vacuum maintained by the usual liberal applications of sealing wax. A breakthrough finally occurred in the late summer of 1931, while Ernest was on the East Coast to meet with the Research Corporation's board in New York and to propose to Molly Blumer, who was soon to begin work at Harvard for a master's degree in bacteriology. ("I am beginning to realize I have two consuming loves—Molly and research!" he confided effusively to his Yale friend Donald Cooksey before leaving for the East.)

Livingston took the opportunity of Ernest's absence to make a subtle change in the vacuum chamber's design on his own. Lawrence had decreed that the electrical field that delivered the sequential jolts to the spiraling protons should be present only in the gap between the two dees. The dees' interior was to be free of any electrical field, which he thought would interfere with the magnetic field keeping the particles in their spiral. They had fenced off the parallel faces of each dee with a grid of fine tungsten wires to prevent the electrical field from impinging within the dees while allowing the particle beam to pass through. Both men were flying blind, neither having had the slightest training in electrical field theory. But now Livingston, frustrated by the low current and energy of the beam, guessed that the grid was blocking more than they thought. He broke open the vacuum chamber and pulled off the tungsten grids by hand. This was a perfect expression of what would become the lab's practice of "cut and try" in its early years: while theory was still so rudimentary, often the only way to test one's hunches was to put them into practice and see what happened. In this case, Livingston would recall, he was operating "more or less intuitively—I didn't have any reason for doing it except that I had an urge to get something out of the way." The current and energy leaped higher instantly. On August 3, Livingston dictated a wire for the Physics Department secretary to send to Lawrence, care of the Blumer household in New Haven:

"Dr. Livingston has asked me to advise you that he has obtained 1,100,000 volt protons. He also suggested that I add 'Whoopee.' " Lawrence read out the message to the Blumer family circle that night, and then escorted Molly outside to the porch and proposed. She accepted, with the proviso that the wedding take place in the spring, after she received her degree. With her promise in hand and the prospect of further breakthroughs awaiting him in the lab, he returned to Berkeley in euphoria.

Chapter Four

Shims and Sealing Wax

Livingston's breakthrough did not conclude the effort to improve the eleven-inch cyclotron's performance. Over the following months, he and Ernest sweated to produce an even more energetic particle beam. This involved long hours, late into the night, with barely any respite. Every component of the machine was disassembled, rebuilt, and reassembled, under the impetus to reach a new voltage threshold. "We were heading for something big, definitely," Livingston remembered. "Lawrence said, 'We're making history.' He wouldn't let me take a minute's time off for anything else." Livingston, habitually a gloomy personality, was beginning to chafe under his mentor's relentless prodding. But Ernest's extroverted enthusiasm and the opportunity to be in on a groundbreaking discovery carried him along, for the moment.

After Ernest returned from the East and examined Livingston's grid-free dees, he realized that allowing the electrical field to leak into the dees actually focused the particle beam, thereby protecting the spiraling protons from obliterating collisions with the walls of the vacuum chamber. He explained the phenomenon to a surprised Livingston by sketching the lines of electrical force on a blackboard. "It was Lawrence's genius for understanding a new phenomenon when he had only a glimpse of it," Livingston marveled years later. Livingston had discovered electrical focusing by sheer ingenuity and serendipity; Lawrence recognized the underlying principle, which enabled him to incorporate it into subsequent designs.

Livingston also was perplexed about his inability to coax the protons to more than seventy-five accelerating steps, a mysterious constraint on the machine's effectiveness. He suspected that the obstacle was the relativistic limit predicted by Einstein's theories: as a particle gained velocity, it also increased in mass, which eventually destroyed its ability to remain synchronized with the oscillating electrical field. If that were so, the reign of the cyclotron as science's most powerful and effective atom smasher might be a brief one. Lawrence acknowledged the presence of some obstacle, but instinct told him they were nowhere near the relativistic limit. Instead, he concluded, irregularities in the magnetic field were driving the proton beam out of resonance. The solution he and Livingston invented was to insert strips of metal, or shims, at certain points in the gaps between the vacuum chamber and the magnet poles to compensate for those irregularities and "shape" the field.

Whether it was Lawrence or Livingston who first came up with shimming is unknown, for both would claim credit. But they spent long hours together testing different shapes, sizes, and placements of the shims in an endless process of cut and try: randomly inserting circles, squares, rings, polyhedrons of metal in the gap, like auto mechanics trying to balance a wheel with lead weights by sight. Eventually they discovered that an elongated teardrop shape, with the wide end oriented toward the magnet's center, quadrupled the maximum potential acceleration to 300 steps, or 150 complete circuits. This milestone was reached on January 9, 1932, when they placed 4,000 volts on the dees and, after 300 kicks, ended up with protons at 1.22 million volts. "Here again, experiment preceded theory!" Lawrence declared. His exuberance almost matched the moment when he had hatched the cyclotron idea itself. "As the galvanometer spot swung across the scale, indicating that protons of 1-MeV energy were reaching the collector," Livingston recalled, "Lawrence literally danced around the room with glee."

News of the discovery spread campus-wide. "We were busy all that day demonstrating million-volt protons to eager viewers," Livingston recalled, leaving it unsaid that what the visitors witnessed was only indirect evidence of the achievement—a needle swinging from one end to the other on an electrical meter. The news created a sensation off campus, too, with stories in all the local newspapers and not a few national publications, reinforcing the impression taking hold in the international physics community that something exceptional was happening in this former backwater on the West Coast. The very day before Lawrence and Livingston breached 1 million volts, the Princeton physicist Joseph Boyce had reported to his colleagues: "The place on the coast where things are really going on is Berkeley. Lawrence is just moving into an old wooden building . . . where he hopes to have six different high-speed particle outfits." He mentioned the eleven-inch, the new twenty-seven-inch accelerator soon to make use of the Federal Telegraph magnet, Sloan's linear accelerator for heavy ions, a Tesla coil, and a Van de Graaff generator (which was never actually erected, though a room was reserved for it). "On paper," Boyce continued, "this sounds like a wild damn fool program, but Lawrence is a very able director, has many graduate students, adequate financial backing, and in his work so far with protons and mercury ions has achieved sufficient success to justify great confidence in his future." The building blocks of Big Science were being moved into place.

But the euphoria proved to be short lived. In late April word arrived of another striking achievement from the tatty corridors of the Cavendish: physicist John Douglas Cockcroft, working with a young graduate of Dublin's Trinity College named Ernest Walton, had disintegrated a lithium nucleus by bombarding it with protons at energies a fraction of those Lawrence had achieved, and a fraction of those Lawrence thought necessary to do the job. Lawrence had invented a superb instrument for smashing atoms, but while he was preoccupied with making the tool better, the quintessential small-science lab had stolen the prize from under his nose.

Cockcroft was thirty-four, a man with the mild eccentricities that American physicists had come to expect in their cross-oceanic colleagues. He was so intensely devoted to experimental physics that even his laboratory fellows considered him hopelessly aloof: "To a superficial acquaintance like myself," remembered a Cambridge experimentalist, "it appeared that when the ice had broken, a lot of cold water would be found underneath." But Cockcroft was a thoughtful physicist in the style treasured by Rutherford, especially in his resourcefulness with equipment, the Cavendish's hallmark. It was Rutherford who had assigned Cockcroft the task of splitting the atom and personally paired him with Walton.

Cockcroft's approach differed diametrically from Lawrence's. Instead of producing beams with high energies but a low current like the cyclotron's—that is, protons that were speedy but few in number—his goal was to produce protons with moderate energies but in great profusion. This method derived from the theories of the Russian physicist George Gamow, who deduced from quantum mechanics that the occasional particle from even a moderately energetic beam could penetrate the nucleus. Produce protons in sufficient quantity, Gamow suggested, and sooner or later a lucky bullet would find its mark.

Cockcroft and Walton aimed to generate a mere 300,000 volts, which they considered attainable with an array of electrical capacitors that would charge up in series and discharge in parallel, like a giant weight winched up in steps and then dropped abruptly to the ground. Their voltage multiplier was arranged as several vertical tubes in an upper story of the Cavendish, with the main accelerator tube inserted through a hole drilled in the floor and terminating in a wooden hutch in the basement. The experimenter folded himself into that cramped space and fixed his eye on a scintillation screen capturing flashes from a bombarded lithium target. On April 16 Cockcroft and Walton nervously summoned Rutherford to the basement so he might see for himself a scintillation pattern they thought heavily suggestive of alpha particles. After a few minutes of viewing, the discoverer of alpha rays crawled from the hutch and declared, "I know an alpha when I see one." The conclusion was inescapable: Cockcroft and Walton had fired a proton into lithium, which in its most common form has a nucleus of three protons and four neutrons. The yield was two alpha particles, each with two protons and two neutrons.

More than a decade earlier, Rutherford had fired an alpha particle at nitrogen and knocked a proton free; his associates had attempted the converse by firing a proton at lithium, and for the first time had split the atom. It was another instance, and not the last, of Cavendish intuition and resourcefulness outplaying labs with more soaring ambitions, superior resources, and, perhaps, a touch less vision.

Lawrence declared himself pleased for his fellow inventors. But plainly he was rankled at having been outmaneuvered on a quest so easily within reach. "We weren't ready for experiments yet," Livingston recalled mournfully. "I had built the machine but had not included any devices for studying disintegrations."

Lawrence moved rapidly to recover from the defeat, albeit at long range. He had wed Molly Blumer on the Yale campus on May 14, a day before her twenty-first birthday and only a few days after the Cockcroft-Walton report reached the United States in the pages of the journal *Nature*. From his honeymoon cabin on Long Island Sound, he wired his graduate student Jim Brady to help Livingston bombard a lithium crystal in the eleven-inch accelerator. Brady had earned his PhD and already accepted a faculty post at Saint Louis University. But Lawrence artfully persuaded Brady's new dean to let him stay on at Berkeley through the summer to acquire a last layer of polish while collecting his first paychecks from Saint Louis—a canny arrangement that gave the Rad Lab an extra pair of hands, courtesy of the SLU budget. Eager to jump into what was now the hottest line of research in nuclear physics, Brady checked out a lump of lithium from the Chemistry Department storeroom. Two of Lawrence's friends visiting from Yale for the summer, Don Cooksey and Franz Kurie, offered to help, and since Cooksey also possessed an expertise in hand-manufacturing Geiger counters, Brady readily accepted.

While the Rad Lab staff worked at reproducing the Cockcroft-Walton results, the newlyweds made their circuitous way back to California, visiting with friends in New York and Chicago and eventually stopping at the Lawrence family homestead in Canton. There Molly came face-to-face with the difference in lifestyles between the cosmopolitan Blumers of New Haven and the Lawrences of pastoral Canton. When she pulled out an after-dinner cigarette, a habit unremarked in the Blumer household, the scandalized Gunda Lawrence hastily drew the window shades, lest her neighbors remark on the sight of a woman smoking in her home. Carl joined Molly with a cigar, meeting his wife's scowl with his customary defense, "Everyone has to have some bad habit."

Ernest and Molly moved into a rented house on Berkeley's Keith Avenue, in a hilly neighborhood just north of the campus. Brady had decamped for his own wedding and his new post in St. Louis; Cooksey and Kurie had returned East. The lithium experiment was in the hands of a graduate student named Milton White, who now got his first exposure to Ernest Lawrence in the throes of a mania.

"The place was beginning at this point to catch fire," White would recall. Lawrence was already occupied with the twenty-seven-inch cyclotron rising in the wooden Rad Lab. But he would appear every day in LeConte Hall, sometimes late at night, to peer over White's shoulder and pore over his results. "He'd come in at two or three in the morning wanting to know why we hadn't gotten more data, what's holding us back, and he was really putting the pressure on." Adding to the tension was what White perceived as "a certain amount of shamefacedness that we hadn't been the first" to disintegrate a nucleus—the more so because the overlooked pieces of the Rad Lab system had been those easiest to design, the counters and detectors. The accelerator was the harder technology, yet *that* they had mastered. Had the lab paused to direct itself at this obvious experimental goal, it would have been the first to disintegrate lithium, not a sheepish second.

The prickliness of the machine's behavior did not help anyone's mood. The eleven-inch accelerator's magnet, which was not water cooled as the new magnet would be, could run for only about an hour at maximum energy before overheating. Then it needed thirteen hours to cool down fully. White would run the machine for the permitted hour starting at one in the afternoon, and then power down, returning for another hourlong run at three in the morning. The process would resume at five o'clock the next afternoon. Beleaguered by sleeplessness, White staggered around campus like a zombie, not knowing day from night.

Still, once yoked to a concrete experimental goal, the eleven-inch proved itself worthy of the task. After only three weeks of bombardments under Lawrence's intent gaze, White had enough data to publish. The Rad Lab's letter reporting its own disintegrations of lithium went out to the journal *Physical Review* on September 15 and appeared in print two weeks later over the names of Lawrence, Livingston, and White. Their report downplayed the fact that they had not really broken new ground but largely had confirmed the Cockcroft-Walton discovery; but they did emphasize that by doubling the Cavendish's energy levels, they had shown that the emission of alpha particles continued to increase with the energy of the bombarding protons, a not-inconsiderable extension of the Cavendish results.

Lawrence's move into the wood clapboard Rad Lab in 1932 marked an administrative, financial, and intellectual break between the lab and the Berkeley Physics Department, its putative parent. The gulf would only grow wider as his stature rose. Raymond Birge, who became department chairman upon Elmer Hall's passing that same year, was fond of remarking with a wry jocularity, "I don't know what goes on over there in that Radiation Lab." Still, he retained a deserved sense of pride in having brought Lawrence to Berkeley in the first place.

The most obvious sign of the difference between the Physics Department and the Rad Lab was the divergence in their research budgets. The department's budget from 1931 through 1933 averaged about $11,000 a year, mostly for apparatus and supplies, falling to $8,000 in the Depression year of 1934. By contrast, the Rad Lab's spending kept rising—from

$17,670 in 1933 to $22,000 in 1936. Those figures did not count the salaries of Lawrence ($5,000 in 1932) and Donald Cooksey, who emigrated from Yale in 1932 to become functionally (if not yet formally) the lab's associate director, at $3,000. Their pay was billed to the departmental budget, as were teaching assistantships and other small stipends paid out to the platoon of graduate fellows who were now tending, in shifts, the cyclotron humming away on the new Rad Lab's ground floor.

Lawrence exploited to the limit the free labor of graduate students, an important key to low-cost operation of the enormous piece of capital equipment he was erecting. In 1937, a typical year, the Rad Lab listed seventeen postdoctoral physics fellows on its staff but paid the salaries of only two. The others were sustained on stipends from donors such as the National Research Council and the Rockefeller Foundation, or were paid out of unallocated grants made to the Rad Lab by the same bodies. Lawrence's ability to run a program costing tens of thousands of dollars by juggling contributions from a dozen sources made the Rad Lab virtually immune to the Depression-related budget cuts afflicting every other part of the university. From 1932 through 1939, the lab's overall staff shot up from ten to sixty, but the number of Rad Lab personnel paid out of state funds never exceeded ten. After 1933, Lawrence even tapped federal New Deal programs such as the Works Progress Administration and the National Youth Authority, which supported as many as fifteen researchers a year.

From this raw material, Lawrence was creating a cohesive research organization. His genial personality provided some of the glue, but so did his single-minded devotion to improving the accelerator and his receptiveness to all varieties of scientific contribution; the Rad Lab was soon populated with chemists, biologists, medical scientists, and engineers in what was, for academic institutions of the time, a uniquely interdisciplinary atmosphere.

The give-and-take of scientific discussion was fostered by another Lawrence innovation: the Journal Club, a weekly colloquium to which all the Rad Lab staff and visitors from other departments were invited. Every Monday evening at seven thirty sharp, the accelerators were turned off and the staff convened in the LeConte Hall library. Ernest managed the agenda, which he did not announce in advance, from a big red leather chair. The evening's speaker might be a graduate student assigned to explicate a recent paper from Europe, or a visiting luminary discussing his own work. The Journal Club proved to be an ideal way to keep the lab abreast of the latest advances in physics, but over time the agenda reflected the lab's own expanding prominence. At its inception, the topics for discussion almost always concerned research done elsewhere; by 1936, they almost always concerned research carried out at Berkeley.

Meanwhile, the wealthy and influential personages of Northern California, many of whom had been educated at Berkeley and supported the university generously, began to eye a new, high-profile candidate for their philanthropy. Bay Area industry began to perceive the value of a relationship with Lawrence and his increasingly famous machine. To obtain radio tubes for David Sloan's X-ray machine, Lawrence appealed to Federal Telegraph, which allowed itself to be bargained down to a steeply discounted price of $225 per tube, deducting the discount as a charitable contribution. Only a few commercial enterprises showed the fortitude to resist Ernest's importuning. One was the giant utility Pacific Gas and Electric Company. Asked in September 1931 to donate 120,000 kilowatt-hours to run the cyclotron for a year, the company's president, August F. Hockenblamer, took a firm line. "This company is the second largest taxpayer in the state of California and as such contributes very largely to the funds of the university," he lectured Berkeley's research dean, A. O. Leuschner. "It seems to me the cost of experiments coming in the category of 'pure science' ought to come out of the funds of the university." PG&E was willing to put up extra money for research it might be able to exploit commercially, such as studies of "the utilization of electricity on the farm," he wrote. The cyclotron did not qualify.

Despite the inflow of cash and capital assets, Lawrence kept a parsimonious grip on his kingdom. The sources and volume of funding were concealed from the staff, who were under constant pressure to work frugally: one day, Jack Livingood, who came from Princeton in 1932, was instructed by Lawrence to collect the spare pieces of solder he had been routinely discarding, which prompted him to spend a full morning on his hands and knees picking up nuggets of scrap metal from the concrete floor. Don Cooksey, who settled into the role of green-eyeshaded keeper of the lab budget, was known to conduct hourlong inquiries over such matters as which heavy-duty battery to purchase, a decision in which the expenditure of a single dollar might be at stake.

As was common in academia, Rad Lab researchers lived a poverty-row existence; Molly Lawrence, who considered her lifestyle "quite affluent" on Ernest's annual salary of $5,000, was appalled to discover that Milton White was sharing a $40-a-month apartment with three other young men in order to stretch the $600 he received as a one-year teaching fellowship. The roommates kept a stewpot simmering perpetually on the stove, tossing in a scrounged vegetable when they had it or a scrap of meat when flush, and squabbling over spare potatoes. So every Monday night before Journal Club, Molly invited members of the penurious staff for dinner (on a rotating schedule), serving a standard menu of roast beef and potato, salad, and apple pie. The staff's living standards would improve in time, but not until the appearance on the scene of an especially well-heeled patron, the United States government.

It was during this period that Ernest acquired the reputation of a relentless taskmaster in command of a growing brigade; the recurring term in personal reminiscences of the Rad Lab is "slave driver." Those who were entranced by his energy and vision thrived under the pressure; everyone else could only submit and simmer—or go. At home Ernest kept his bedside radio tuned to the frequency of the oscillator, which was so poorly shielded that its interference easily reached the quarter mile or so to his home; if it fell silent for even a few minutes, he would be on the phone to the lab—or even materialize in person—demanding to know whether the machine was out of commission or simply had been left unattended while the late shift slipped out for a beer. Molly became accustomed to having their Friday-evening movie dates hijacked by the machine. The first time, Ernest said casually as they left the house, "Do you mind if I drop by the lab to see how the boys are doing?"

"Of course," she remembered later, "the boys weren't doing so hot; there was another of those pesky leaks in the vacuum, and we never did get to the Paramount. Finally, I found an old chair in the corner and sat down."

There would be many evenings like that. But Molly, as the daughter of a physician, understood that she was witnessing the birth of team science; a new paradigm of a research lab. "There appeared to be no 'pecking order,' nobody really in command—not even Stan Livingston, the pioneer member of the crew and undoubtedly the most experienced operator," she recalled. "Every one of the five or six men present, including Ernest, was busy getting his hands and his lab coat dirty. Every now and again they would go into a huddle, like a football team, to call the signals for the next play. But every man was a quarterback; a suggestion from the newest and youngest recruit would be considered and acted upon with as much respect as one proposed by Stan or Ernest. The whole scene was a far cry from my idea of scientific research."

This egalitarian atmosphere helped to moderate the demanding routine of the average Rad Lab staff member. So too did Lawrence's bracing personality. "He was warm and friendly, and he had a feeling of drive and latent accomplishment I've never seen anywhere else," recalled Malcolm Henderson. The son of a Yale Medical School professor, Henderson had grown up with the Blumer girls and earned his doctorate at the Cavendish before joining the Rad Lab in 1932.

Ernest's imperturbable demeanor completed the picture. Displays of temper were almost unheard of; no utterances of profanity are recorded in the extensive recollections of his students and colleagues. A very large frustration—the failure of the machine during the visit of an important donor, for example—was required to provoke him to a maximal "*Oh, sugar!*"

• • •

After his return from the East Coast with his bride, and the completed installation in the new Rad Lab of the magnet and vacuum chamber for the twenty-seven-inch machine, Ernest began to leave a larger share of the manual work to his grad students. A few months earlier, while still a bachelor, he had described his routine as "working day and night with Livingston and Brady on our big magnet and associated apparatus," and confessed to having "neglected everything else, even my fiancée." Now married life prompted him to place a higher value on his free time from fund-raising and managing the lab. Sundays customarily meant a competitive game of tennis. "He'd come in, and the machine would be busted on a Sunday morning," recalled Livingood, "and he'd say, 'Well, I'm going out to play tennis, and you people have to fix it.' "

Yet under his leadership, the lab was coalescing into a unit. There were shared trips to Yosemite National Park or down to the shore at Carmel, sometimes in Lawrence's sedan, graciously loaned out if he was staying home. The entire lab convened monthly for staff dinners at DiBiasi's, an eatery just over the Berkeley city line where the food was of middling quality but tolerance high for a boisterous group that grew larger with every appearance. The lab itself was always bustling: David Sloan's X-ray tube working in one room, a linear accelerator in another, and in a third room the proton accelerator—which by 1935 was commonly referred to as the cyclotron, in what Ernest described as "a sort of laboratory slang." (By 1939, the name was official enough to be used in the citation for Lawrence's Nobel Prize.) But the quirkiness of the machinery produced a lot of downtime. "Everyone had to wait for sufficient vacuum in the cyclotron tank," recalled Henderson, whose own peculiar mode of relaxation was to play the bagpipes. "If it were nice weather, then there might be touch football outside or time for a run over to the Faculty Club for billiards or pool." Yet once there was work to do, the horseplay stopped.

One thing that did not change was the lab's slapdash approach to scientific procedure, including safety procedure. For scientists who should have been familiar with the behavior of high-tension electrical fields and with the possible effects of radioactivity on living tissue, the staff worked with surprising carelessness while in close proximity to these unforgiving physical phenomena.

Jack Livingood had the closest call. One morning in 1934, he was perched atop a ladder, testing the electrical tuning coil of Sloan's X-ray machine by poking at it with a wooden yardstick fitted with a nail at the end. It was routine procedure for the X-ray team, but this time a crackling spark jumped from the coil to the stick, sending 16,000 volts into Livingood's thumb, through his body, and out to the ground through the nails in his shoes. He was jolted off the ladder onto the concrete floor, where he lay writhing on his back.

Malcolm Henderson, working on the cyclotron in the next room, was used to hearing the X-ray workers shout to make themselves heard over the thrum of the tube's oscillator. This time the sound had "a different color," alarming enough to bring him running. He saw the prostrate Livingood surrounded by a crowd that included a dumbfounded and fretful Ernest Lawrence. A doctor's son, Henderson checked Livingood's pulse and packed him into a car for a trip to the hospital. Livingood stayed there for ten days, recovering from severe burns on his hands and the soles of his feet.

Ernest was severely rattled, reporting to Cooksey that he had been certain that Livingood was "a goner." He hoped the experience would "impress the fellows with the ever-present dangers of the high-voltage equipment," but allowed that "it isn't practical to make the equipment foolproof when it is being rapidly altered and the only procedure is one of intelligent caution." He claimed to have consistently warned the staff of the danger—"so much so that I see myself as an 'old crab,' but now maybe warnings will be given more respect." The truth, of course, was that any warnings he may have issued about the dangers of the lab's physical environment were drowned out by his endless demands and tight deadlines for improvements to the machine.

The staff's indifference to the energy fields produced by the cyclotron and X-ray tube was even more remarkable. They took only minimal precautions against the copious alpha and gamma rays emitted by the equipment; David Sloan enveloped part of his X-ray tube in thin sheets of lead, but the ease with which its rays penetrated the shield was interpreted as a gratifying sign of the tube's great efficiency rather than as a reminder of its dangers. The effect of radioactivity on health was hardly unknown—not least because of widespread press coverage of the poisoning of working women at U.S. Radium Corporation in the 1920s. The workers, who marked watch dials with luminous radium paint, customarily licked their brushes to a fine point, ingesting the radioactive substance. The interest shown in the Sloan apparatus by medical researchers in San Francisco and New York arose, after all, because X-rays were known to destroy cancer cells, as long as they could be focused precisely enough to leave normal cells unharmed.

These issues were typically treated with graveyard humor in the Rad Lab. Molly Lawrence tried to remain silent about her own fears—one of the guests at her own wedding, a physicist named Joe Morris, had arrived with his hand enveloped in a black glove to conceal radiation burns—until the night a breakthrough with the Sloan tube kept Ernest out until after three in the morning. Unable to reach him by phone and confined at home because Ernest had the car, she let her imagination magnify "all the horror stories about what excess radiation might do to the men." When her husband arrived home, she met him at the door and demanded, "Are we going to have a family or aren't we?" Her fears were not fully assuaged until she became pregnant early in the new year and delivered a healthy boy, John Eric Lawrence, in mid-October. There would be five more Lawrences, all perfectly healthy at birth.

The men continued to work heedlessly in a miasma of radioactive emissions, including neutrons, which Lawrence recognized as especially active on human tissue. "We have been giving ourselves undesirably great exposure to the neutrons," he informed Cockcroft in 1935, acknowledging that the Rad Lab had determined that the particles "are about ten times as lethal as X-rays" and that he had decided to move the cyclotron's control panel "further from the magnet with suitable material interposed." The "material" was cans of water.

The staff's insouciance may have reflected the infrequency of acute injuries. A researcher occasionally might be winged by a metal tool sent airborne by the magnet's powerful field—one piece of flying shrapnel even nipped off the end of Ernest's finger—but more serious physical effects remained latent, not to emerge until many years later, when exposure to Rad Lab equipment was impossible to pinpoint as the cause. Thus when Dean Cowie, who had arrived at the Rad Lab as a twenty-year-old graduate student in 1935, developed cataracts at the alarmingly young age of thirty-two, there was no way to determine if the condition resulted from his frequently scrutinizing the Berkeley cyclotron's beam with the naked eye; or from his later cyclotron work with Merle Tuve at the Carnegie Institution of Washington; or even from an undiagnosed predisposition to eye disease. His Rad Lab colleagues could only say that they recalled Cowie as "neither particularly careful nor careless, but just average." The cost of operations to restore Cowie's eyesight was eventually covered by the Carnegie Institution.

In assembling his staff, Lawrence was organically developing a new paradigm of the research lab, compounded from his insatiable demand for brainpower and manpower, and his willingness to employ anyone, no matter his formal scientific discipline, who might be willing to keep the accelerator running.

Some of his new employees were distinctly oddballs for an academic laboratory. One was Commander Telesio Lucci, a courtly fifty-six-year-old who owed his title to service in the pre-Mussolini Italian navy. Lucci had come to America to serve his nation as a consular officer in Pittsburgh (among other postings), and presently fetched up in Berkeley with his American wife, Winifred. An amiable gentleman with an endless store of personal tall tales and a vehement hatred for *Il Duce* that he would happily air at the slightest prompting, he was serving Leonard Loeb as a general dogsbody when he became enthralled by the goings-on at the Rad Lab and offered his services as a spare pair of hands for free, which Ernest considered an irresistible price. Lucci was adept with a screwdriver but innocent of any physics knowledge; his particular pleasure was to throw the heavy knife switch that turned on the accelerator, an operation that delivered an electrical surge occasionally strong enough to knock out power to the lab, the campus, or even to the entire city of Berkeley. Convinced that the problem was a too-abrupt closing of the circuit, Lucci would move the switch with elaborate delicacy, to the amusement of his colleagues.

But it was the presence of scientists from outside physics that marked the real departure from custom. Like most other major universities, Berkeley tended to think of chemists, biologists, physicists, and engineers all as inhabitants of discrete sandboxes. That began to change as Lawrence embraced these strangers within the Rad Lab. To a great extent, the outreach was dictated by necessity. James Chadwick's discovery of the neutron in early 1932 posed questions about atomic structure and behavior that could not be fully processed without an understanding of chemistry in addition to physics. As for the biological sciences, it was an inescapable fact that research foundations were far more eager to fund medical studies than basic scientific research; this was demonstrated to Lawrence by the interest shown in Sloan's X-ray tube by two major cancer research institutions: Columbia University's Institute of Cancer Research and Berkeley's own medical school, located across the bay in San Francisco. Financial backers of both institutions—the Chemical Foundation in Columbia's case and William Crocker, a banking and railroad magnate, in Berkeley's case—seemed eager to pony up funds to cover Sloan's further research. How eager was demonstrated by Crocker during a visit to the Rad Lab. A regent of the University of California, Crocker had offered $12,000 to fund Lawrence's construction of a Sloan tube for the medical school. But the tube failed on the day that Crocker came out to the Rad Lab to see it in action. ("Oh, for heaven's sake!" Ernest exclaimed.) As a substitute demonstration, Ernest ordered the device to be dismantled so he could explain the function of each part to the wealthy donor. Crocker got the point. "How much is needed to make it work?" he asked.

Lawrence had spent a difficult year cadging money from philanthropists and foundations for a proton accelerator, the purpose of which had to be explained to them in the abstract terms of basic science. He was stunned by Crocker's question, which hinted at the ease with which he could extract funds for a device with presumed, but far from certain, medical applications. His insight was reinforced by a demand from the Research Corporation's Howard Poillon that he bring the X-ray tube to a patentable stage without delay and without disclosing his design to outsiders. "I have warned [Sloan and Lawrence] about directing the attention of others to it lest . . . the patent situation becomes very cloudy," he informed his patent lawyer, Arthur Knight. Ernest did not object, now that the tube's capacity to generate research funds appeared to outstrip that of the cyclotron. "I am told there is a very big market for such deep therapy outfits," he wrote Poillon. "Go ahead as rapidly as possible with the commercial development, for I am afraid that the General Electric Company and others will be entering the field and harass our progress."

Crocker's donation for the medical school project solved one of Lawrence's immediate problems: keeping Stan Livingston on hand after the expiration of his one-year instructorship at Berkeley. Ernest simply assigned Livingston to help Sloan install the X-ray tube across the bay. In effect, Livingston was slotted into the same role on the X-ray tube that he had played on the eleven-inch and twenty-seven-inch cyclotrons. Although his salary would be paid for the coming year from the medical school budget, he would be near enough to be summoned for work on the cyclotrons when needed.

Livingston threw himself into his new task with characteristic single-mindedness, even training himself to operate the tube on some early patients. Many years later, he recalled his delight at having "had it running in six months and producing one-million-volt X-rays, the first in history." He surely appreciated the assignment for another reason: it was his first chance to move out of the shadow of Ernest Lawrence.

Stan Livingston was one of the first scientists, though not the last, to discover that the teamwork demanded by Big Science might not suit them. Although it was not unusual for even talented scientists to become subservient to those of greater talent or stronger personality—as were the assistants, say, to Ernest Rutherford—researchers were still imbued with the self-image of "the lone scientist in pursuit of the truth." That romantic phraseology came from Robert R. Wilson, whose preference for solitude sprang from his childhood on the Wyoming range. At the Rad Lab, Wilson contributed ingenious improvements to the cyclotron, among them a rubber gasket allowing probes to be inserted and withdrawn from the chamber without compromising the vacuum. But he bristled at subordinating his personal quest for the truth to the priorities of the group, which included, tediously, the maintenance of the machine. "As the youngest member of the laboratory, it was I who had to yield and to watch the team go its way rather than my way," he reflected later. "It was [Lawrence] who was independent and creative. In some degree, the members of the team just carried out his ideas." The only way he could maintain his own independence and creativity was by working very late at night, when he could have the run of a great laboratory almost to himself. After receiving his doctorate at Berkeley, Wilson fled to Princeton, which then was more accommodating to solitary inquiry in the old style.

It was Stan Livingston who chafed most deeply under the Lawrence regime. Despite his stature as a charter member of Ernest's inner circle—after all, he was one of Lawrence's first PhDs and the designer of his first working cyclotron—Livingston remained aloof to the camaraderie of the Rad Lab, seldom a participant in the off-hours horseplay. Having come to Berkeley at the age of twenty-five, he was the other researchers' contemporary, yet he struck them as older, more restless. "I don't know what makes him so prickly," Malcolm Henderson would remark later.

Livingston was an indefatigable worker, keeping such long hours while Lawrence was on his honeymoon in 1932 that physics chairman Elmer Hall, alarmed at his bedraggled appearance, ordered him to "get away from Berkeley and from his problem" for a spell. Hall informed Lawrence: "Livingston looks tired, and I suggested to him yesterday that he ought by all means to take at least two weeks' vacation. I fear he will go 'stale' . . . if he plods on without intermission." But Livingston had just had his brainstorm about removing the grids from the dees, and taking a respite at such a moment fit neither his personality nor Lawrence's calendar. He worked hard through the summer, and harder when Ernest returned. They labored together so closely for such long, unbroken hours that it was understandable that their contributions to the finished product blurred into one another.

Livingston considered himself an equal partner with Lawrence in the cyclotron's early development. Was it not his name next to Ernest's on the very first article ever published describing a working cyclotron (in the *Physical Review* of April 1932), the device pictured in its pages his own handmade eleven-inch brass box? Whether the success of the project owed more to Lawrence's vision or to Livingston's design technique sometimes seemed debatable, even if in retrospect it was self-evident that without Ernest Lawrence, the Berkeley cyclotron could not have been born. About that, there was never any doubt in Lawrence's mind, at least.

To most of his students and colleagues, Ernest seemed generous with credit to a fault. He often allowed their names to come first in journal articles announcing new Rad Lab discoveries, sometimes even refusing any byline whatsoever—both practices almost unheard of in major scientific laboratories led by an eminent figure. Most of the staff seemed to think the distribution of credit at the Rad Lab was fair. "Ernest had enough credit to go around, and we all got it," reflected Henderson. "I know I got fully as much as I deserved."

Not so Stan Livingston. As the cyclotron attracted more attention nationwide, and Lawrence spent more time escorting luminaries through the Rad Lab, basking in their compliments and sometimes obliviously walking them silently past the laboring Livingston, the matter of credit began to grate. One day he broached the issue with Ernest face-to-face, explaining that he considered his contribution unappreciated.

The coldness of Ernest's reply was shocking. "I'm running this myself," he said. "If you're unhappy, why, feel free to go on any other project. Because I can get any number of graduate students to do what you are doing."

Staggering out of Ernest's office, Livingston ran into Jim Brady, whose seniority at the lab matched his own. His face ashen, he recounted the interview in all its excruciating detail. Brady mouthed some sympathetic words, but privately, he agreed with the boss. Lawrence provided the vision, he reflected later, and "Livingston was a pair of arms." This was perhaps unduly harsh; Lawrence's effort to secure Livingston another year's stipend through Crocker's medical school donation attested to the esteem he held for this particular pair of arms.

But Stan Livingston plainly could no longer survive in the psychological environment of the Rad Lab, where the boss's vision was realized through the resourcefulness of the staff—but where no one was permitted to forget that Ernest's vision was the animating force. Leonard Loeb, an outsider secure in his own self-importance who had the luxury of pondering the dynamics of the Rad Lab from a safe remove, observed: "Certain things will occur when a man like Ernest Lawrence with his . . . unconscious and clean enthusiasm and leadership in general draws into his orbit the more suggestible and the weaker members of the scientific community." The most brilliant researchers drawn into the Lawrence orbit—Nobel-caliber scientists such as Luis Alvarez, Edwin McMillan, and Glenn Seaborg—would find their own methods to exploit Lawrence's laboratory resources to make their own reputations independently. Lawrence recognized that their accomplishments added luster to the Rad Lab, and, accordingly, gave them as much freedom as they needed. Others, like Wilson, absorbed what knowledge and experience they needed from the Rad Lab and launched successful careers elsewhere. Still others, like Donald Cooksey, settled into long and fruitful careers as acolytes to the leader. The research paradigm Ernest was creating was novel for everyone.

By Loeb's reckoning, Livingston was a personality type destined to occupy an uneasy limbo within this continuum. "Livingston unquestionably belonged to the fringe of the suggestible," Loeb observed to Lawrence's authorized biographer, Herbert Childs, "but had sufficient individualism in himself to desire to escape . . . The bitterness came in that Livingston had worked too hard and for too many years as a cooperator . . . and became acutely aware of his contributions. Being on the fringe of belonging to the leadership group and unquestionably under Lawrence having achieved beyond his normal capabilities, he suffered a serious setback to his morale when he . . . realized that after all the credit went to Lawrence. As society is constituted, it could not have been otherwise." It is telling that even after Livingston left Berkeley in July 1934 for a faculty post at Cornell University—another American physics backwater that was about to leap up in stature with the appointments to its faculty of Hans Bethe and Robert Bacher, both physicists of the first rank—he remained a devoted member of what became known as the Cyclotron Republic. Cornell had hired him primarily to build its cyclotron; its eleven-inch dimension was dictated by the school's meager resources, but Livingston brought it the distinction of building the first successful cyclotron in the United States outside Berkeley. More broadly, he became the unofficial curator of cyclotron history and an assiduous chronicler of the spread of the technology, keeping Ernest supplied with frequent reports from the field.

Yet he never lost the conviction that his role in the machine's development had been overlooked. For this, he chose to blame Donald Cooksey, who, as the longtime associate director of the Rad Lab, served as the official keeper of the Lawrence flame. Cooksey "brought a certain idolization of Lawrence into the laboratory," Livingston observed more than thirty years after his departure. "Cooksey's handling of the story of the early days caused the new generation to be unaware of what had been done in those early days and to think that Lawrence must have done all of it with his own hands . . . I think he did harm to history."

Livingston may not have recognized that the Rad Lab already had acquired unique prestige at the moment that he challenged Lawrence for credit. If Lawrence's attention was still devoted more to perfecting the machine than to exploit its powers, both its performance and its results already were being noticed around the world. Physicists of international standing were showing up in Berkeley to see the marvel in action and contemplate what they might accomplish with a cyclotron of their own. Berkeley and its remarkable style of scientific inquiry was on the rise, and that was Ernest Lawrence's doing—mostly.

For there was one other factor: Ernest had forged a partnership with an outstanding young scientist who was building his own international reputation as a theoretical physicist. He was Lawrence's great friend, yet as different from the South Dakota boy as the moon is different from the sun. His name was Oppenheimer.

Chapter Five

Oppie

It is the emblematic photograph of Ernest Lawrence and Robert Oppenheimer together in the full bloom of their friendship, long before their personal relations became soured by rivalry, suspicion, and politics. Snapped at Perro Caliente ("Hot Dog"), the New Mexico ranch Oppenheimer leased with his brother, Frank, it is undated, but the time must have been the early 1930s. The friends are both wearing riding boots encrusted up to the calf with desert sand from a recent outing on horseback. Ernest stands evenly balanced on the balls of his feet, like a youthful Mark Antony, in command of his surroundings; he wears a neat checked jacket over a V-necked sweater and a knotted tie, grinning broadly at the camera. Robert slouches against the fender of his Packard automobile, his shapeless dark jacket covered with dust, his hair an unkempt mop, his eyes glaring mistrustfully at the lens from under hooded brows.

What was it that united these men from irreconcilably divergent backgrounds? To those who knew them both during the quarter century in which they joined to create Big Science and dominated American physics, Ernest Lawrence and Robert Oppenheimer were a most enigmatic pair: Ernest, the offspring of Lutheran schoolteachers, raised in the upper Midwest and educated at a land-grant college; Robert, the scion of a Jewish merchant family, the product of Harvard and the great European temples of learning. Lawrence was broad shouldered and athletic (Robert would marvel at his "unbelievable vitality") and always neatly groomed; Oppenheimer was alarmingly thin and permanently disheveled, a cigarette almost invariably drooping from his lips. Even their personal inconsistencies seemed like photographic negatives of each other's. Ernest projected the air of a worldly bon vivant, but, in truth, the work of his lab always came first; Robert projected the air of an ascetic, but his indulgences were manifold and libertine: wine, women, food, music, and politics. Around the time they first met, the extroverted Lawrence was preparing to become engaged to the woman to whom he would remain married all his life; the introspective Oppenheimer arrived in Berkeley with several love affairs under his belt and with more yet to come.

The common force in their lives was physics. But that is an incomplete answer, for their approaches to science were also divergent: Oppenheimer was a theorist who could barely turn a bolt with a wrench; Lawrence an experimentalist whose inspired gadgetry transformed how physics—including Oppenheimer's physics—was done. Perhaps that was the secret. They seemed to be complementary pieces making a whole, the way particle and wave manifestations together defined a photon. "Lawrence the experimentalist and Oppenheimer the theoretical man formed about as strong a team as you could imagine in physics," James Brady told an interviewer years later. "And they were always together."

They felt an indentical compulsion to ride the dramatic new developments in their chosen field to their logical destination—to a Big Science focused paradoxically on the infinitesimally small—and to turn their academic home into the dominant center of learning and discovery in that field. Lawrence would provide the instrumentation and nurture the new sources of money and patronage needed to make it ever more powerful; Oppenheimer would provide the intellectual bedrock on which Lawrence's machinery would stand. Neither could have achieved his goals without the other.

This would be the most important and lasting professional relationship in either man's life. And it would resonate worldwide. The bond between Ernest Lawrence and Robert Oppenheimer would influence the development of nuclear physics itself, allied strategy in World War II, and civilian and military nuclear policy through the postwar years. There could be few relationships between any two men that left so profound a legacy for the world we live in today.

Julius Robert Oppenheimer had arrived on the Berkeley campus almost exactly one year after Ernest Lawrence, and with every bit as much thunder.

It was the summer of 1929, a few weeks before the beginning of term. Oppenheimer had received his doctorate only two years earlier under Max Born at Göttingen, where he had communed with such rising stars of quantum mechanics as Werner Heisenberg and Paul Ehrenfest. He seemed to absorb the baffling paradoxes of quantum theory with ease. Following the oral examination for his doctorate, one of the examiners who had grilled him, the freshly minted Nobel laureate James Franck, told a colleague, "I got out of there just in time. He was beginning to examine *me*."

From among ten job offers awaiting Oppenheimer upon his return to the United States, he selected two, reaching an unusual joint arrangement with the California Institute of Technology and Berkeley that would allow him to teach in each place in alternating semesters. That served both universities well, but served him better: He could build a new school of theoretical physics in the "desert" that was Berkeley, while remaining au courant with the latest work in the field via Caltech's more traditional Physics Department. Berkeley, he would recall, "had no theoretical physics. Its experimental physics was pretty old-fashioned and sleepy . . . And that was a nice climate and a challenge. And I regarded Caltech as so much more in touch with physics that I wouldn't get completely isolated." The willingness of the two universities to share Robert Oppenheimer attested to the scarcity of qualified theoretical physicists in American academia, especially theoreticians of Oppenheimer's distinction. Within the decade, Oppenheimer's theoretical teachings and Ernest Lawrence's cyclotron would transform Berkeley into very much Caltech's superior—no longer a desert but the preeminent center of nuclear physics in the world.

Oppenheimer's arrival at Berkeley was rather less dramatic than his arrival a few months earlier at Caltech, which he had reached after a breakneck automobile drive through the desert punctuated by two serious crack-ups. Fresh from the journey, he materialized in Caltech's physics lab with his arm in a sling, and declared, "I am Oppenheimer." At Berkeley, he moved into the Faculty Club, a haven for bachelors where Ernest Lawrence—at twenty-eight, Oppenheimer's senior by nearly three years—was his neighbor. The two became instant friends.

While it was not unusual for a scientist with an advanced degree to have read widely outside his or her chosen specialty, Oppenheimer's range of interests was unusually broad. At Harvard, a fellow student marveled, "he intellectually looted the place," drinking deeply of physics and chemistry, of course, but also mathematics, philosophy, and French literature. He learned Greek in order to read Plato in the original and later took up Sanskrit to study the Bhagavad Gita. During a European sojourn, he surprised a colloquium at the University of Leiden in the Netherlands by delivering a lecture in Dutch, which he had taught himself. "I don't think it was very good Dutch, but it was [appreciated]," he recalled. Years later he would ask Leo Nedelsky, one of his Berkeley graduate assistants, to substitute for him in delivering a lecture. "It won't be any trouble," he told Nedelsky. "It's all in this book." When Nedelsky pointed out that the book was in Dutch, Oppenheimer replied, "But it's such easy Dutch."

Oppenheimer's compulsive polymathy, however, pointed to his one outstanding intellectual flaw: he lacked the patience to make a single subject his own. That was surely a factor in his becoming perhaps the most accomplished American physicist not to receive a Nobel Prize, but it does not mean he failed to do pioneering research. During a period of unexampled intellectual ferment in physics, there were few topics on which Oppenheimer failed to publish a letter or paper. These were always original and often influential, even seminal. In 1930 he predicted the existence of the positron, a positively charged electron. But having "argued himself into a correct conclusion," as a fellow theoretician put it, he lost interest in the topic; it would be Carl Anderson, his own student at Caltech, who would discover the particle—and collect a Nobel Prize for doing so. Oppenheimer's study of astrophysics during the 1930s predicted the existence of neutron stars and, more astonishingly, of black holes: massive stars that collapsed into objects of such enormous gravitational force that not even light could escape their pull. Neutron stars would not be detected until 1967 and hard evidence of black holes not found until the twenty-first century, underscoring Oppenheimer's achievement and the tragedy of his curiously unfulfilled career.

"Oppie was extremely good at seeing the physics and doing the calculation on the back of the envelope and getting all the main factors," recalled his colleague Robert Serber. "As far as finishing and doing an elegant job . . . that wasn't Oppie's style." Quite the contrary: some of his most famous papers are marred by rudimentary mathematical errors, occasionally leading to erroneous conclusions. Serber again: "His physics was good, but his arithmetic awful."

Oppenheimer's true gift was for synthesis. His grasp of physics enabled him to establish a theoretical foundation for almost any new experimental finding. Luis Alvarez, one of the most distinguished members of Lawrence's Rad Lab, witnessed this talent in action one afternoon in 1939 when he burst in on Oppenheimer with the startling news that the German chemist Otto Hahn and his assistant Fritz Strassmann had announced the discovery of nuclear fission, which split the heavy uranium nucleus in two. Standing by the blackboard in his office at LeConte Hall, attended by his ever-present students, Oppenheimer declared promptly, "That's impossible," and proceeded to demonstrate mathematically why Hahn and Strassmann must have been mistaken. This was Oppenheimer's intellectual arrogance, his least endearing quality, in action. But the very next day, he visited Alvarez's lab to witness a demonstration of the phenomenon. "In less than fifteen minutes, he not only agreed that the reaction was authentic," Alvarez recalled, "but also speculated that in the process, extra neutrons would boil off that could be used to split more uranium atoms and thereby generate power or make bombs." It was an extraordinary demonstration of scientific percipience, and classic Oppenheimer. Abandoning his own initial error, he promptly apprehended the underlying physics and, more impressively, envisioned the extended implications like a chess master thinking dozens of moves ahead.

At Berkeley, Oppenheimer displayed the peculiar combination of personal charm and intellectual magnetism that would make him so effective a leader of the atomic bomb program at Los Alamos. On campus, he emerged as the "Pied Piper of theoretical physics," Serber later remarked. His "entourage," as Alvarez labeled it with perhaps a hint of jealousy, adopted his every tic and peculiarity. They chain-smoked his brand of cigarette, Chesterfields, and mimicked his long-legged gait and his scarcely audible mumble. His artistic taste became their own: "We weren't supposed to like Tchaikovsky, because Oppenheimer never liked Tchaikovsky," objected one student, Edwin Uehling. At the semester break every spring when Oppenheimer decamped for Caltech, the entourage followed him south in a caravan of rickety vehicles. Come August, they would migrate north again together.

Oppie's classroom style was unusual. He would stand with his back to the class, scribbling complex formulas at random spots on the blackboard, sometimes erasing them to make room for more before the class had a chance to copy them down. "I still visualize him in his characteristic blackboard pose, one hand grasping a piece of chalk, the other hand dangling a cigarette, and his head wreathed in a cloud of smoke," recalled Edward Gerjuoy, who got his Berkeley PhD under Oppie. He would mumble sotto voce, pausing now and then to emit a murmur that students caricatured as "nim-nim-nim." Visiting Caltech, Paul Ehrenfest, his friend from Europe, strained during one lecture to make out Oppenheimer's words from the front row and finally exclaimed, "Oppie, is it a *secret*?"

Not only his mumbling made his lectures incomprehensible. The subject matter was beyond obscure: a challenge even to the most experienced theoreticians in the world. As a Caltech graduate student, Carl Anderson spent several days in a packed lecture hall straining unsuccessfully to make sense of Oppenheimer's class on quantum mechanics. Finally, he confessed to Oppenheimer that he was so thoroughly at sea he would have to drop the course. Unnerved, Oppenheimer confided that every other registered student had already done the same—the hall was filled with students auditing the course for no credit, struggling to understand the subject matter without risking a grade. He pleaded with Anderson to stay, for without a single student, the course could not be counted as part of the Caltech curriculum. Anderson did so and received an A, even though the material "was over my head, all the way through," he recalled.

Unlike Lawrence, who was skilled at communicating concepts but impatient with the burden of classroom teaching, Oppenheimer liked teaching; he just was not very good at it. The opacity of his classroom manner reflected his own insecurity with newly discovered phenomena that require a novel cast of mind to understand even today. "In those days," he explained sheepishly to a former graduate student who reproached him for talking over his students' heads, "I was just trying to educate myself."

Those outside Oppie's charmed circle found it all rather bewildering—even Enrico Fermi, who should have recognized the behavior, since he was the object of similar veneration by his own students. Sitting in on a Berkeley seminar given by a group of Oppenheimer students in 1940, Fermi found himself unable to follow the mumbled discussion. Afterward he lamented to his friend and colleague Emilio Segrè: "I went to their seminar and was depressed by my inability to understand them. Only the last sentence cheered me up. It was: 'And this is Fermi's theory of beta decay.' "

But Oppenheimer was building the leading school of theoretical physics in the country. The two dozen doctoral theses he supervised at Berkeley between 1929 and 1943 (when he left for Los Alamos) comprised a large share of all the American doctorates awarded in the field during that period. One reason is surely that he was one of the few teachers in the United States whose grounding in quantum mechanics came directly from the Europeans who originated the theory, and one of the few determined to impart his knowledge to a new generation. Another reason is that until the late 1930s, when the clouds of dictatorship and war in Europe inspired the great emigration of European physicists to the United States, Oppenheimer had the American market virtually to himself. As late as 1937, when Gerjuoy asked his advisors at New York's City College where he might continue the study of theoretical physics he had begun there as an undergraduate, "the only established group they could point me to was Oppie's."

In those first years of joint bachelorhood, Lawrence and Oppenheimer were virtually inseparable. They socialized together, played together, shared a few of the same habits, though each in his own way: they both smoked, but Oppenheimer ceaselessly and Lawrence sporadically (and, as though in deference to his puritan upbringing, furtively). The effort to define, or merely understand, the relationship between Lawrence and Oppenheimer consumed the idle hours of many who knew or worked with them. The chemist Martin Kamen left a succinct effort: "Oppie—highly cerebral and introspective, by turns arrogant and charming—was continually plagued by a sense of insecurity. He possessed extraordinary analytic powers, but little manual ability. E.O.L.—less cerebral and highly intuitive—showed practically no self-doubt and remarkable mechanical skills . . . With the theoretical acumen of the one complementing the experimental skills of the other, there was a basis for an intimacy that minimized the gulf separating them intellectually and culturally." In one respect, they matched each other precisely, Kamen observed: "They shared a common drive to be center stage."

What made their partnership extraordinary was that during this nascent period of nuclear physics, theorists and experimentalists customarily regarded each other with mutual condescension and suspicion. They had stereotypically different personalities, different worldviews, and even different politics. "Theorists tend to be more liberal in their politics, liberal ranging on into radical," the Nobel laureate Edwin McMillan,

Lawrence's laboratory associate and brother-in-law, observed years later. "Experimenters . . . were more on the political right." McMillan was looking back at a postwar era when the war in scientific politics had raged out of control, ruining careers and reputations, but there was no question that even in the twenties and thirties, the way that one approached physics reflected, and determined, the way that one viewed the world.

But Lawrence and Oppenheimer understood that they could not have achieved what they did without each other. "Lawrence leaned heavily on Oppie," Brady said. Let the cyclotron produce a perplexing result, and Lawrence's reaction was invariably, "Let's ask Oppie."

"So we went to Oppenheimer," Brady recalled of one such occasion, "and Lawrence hardly finished the first sentence when Oppie said, 'No, no, no, no, no. This can't be. It would be a violation of the first law of thermodynamics. Can't be.' And Lawrence just said, 'Okay, forget that.' They worked this way all the time."

For his own part, Oppie was profoundly stimulated by the torrent of experimental results from Lawrence's magnificent machine. "Very often the things they found [via the cyclotron] were so astonishing that I said I just don't see how that could be," he recalled late in life. "Sometimes I'm sure I was astonished. I was wrong." He was equally enthralled by Lawrence's innovative Journal Club, that weekly free-form exchange of scientific data and news for graduate students, physics faculty, and the occasional visiting eminences that Oppie called Lawrence's "other great invention." While he was in residence at Berkeley, Oppenheimer rarely missed a session and not infrequently chaired them, although on those occasions, Lawrence, having ceded the floor, could be spotted in the audience straining to comprehend Oppenheimer's mumbling, like everyone else.

Nor did Oppenheimer share the hauteur that other theoretical physicists displayed toward Lawrence and his crew of preoccupied tinkers. Lawrence's achievement, he judged, "wasn't in the realm of understanding of nature, but it was in the realm of understanding the problem of studying nature. And he as much as anybody contributed to the whole style of physics." Oppenheimer was rare among theorists in conceding value to the uncomplicated perception of nature, free of abstractions, that happened to be Lawrence's approach. Abstraction, he related, "wasn't primarily [Lawrence's] dish. His dish was to build and to expand a technique. This is an instrumental approach, something without which astronomy and physics would not amount to much."

On the surface, the warmth of their friendship endured well into the 1940s, surviving the kinds of changes in personal lives and careers that can drive a rift between the closest of friends. Ernest was a regular visitor to Perro Caliente, where he would don "very proper riding clothes" and perch on a English saddle. At home, he and Oppenheimer would take long walks around Berkeley and into the bucolic woods of Northern California. "We talked about physics," Oppenheimer recalled, recognizing that Ernest's intellectual interests were narrower than his own and that he might not be especially receptive to wide-ranging symposia on Eastern philosophy and Western art.

There was something endearing about their mutual trust and solicitude. In October 1931, when the mortal illness of Oppenheimer's mother called him away to New York, Lawrence wrote him every couple of days, plainly aware of how deeply Oppenheimer would be affected by his mother's death. "I feel pretty awful to be away so long," Oppenheimer responded to one such expression of solicitude, and even went so far as to ask Lawrence to "do what you can for the fatherless theoretical children, won't you"—though what he thought the arch-experimentalist could do for Oppenheimer's students of theory remained unsaid.

Toward the end of that year, the friends reunited at the American Physical Society meeting in New Orleans, where they basked in their rising eminence and remained, evidently, inseparable. One colleague's wife witnessed them holding an extended conversation on either side of an elevator door, Lawrence stepping in and out to elucidate one more point for Oppenheimer, until the elevator operator put an end to the dance with the words "Break it up, sweethearts."

The same meeting showed Ernest in another light, one that would become more pronounced as their relationship matured: as counsel and mentor, very much the role that Merle Tuve had played for him at that critical moment before his departure for Berkeley. At the presentation of his conference paper, Oppenheimer was run through the intellectual wringer by Robert Millikan, the notoriously prickly Caltech president. Millikan may have been irked by Oppenheimer's challenge to his theory of the origin of cosmic rays, which he had vehemently defended for more than a decade (and which eventually was proven wrong). Oppie deeply appreciated Ernest's moral support at the event. "It was like you, Ernest, and very sweet, that you should whisper to me so comforting words about the Wednesday meeting," Oppenheimer wrote him a few days later. "I was pretty much in need of them, feeling ashamed of my report, and distressed rather by Millikan's hostility and his lack of scruple." As if to repay Lawrence for the consoling words, Oppie revealed that Millikan's behavior had pushed him to start cutting the cords with Caltech and commit himself more to Berkeley, as long as that could be achieved "without complete rupture with Tech."

But it was around this time that their relationship began to change subtly. Oppenheimer's repute as a theoretician was growing, but the rise in Lawrence's professional stature was on a higher plane entirely. Lawrence "became a relatively prominent guy during the thirties, and rightly so," Oppenheimer would recall. "He regarded me as potentially a very good physicist . . . But in a certain sense, not worldly, not experienced, and not very sensible."

Oppenheimer recognized that their social milieus were fated to remain distinct. Robert and his wife, Kitty, were happiest among intellectual bohemians and political leftists; they chose their physicist friends from among those with catholic tastes in music and art, people like the Serbers and Caltech's Linus Pauling. As Lawrence rose in professional esteem, his social circle came to encompass the banking and oil magnates who had become his financial patrons. As early as 1932, Sproul sponsored him for membership in the Bohemian Club, the most august organization of prominent citizens in San Francisco. For Oppenheimer, a genuine bohemian and a Jew, membership was out of the question.

Indeed, throughout the thirties, Oppenheimer swam upstream against a genteel anti-Semitism in the professions and academia. In 1936 he had struggled to obtain an appointment for Robert Serber as his research assistant at Berkeley, where Birge refused to contribute more than a cheeseparing $1,200 for Serber's salary. (At Oppie's request, Lawrence chipped in another $400 from Rad Lab funds.) But his efforts to secure an assistant professorship for Serber came to naught; only years later did Serber discover that the roadblock was Birge: as he had written to a friend at the time, "One Jew in the department is enough."

Yet what really drove a wedge between Lawrence and Oppenheimer was politics. Through the 1930s, Lawrence considered himself a New Deal Democrat. But he was intent on keeping politics out of the lab. Indeed, he thought it inappropriate for scientists to engage in political activity of any kind—"political fiddling around," as he called it in a conversation with Oppenheimer. "Why do you fool with these politics?" he once asked Oppenheimer's younger brother, Frank, a member of the Rad Lab. "You don't have to—you're a good physicist." It was a prescient question, for Frank Oppenheimer's association with Lawrence's lab would end abruptly, following the disclosure of his earlier membership in the Communist Party.

Robert Oppenheimer could no more divest himself of political concerns than he could give up music and wine; they were all essential to his method of engaging with the outside world. What would irk Lawrence most about his political activism was not merely its tinge of radicalism, to which Ernest would grow increasingly hostile, but Robert's insensitivity to the damage it could do to the university in general and the Physics Department in particular, as the breadth of acceptable political discourse narrowed in the immediate postwar years. At first Lawrence dealt quietly with Oppie's thoughtlessness, as on the day when Oppie scribbled an announcement of a cocktail party for relief of the antifascist forces in the Spanish Civil War on the Rad Lab's blackboard. Spotting it on his daily rounds of the lab, Lawrence mutely obliterated it with the violent strokes of an eraser. As time passed, he found it harder to remain silent about his friend's politics, eventually upbraiding Robert for what he termed his "left-wandering activities" and counseling him that they were likely to constrict his opportunities at the university, in industry, and, as war loomed, in government service.

For a few years, the tension remained submerged. After Lawrence brought his bride, Molly, to Berkeley, Oppenheimer remained an intimate part of the family circle. The Lawrences' second son, born just after New Year's Day 1941, was christened Robert, after Oppenheimer. And the day after Oppenheimer brought his own new bride, the divorcee Katherine "Kitty" Harrison, home to Berkeley in November 1940 (Kitty already in maternity clothes), the Lawrences were the first couple in town to welcome them with a dinner.

But as their careers and their politics began to pull them apart, something even more fundamental seemed to come between them. Years later, when J. Robert Oppenheimer was facing the great public crisis of his life and a word from Ernest Lawrence might have spared him from a politically motivated ordeal, Jim Brady asked Lawrence why he and his group at Berkeley had not uttered a word in Oppie's defense.

"There's a very good reason for it," Lawrence replied. "We're the only ones who really know that man."

"It seemed to me," Brady reflected, "to be almost personal." Only time would clear up the mystery.

But the split lay far in the future. By 1933, their collaboration had poised the University of California to take its place as one of the great academic centers of the world, one of the richest institutions and most ambitious. Berkeley attracted the most promising young graduate students, hosted the most eminent visiting lecturers, pocketed the largest contributions from research foundations and garnered the lion's share of public interest and acclaim. Robert Oppenheimer was the leading theoretician in the nation. Ernest Lawrence's renown as an experimentalist and the father of the most productive instrument for researching the atom had spread coast to coast. A thrilling sign arrived that year that his fame had spread to Europe. It was an invitation to the Solvay Conference, an elite triennial international convocation in Brussels, Belgium. Lawrence was the only American invited to an event at which he would be rubbing shoulders with twenty-one other present and future Nobel laureates.

Asked by the conference chairman, the French physicist Paul Langevin, to outline his proposed presentation, Lawrence replied that it would be his dramatic new theory that nuclei of deuterium, a heavy isotope of hydrogen, disintegrated upon striking another nucleus. This was an extraordinary claim that promised to rewrite fundamental laws of physics, backed up by extensive results from the cyclotron. Reassuringly, the results had been validated theoretically by Oppenheimer, who was gratified that they undermined certain European theories of quantum dynamics with which he heartily disagreed. As Oppie wrote his brother cheerily on the eve of the conference, Lawrence "has definitely established the instability of the H_2 nucleus. It decomposes upon collision into neutron and proton . . . That makes as far as I can see a hopeless obstacle to Heisenberg's pseudo qm [quantum mechanics] of the nucleus."

It could have been the greatest moment in the partnership of Lawrence and Oppenheimer; the unveiling of earthshaking experimental findings with solid theoretical underpinnings. Instead, they committed a world-class blunder on an international stage. Big Science, as it happened, was not quite ready for its time in the sun.

Part Two

THE LABORATORY

Chapter Six

The Deuton Affair

Ernest's invitation to the Solvay Conference was more than an acknowledgment of the Rad Lab's standing in international science; it was the lab's coming-out party.

The October 1933 conference, which was devoted specifically to "the structure and properties of the atomic nucleus," was the seventh in the series founded by the Belgian chemist and industrialist Ernest Solvay in 1911. Ernest's appearance would be only the eighth by a US physicist since the inception. He would be the lone American in attendance this time, rubbing shoulders and debating the fine points of nuclear physics with the most glittering stars of European science. Among the other guests would be Einstein, Heisenberg, Bohr, Erwin Schrödinger, and Marie Curie and her daughter and son-in-law, Irène and Frederic Joliot-Curie, none of whom had anything like the resources at Ernest's disposal when they did their groundbreaking work. There would be eight delegates from Cambridge alone, including Rutherford, Chadwick, and Cockcroft.

Ernest lost no time spreading the news of his invitation, declaring himself to be "surprised and tremendously pleased" in letters to Swann and other friends, and cadging $300 from the university to help cover travel expenses. In the weeks before the event, he prepared detailed comments on papers submitted by Cockcroft and Chadwick—laying out, in effect, a direct challenge to the work of the revered Cavendish Laboratory. But Lawrence was about to step onto thin ice. When he plunged through, he would almost drag the reputations of the cyclotron and the Rad Lab down with him, sullying the perceived promise of Big Science.

The subject at hand was a particle known then as the "deuton." (Today the accepted term is "deuteron.") A nucleus of heavy hydrogen, or deuterium, the deuton and Ernest Lawrence made each other famous.

The neutron and deuterium were both discoveries from that miraculous year of 1932, when nuclear physics gave up some of its greatest secrets. Deuterium had come first. Harold Urey, a Berkeley chemistry PhD working at Columbia, had set out to identify a heavy isotope of hydrogen that had been postulated by Raymond Birge, among others. Urey's quarry was an atom of mass 2 that by Birge's reckoning appeared in hydrogen gas at a concentration of 1 for every 4,500 atoms of hydrogen-1. His discovery was a triumph of scientific deduction, for the neutron, which gives deuterium its additional weight, was not discovered until many months after he identified the isotope itself. When James Chadwick found the neutron, that uncharged nuclear particle that Rutherford had been seeking for a decade, the puzzle of deuterium's atomic structure was complete: while the nucleus of common hydrogen comprises a lone proton, deuterium's nucleus—the deuteron—comprises both a proton and a neutron.

Urey's discovery inspired Gilbert Lewis, his former Berkeley mentor and the legendary head of its Chemistry Department, to search for a way to produce large quantities of heavy water—that is, water molecules with deuterium in place of common hydrogen—to function as a medium for experimentation on the new isotope. Living up to his reputation for experimental resourcefulness, Lewis conceived an electrolytic process involving distilled battery acid; soon he was turning out samples with a 50 percent concentration of deuterium, a higher purity at greater volume than anyone else. Lewis was so confident of his process that he was profligate with what was still a rare substance. "He liked to tell how he fed some of his first heavy water to a fly," recalled one student, "and it rolled over on his back and winked at him." More plausibly, he told of feeding his initial sample by eyedropper to a mouse, which showed no ill effects after ingesting what was then "the world supply" of heavy water.

Lewis soon was providing heavy water in quantity to Ernest Lawrence, who vaporized it into gas to pump into the cyclotron. Lewis was an avatar of old-school solitary research, but he was delighted to play a role in stretching science's boundaries. Now he became a fixture in the Rad Lab, wreathed in smoke from his ever-present black cigar, as he perched on a stool and watched Ernest's assistants bombard every element they could find with this new, startlingly effective projectile.

Lawrence's excitement about the possibilities of the deuton matched Lewis's. Any other ion twice as heavy as a simple proton—a proton-proton pair, for example—would have packed an extra wallop when aimed at a target nucleus, but because of its doubled charge also would have been more strongly repelled by the target's own positive charge. The deuton, however, had twice the heft of the proton but not the additional charge, so it should be better at penetrating a target's electromagnetic field. But even Ernest was unprepared for how effective the deuton turned out to be. "As soon as we used deuterons," Livingston recalled, "we got enormous yields of reactions that had never been seen before." Aimed at lithium, the deuton produced ten times as many disintegrations as did mere protons (measured by the emission of alpha particles); aimed at beryllium, the yield jumped a hundredfold.

"Ernest's love affair with the deuteron beam was legendary," Luis Alvarez would observe later. Eventually the Rad Lab was able to manifest the projectile's power visibly by deflecting deutons out of the vacuum chamber and into the air via a platinum "window": Lawrence would never tire of displaying for visitors the eerie purple glow produced by the deuton beam as it ionized nitrogen in the air. But there was much more to it than a purple glow. Every element bombarded with deutons yielded copious alpha particles, signifying their disintegration.

The discovery returned Lawrence to a round-the-clock schedule at the

Rad Lab, supervising the bombardment of dozens of elements with his magic bullet. For light elements like lithium and beryllium, the high yield was not particularly surprising, but the same thing happened with heavier atoms like gold and platinum. At last, the Rad Lab was humming with excitement over experimental results, not merely from the achievement of building a bigger machine. Finally, the lab was exploiting the advantages that made it unique: in this case, Lewis's large supply of deuterium combined with the high energies produced by the cyclotron. No lab in the world could match it in either respect. Suddenly the Rad Lab was looking like a respectable rival to the Cavendish, which was still basking in the glow of Chadwick's discovery of the neutron. That was especially true of the Rad Lab's work with heavier elements, which required bombardments at energies well beyond the relatively meager powers of the Cavendish's Cockcroft-Walton apparatus.

"All of a sudden we were flooding the world with papers on nuclear physics," Stan Livingston recalled, "in a field that no one else could enter because they didn't have the deuterium and they didn't have the high energies." The deuton, Livingston added, "was what made the Berkeley laboratory famous. We were opening up a whole new field of science." Before the end of May, Ernest authorized Berkeley's publicity department to issue a release detailing the "transitions" seen in lithium, beryllium, boron, nitrogen, fluorine, aluminum, and sodium, and declaring that "at this rate of progress, one dares not guess what will be achieved in nuclear physics within a few years."

That was only the beginning. As they prepared for the formal unveiling of their deuton results to the world in the July 1933 issue of *Physical Review*, the Rad Lab team noticed that every bombardment, no matter the target, emitted protons of identical energy and range—eighteen centimeters, or about seven inches, in air. According to conventional nuclear physics, this was extraordinary, even bizarre: nuclei of different weights would be expected to emit disintegration products of widely variable energies, with the heavy elements producing the more energetic recoils. "I am almost bewildered," Lawrence wrote Cockcroft.

His perplexity did not last long. Within a few days, he proposed a solution: the protons were disintegration products not of the targets but of the deutons themselves, which were "exploding" upon contact with the atomic nuclei. This conclusion led to another equally astonishing hypothesis: if the shattering of the deuton imparted equal energy to its two constituents, proton and neutron, then simple math yielded a weight for the neutron of one atomic mass unit, or "unity." (The mass unit was then pegged at one-sixteenth the weight of the oxygen atom.) This was a neutron much lighter than any other lab had postulated.

Lawrence's results bore important implications for nuclear physics. His neutron's weight directly contradicted the value proposed by the Cavendish, which was in the range of 1.0067 to 1.0072 units. The Cavendish understandably harbored a proprietary interest in the characteristics of the particle it had discovered, not to mention a native pride in its ability to extract precise measurements from its meticulously hand-built equipment. It was not about to take Lawrence's challenge lying down. A major battle between small science and Big Science was taking shape in the run-up to the Solvay.

Initially, Lawrence's brash salesmanship held the stage. During a May conference at Caltech honoring the visiting Niels Bohr, he alluded to the lightweight neutron while describing the disintegration of eight heavy targets, up to aluminum, by deuton bombardment. Bohr pronounced the results a "marvelous advancement." Caltech president Robert Millikan suppressed his institution's feelings of rivalry with Berkeley to compliment Lawrence on his "altogether extraordinary" discoveries.

Then it was on to the annual meeting of the American Physical Society, held in Chicago in the glare of the 1933 World's Fair. This was Ernest's debut on the national stage. He proved to be fully up to the challenge, reigning as the star of two front-page articles in the *New York Times* and a prominent feature in *Time* magazine. The *Times*'s science correspondent, William L. Laurence, labeled Berkeley's deuton "a new miracle worker of science . . . The most powerful cannon yet found for liberating relatively enormous stores of energy locked up in the inner core of the atom." Cribbing a metaphor from Francis Aston of the Cavendish, Laurence reported that the energy unleashed from a glass of water could power the ocean liner *Mauretania* "across the Atlantic and back again." He introduced Ernest Lawrence to his readers as the leader of a "scouting party" of Berkeley wunderkinder, most of them "still in their early thirties."

Lawrence's talent for communicating with a lay audience was the theme of *Time*'s report. The article began by describing Bohr's becoming entangled in his microphone cord, drawing a pained high-pitched squeal from the loudspeaker. "It was much easier, and more pleasant, to understand round-faced young Professor Ernest Orlando Lawrence of the University of California tell how he transmuted elements with 'deuton' bullets." *Time* had detected the transition taking place between the "philosophizing" old guard of small-scale physics represented by "Theorist Bohr," whom the audience "tried hard to understand," and the new breed of strapping young experimentalists such as Ernest Lawrence, who skipped blithely over the thickets of theory to fire deutons at atoms of lithium "like a boy with a sling shot." Lawrence had been talent-spotted by *Time*; his elevation to the magazine's cover boy as the symbol of modern American science would soon follow.

The Cavendish, meanwhile, had not been idle. Ernest Rutherford was as quick as Lawrence and Lewis to divine the virtues of the deuton as a nuclear projectile. Unable to produce deuterium in his own lab, he acquired a tiny supply from the visiting Gilbert Lewis that May—about half a cubic centimeter of pure heavy water, or a tenth of a teaspoon, sealed in three delicate glass ampoules. Mark Oliphant developed a method of converting the water to gas with virtually no loss, allowing the lab to recycle the precious supply over and over again.

Generously, Rutherford sang Lawrence's praises to Lewis, offering congratulations to "Lawrence and his colleagues for the prompt use they have made of the new club to attack the nuclear enemy . . . These developments make me feel quite young again." But Rutherford's colleagues found reason to doubt Lawrence's results. Knowing that the immense energies available to Lawrence via the cyclotron were hopelessly out of reach of their equipment, they opted for lower energies but a higher proton current—less power but more particles. They had no trouble finding Lawrence's 18-centimeter protons, but only in the lighter elements lithium and beryllium. At the Cavendish, gold resisted the disintegration effect, except for a modest result the scientists traced to contamination of the heavier targets by light impurities such as boron. Tellingly, Oliphant detected proton emissions even from a clean steel target, which should have been all but inert. Having also determined that the emission rate increased with the length of the bombardment, he concluded that deutons were "sticking to the target," and therefore that what Lawrence interpreted as the disintegration of the bombarding deuton was, in fact, deutons merely striking other deutons on the targets' surfaces. In other words, Lawrence's targets were contaminated—his supposed great discovery the product of poor technique. The difference in experience between the two labs told the tale: to the veteran experimentalists of the Cavendish, contamination was a familiar and well-understood phenomenon, but not so to the brash bombardiers of Berkeley.

As the Solvay drew near, interest in the deuton spread beyond Berkeley and the Cavendish, as did doubts about Lawrence's disintegration theory and, consequently, his calculation for the neutron's weight. From Merle Tuve at the Carnegie Institution came a sharply worded warning to his boyhood friend "Ernie" about his tendency to jump to unwarranted conclusions before all the facts were in. Lawrence wrote back, at once chastened and defensive, on the eve of his departure for Brussels. "I quite agree that the production of neutrons from beryllium is no evidence for the disintegration of deutons and a low mass for the neutron," he wrote. "However, I think we have now pretty conclusive evidence on the point . . . We have observed neutrons from targets other than beryllium in just the amount we expected." He set sail brimming with confidence. To his friend Henry Barton, director of the American Institute of Physics, he wrote that he was prepared "to convince anybody that the deuton is disintegrated." The toughest audience he ever faced was waiting for him across the Atlantic.

Lawrence arrived in Belgium only a few weeks after he engaged in a minor dustup with Rutherford, at long range, over the prospects for useful energy from the atomic nucleus. At a September meeting of the British Association for the Advancement of Science, Rutherford had thrown cold water on the very idea: "Energy produced by the breaking down of the atom is a very poor thing," he cautioned. "Anyone who says that, with the means at present at our disposal and with our present knowledge we can utilize atomic energy, is talking moonshine."

The "moonshine" quote raced around the world, heartening skeptics. "Perhaps this will at least partially cool the ardor of irresponsible writers who have . . . told their impressionable readers that . . . the energy now locked up within the atoms contained in a mere thimbleful of matter will drive a liner across the Atlantic and back," declared *Scientific American* magazine. When it was brought before Lawrence, however, he took a very different line. He agreed with Rutherford that the energy produced by nuclear reactions was a "poor thing," but he attributed this to "purely a matter of marksmanship. At the present time it is possible to break up the atom by disintegrating its nucleus only about once in a million 'shots' . . . But the fact remains that when a 'hit' is made, the atom gives up about twenty times as much energy as was needed to break it . . . Personally, I have no opinion as to whether it can ever be done, but we are going to keep on trying."

Superficially, the two Ernests seemed to be in direct disagreement, and that was how their statements were taken at the time. In truth, however, their views were not so divergent. Rutherford had carefully specified that he was thinking in terms of "the means at present at our disposal." He certainly was not ruling out that at some point in the future, the "marksmanship" mentioned by Lawrence would improve to the point that the energy required to split the atom would fall well below the resulting yield. The big difference between them was that for Lawrence, the future might well be now. In this, as in many other scientific debates when foresight would eventually trump nearsightedness, Lawrence was right.

In any case, discussion at the Solvay would turn not on predictions of the future but on results nearer at hand. Lawrence knew the Cavendish delegates were skeptical not only of his deuton theory but also of the cyclotron itself as a laboratory tool. He had received copies of the participants' papers in advance and marked up their texts accordingly—angrily crossing out a line in Cockcroft's paper asserting that "only small currents are possible" from the cyclotron and scribbling the marginal note "Not true!"

At the conference, Cockcroft sugarcoated the Cavendish's doubts about the exploding deuton by labeling Lawrence's contention not necessarily erroneous but certainly premature: "It is rather superfluous to discuss further the nature of the transformations . . . until we have more experimental information"—preferably from his own accelerator, its current, if not its energies, so much stronger than that of the cyclotron. "Our present information does not suffice." Rutherford and Chadwick stated flatly that they had not found Lawrence's suspiciously light neutrons in any of their bombardments and saw no reason to revise their calculation of the neutron's weight to conform to Lawrence's.

Lawrence was more surprised to hear his deuton theory come under sustained attack from other European delegates. Werner Heisenberg argued that the Rad Lab's experimental results simply could not be reconciled with nuclear theory, and showed his contempt for American science by declaring flatly that it was Lawrence's results, not established theory, that must yield. Theory dictated that if the supposed disintegration were occurring within the nucleus's electric field, the production of protons and neutrons would diminish as the targets got heavier. Niels Bohr, who had so fulsomely lauded Lawrence at Caltech, now backed his friend and student Heisenberg, positing that even if the deuton did split after entering a target nucleus, the speed and range of the ejected proton would increase with the weight of the target—not remain constant, as Lawrence had found. Marie Curie and the Joliots proposed a neutron even heavier than Chadwick's, at a value that solved numerous riddles of nuclear activity in one stroke. They tried to let Lawrence down gently, speculating that perhaps there were different types of neutrons with varied weights; but their posited neutron weight ultimately would prove to come closest of all to the right answer. Lawrence did his best to field all these challenges, but plainly he was overmatched.

"Lawrence both made his reputation with some people and lost it with others at that meeting," Livingston reflected later. Chadwick's disdain was especially irksome: "For years afterward, Chadwick thought all the reports from Berkeley must be literally fabricated." He was not assuaged by the more charitable opinion of his superior, Rutherford, who was delighted with Lawrence's spirit in defending his results. Rutherford lectured Chadwick, "He is just like I was at his age."

Rutherford cheerily tendered an invitation for Lawrence to visit the Cavendish, which Ernest accepted eagerly. During the visit, Chadwick continued to show his sulky side, treating the American so rudely that his colleagues and friends were forced to make excuses. "I gathered that you 'ran up against' Chadwick while in Europe," commiserated Ernest Pollard of Yale, who had received his doctorate under Chadwick the year before. "I think what's wrong with him is he is incredibly overworked. In the two years I worked with him, he seemed thoroughly tired the whole time—he and not Rutherford is the true director of Cavendish research." Ernest acknowledged in reply, "He was a bit abrupt with me . . . Though I was a bit disappointed, I really don't have anything against him because I respect his work so much." Rutherford, however, compensated with his own blustery good will: "He's a brash young man," he told Oliphant after Lawrence departed, "but he'll learn!" That made it all the more curious that it was Rutherford who would continue to resist the installation of a cyclotron in his distinguished old lab, and Chadwick, having moved on to the University of Liverpool, who would launch the first cyclotron construction project in Great Britain and become one of Lawrence's closest professional friends.

But that was years in the future. For the moment, Chadwick's scorn, sounding as it did from deep within the bastion of small science, rankled deeply. Back in Berkeley, Lawrence redoubled the pace of the deuton bombardments. He seemed driven by the urge to show Chadwick up. Bragging to one colleague about the magnitude of the bombardment program at Berkeley, he could not resist adding that "perhaps before long, the evidence will be such as to convince the most skeptical . . . even Chadwick." Informed by Livingston, who was on sabbatical at Caltech, that Charles Christian "C. C." Lauritsen, a Danish physicist recently appointed to the faculty there, was observing neutrons from deuton bombardments of aluminum, carbon, and copper, he replied, "It seems to me that Chadwick will have to come down off his high horse now." He even poked a stick into the lion's cage with a letter to one Cavendish scientist bragging that he had found "unambiguous proof" of deuton disintegration and adding, "It would seem now that even Chadwick would agree." And in a paper for the *Physical Review*, he attempted to refute directly the Cavendish assertion that his results were caused by contamination: "A series of measurements with several sets of carefully cleaned targets showed the same phenomena." For the Research Corporation, which was funding much of this work, Lawrence erected a lofty scientific beacon on a foundation that almost everyone but he thought shaky: "This first definite case of an atom that itself explodes when properly struck is of great interest, not only as a possible source of atomic energy, but especially because it is not understandable on contemporaneous theories . . . [It] promises to be a keystone for a new theoretical structure."

Yet the tide was distinctly turning against Lawrence's position. Rather than confirm Lawrence's results, Caltech's Lauritsen contradicted them, positing that the neutrons produced by his bombardments had come from disintegrating targets, not disintegrating deutons. His report, which appeared in the *Physical Review* next to another defense of the lightweight neutron from the Rad Lab, accepted Chadwick's neutron weight as a given without even alluding to Lawrence's theory, as though it were not worth any attention.

A worse blow was coming. Its source was Merle Tuve, whom Lawrence had visited in Washington on his way home from Europe. Lawrence had asked his childhood friend to double-check the Rad Lab's results using the Carnegie's accelerator, the only machine in the world that could approach the cyclotron's energies. All that Tuve could confirm, however, was the shocking sloppiness of the Rad Lab's work. "After working up all of our data," he wrote in a "Dear Ernie" letter in February, "we reached the astounding conclusion that we were unable to check a single one of the observations which you have reported." The inescapable explanation, he declared, was contamination. Put bluntly, Lawrence's poorly focused beam was coating the cyclotron's interior surfaces with deutons. Rather than a self-destructing deuton, which would be remarkable, Lawrence was detecting the effect of deuton-deuton fusion, which was also remarkable in its way—just not in the way Lawrence had defended so vehemently all these months.

The evidence from the Cavendish, the Carnegie Institute, and Caltech had to be judged irrefutable. On February 28 Cockcroft wrote to tell Lawrence that painstaking purification of the bombardment targets eliminated evidence of an exploding deuton entirely, providing "very good justification for [our] refusing to commit ourselves to your hypothesis." The gentlemanly Oliphant followed up two weeks later with the observation that as little as a single layer of contaminating deutons would produce the erroneous results. ("Do you think this at all possible?" he asked charitably.) The ultimate explanation of Lawrence's error came, perhaps inevitably, from Rutherford, who demonstrated once again the potency of his theoretical instincts when they were yoked to the experimental precision of the Cavendish. One night he jolted Oliphant awake with a three o'clock phone call to declare that the deuton-deuton collisions produced two reactions with almost equal frequency: one emitting a proton and creating a hydrogen isotope with two neutrons (that is, tritium), and another emitting a neutron and creating a helium isotope of atomic number 3.

A startled Oliphant asked him what reasons he could have for reaching such a conclusion. "Reasons! Reasons!" Rutherford roared back. "I feel it in my water!" Rutherford's conclusion meant that Lawrence, in his myopic insistence on a dubious model, had missed the discovery of two new isotopes—which Rutherford, performing as a one-man old guard, had recognized in a characteristic flash of insight.

Lawrence now faced the task of climbing down from his discredited position without sacrificing his dignity or the Rad Lab's youthful reputation. The process began gingerly, with a candid letter to the *Physical Review* signed by Lawrence, Lewis, Livingston, and Henderson conceding that "alternative and reasonable explanations have been found for those phenomena which originally led us to the hypothesis of the instability of the deuton." They acknowledged that further studies would likely show that contamination "will ultimately account completely for our observations."

This was an exceptionally frank confession, though no less so than was demanded by scientific convention. Lawrence followed it up with personal letters to Cockcroft and Tuve bearing almost identical words of contrition: "I can not understand my stupidity in not recognizing this possibility [that is, contamination] when the experiments were in progress . . . I regret very much that the question of deuton instability involved you in so much work and want to thank you very much for stepping in and clearing the matter up so effectively and so promptly." His letter to Cockcroft crossed in the mail with one from Ralph Howard Fowler, Rutherford's son-in-law and deputy at the Cavendish. Gracious to a fault, Fowler comforted Lawrence with the gratifying assertion, surely untrue, that "for a long time Rutherford and Chadwick were nearly convinced that your explanation was right."

Tuve was not so forgiving. To Cockcroft he grumbled that Lawrence's mistake was "one of judgment and point of view rather than of the errors in technique which can give rise to such a situation." Perhaps he had become frustrated watching his boyhood chum collect accolades and public acclaim for his marvelous engineering while complacently performing atrocious physics. Even worse, Tuve felt, was that Lawrence's disinclination to confirm his own results before publishing them had placed an intolerable burden on scientists at other labs, who had wasted time and money trying to confirm unconfirmable results.

Tuve upbraided Lawrence personally in even a harsher tone. "In the face of the very general interest which has been aroused by your publication," he wrote irritably, "we have decided that the only way to handle the situation was to make a bald statement of the extent to which we have endeavored to check your results and failed. There is no way of evading the question much longer . . . I must say that we here have certainly not enjoyed the position in which we have been placed. Once in a lifetime is once too often." He then disclosed that he had already mailed his report to the *Physical Review*. Tuve's indignation revived Lawrence's defensiveness: "It would seem," Ernest replied, "that you are overstating things a bit with the remark that you are unable to check a single one of our observations."

The final act of this family drama played out at the annual meeting of the American Physical Society, held on Lawrence's home turf in Berkeley in mid-June. Tuve presented his findings, as did Caltech's Lauritsen, neither leaving any doubt about the divergence between his results and the host's. "Intensive discussion" ensued, according to the official report of the meeting in *Science*, written by Ernest's departmental colleague Leonard Loeb. Those mild words failed to do justice to the bitterness of the discussion, which included intemperate attempts by both sides to discredit each other; at one point, Raymond Birge, the easygoing Berkeley physics chairman, had to physically separate Lawrence and Tuve and calm their inflamed feelings. Loeb's report attempted to paper over the differences in the experimental results, stating that the findings by Lawrence, Tuve, and Lauritsen "are not contradictory in the least, but rather supplementary" and assuring readers that the three papers "made a consistent picture."

This incensed Tuve, who shot off a "correction" attacking Loeb's report as "erroneous and misleading." Lest anyone doubt where he placed the blame, he referred to the Rad Lab's "abandonment several months ago of a striking hypothesis . . . which they were notified could not be substantiated in Pasadena, at Cambridge nor here in our laboratory." The idea that the three labs' results were " 'not contradictory' . . . is an optimistic one for which I am not responsible, and to which I do not subscribe."

The deuton affair was a turning point for the Rad Lab. All through the fall of 1933, Ernest Lawrence had been on a soaring high, his fame growing within the physics fraternity and among the public. The cyclotron's celebrity as a technological phenomenon fed on itself, fueling its inventor's preference for engineering and salesmanship over the tedious drudgery of hard science. The thousand-ton behemoth filling the old wooden shed on the Berkeley campus was a device that industrialists and foundation directors could appreciate; it was harder for them to apprehend the abstruse science that was its proper quarry. The failures of Ernest's science seemed not to matter as long as his patrons were willing to write their checks, and researchers and laymen were so eager to be thrilled by the concept of multimillion-volt protons that they did not pause to ask what they were good for. But they mattered to scientists, who would be the ultimate judges of whether the money being spent on Big Science was spent wisely.

Even before the deuton affair, some Rad Lab researchers were questioning whether the cult of the machine had not overwhelmed the drive for basic science. They swallowed their complaints, sometimes because they too were swept up in the excitement of pushing the technical capabilities of the cyclotron further and further. At the Rad Lab, it seemed, one could get away with a half-finished research project, but shirking one's nighttime shift operating the machine was a sin. The harvest of this upside-down approach to nuclear research now lay before them.

For those already skeptical of the cyclotron, Lawrence's error reinforced their scorn. Among them was Rutherford. In the aftermath of the controversy, he told Chadwick, "I'm not going to have a cyclotron in the

Cavendish." Rutherford's attitude arose not merely from the poor showing the machine's results made at the Solvay in 1933 but from his personal approach to research. "Rutherford had a horror of complicated equipment," James Chadwick would recall. Of course, he added, this was quite natural for a scientist who had achieved repeated triumphs with instruments scaled to fit on a laboratory bench.

But the complexity of nuclear physics was rapidly outstripping the capabilities of the laboratory apparatus of small science. Chadwick saw the light well before Rutherford. In 1935, chafing under his mentor's dictatorship much as Stan Livingston chafed under Lawrence, he decamped for Liverpool University, a steep downscale move for such a star of the Cavendish's firmament, but preferable to picking a fight with Ernest Rutherford over research technique. "I was not prepared to quarrel with him," he explained. Small science had not quite emptied its quiver: it would produce a few more spectacular successes, again somewhat at the expense of the Rad Lab's reputation. But Chadwick knew the moment had arrived when only the high energies produced by the cyclotron would serve physics.

Liverpool was a poor institution with an undistinguished physics faculty. Chadwick would transform the place. A few weeks after he arrived, he received word that he had been awarded the Nobel Prize for discovering the neutron. One of the first letters of congratulations came from Lawrence. Ernest disclosed that he had received a visit from Arthur P. M. Fleming, research director of the British industrial giant Metropolitan-Vickers, and persuaded him to put up the money for a Liverpool cyclotron. Chadwick's enthusiastic reaction showed him to be at heart not the crusty malcontent Lawrence had met at the Solvay but a professional devoted to science, eager for any assistance that might propel his penurious university into a higher sphere. "You would be surprised to know what this laboratory has been running on in the last few years," he told Lawrence—"less than some men spend on tobacco."

Liverpool was soon the happy recipient of Lawrence's blueprints and two English-born, Rad Lab–trained physicists, dispatched to help Chadwick build his machine. As part of the vanguard of European cyclotron construction, the university soon would rival the Cavendish as the hub of nuclear science in Britain. As a European cyclotron center, it would be joined by Frederic Joliot's lab in Paris, Bohr's in Copenhagen, and then, miraculously, by the Cavendish, which in 1936 found itself "wallowing in cash" from two windfalls. The first was £30,000 paid by the Soviet Union to purchase the Cavendish lab equipment of Peter Kapitza, a Soviet citizen who had been detained by the regime during a visit home in 1934 and was kept happy by its commitment to duplicate his British laboratory in Moscow. The second was a donation of £250,000 from the automobile magnate Lord Austin, which ended the Cavendish's poverty-row ways for good. All these places were soon populated by Berkeley-trained cyclotroneers, carriers of the DNA of Big Science around the world.

The Rad Lab's recovery from the deuton embarrassment was aided by Lawrence's candid confession of error (at least once the evidence became irrefutable). Inside the lab, he was sheepish: "This was a mistake, and a serious one," Livingston recalled. "He told us we had allowed our enthusiasm to carry us along too fast and that we should be much more careful in the future in analyzing our results before we published." But he also declared that error was an inevitable, even indispensable, part of the scientific method. "I have gotten over feeling badly," he told Cooksey. "We would be eternally miserable if our errors worried us too much, because as we push forward, we will make plenty more." But he resolved to bring theorists and experimentalists into closer contact in the lab, the better to challenge or reinforce one another's judgments. And soon his cadre of skilled cyclotroneers would be supplemented by an infusion of inspired research talent, as scientists like Edwin McMillan, Franz Kurie, and Luis Alvarez would assume the job of turning the cyclotron from an engineering prodigy into a source of genuine scientific achievement. The first opportunity to show what it could do was just around the corner. But there would be one more harsh lesson to show how inattention could trump even the most inspired engineering.

Chapter Seven

The Cyclotron Republic

As the daughter and son-in-law of Marie Curie, Irène and Frederic Joliot-Curie were members of physics royalty. But that had not saved them from treatment as severe as Lawrence received at the Solvay Conference. Their error, according to the unforgiving delegation from the Cavendish, was in proposing a neutron even heavier than Chadwick's. The result of their bombardments of boron and other light elements by alpha rays, moreover, had led them to propose that the proton was composed of a neutron and a positive electron, or positron. This contradicted Chadwick's picture of the neutron as a compound proton and electron, though it raised the same difficult question as his: how to fit an electron, no matter its charge, into the prevailing model of the atomic nucleus.

Lawrence came home battered and bruised from Solvay, and capitulated to the criticism; the Joliots went home to their modest Paris laboratory determined to validate their theory of the neutron, and earned the Nobel Prize in the effort. Their method involved bombarding aluminum foils with alpha rays produced by their usual rudimentary source, a hunk of the inexpensive but vigorous alpha emitter polonium. As they expected, the bombardments drove positrons from the target foils. When they ceased the bombardments, however, the emissions continued in the same pattern as one would expect from a naturally occurring radioactive isotope. But theirs was not a natural isotope, it was an unstable isotope of phosphorus created in their laboratory. As their report in the French journal *Comptes Rendus* made plain, they had discovered artificial radioactivity.

The discovery thrilled physicists around the world, not least Marie Curie, to whose deathbed the Joliots brought a test tube of the first artificially produced radioelement. "I can still see her taking [it] between her fingers, burnt and scarred by radium," Joliot recounted later. "I shall never forget the intense expression of joy which seized her."

The Rad Lab received the news from Ernest Lawrence, who came "roaring into the lab . . . waving a copy of *Comptes Rendus* over his head," as Livingston remembered the moment. The staff did not greet the Joliots' achievement as joyously as Madame Curie, for the discovery was another reproach to their slapdash methodology. No laboratory in the world was as well equipped as theirs to discover artificial radioactivity, for none was as capable of such sustained bombardment. No lab had less excuse for overlooking a phenomenon that indeed had been in front of the scientists' eyes for months. They had bombarded dozens of elements with deuterons and assiduously tracked emissions of alpha particles during the bombardments; but the continued emission of electrons or positrons *after* the bombardments ended had escaped their notice.

It only added to their shame that they were able to reproduce the Joliots' findings almost immediately with the cyclotron. The Joliots had hypothesized that the same radioactivity they had induced in boron with alpha particles would be produced by deuterons acting on carbon. Deuterons being the Rad Lab's stock in trade, Lawrence commanded, "Let's try it out." It took only minutes for Livingston and Malcolm Henderson to discover the explanation for the lab's failure to detect what the Joliots had found: the same switch operated the cyclotron's oscillator and its Geiger counter, so that turning off the cyclotron also turned off the detection apparatus. Livingston and Henderson rewired the switch and trained the deuteron beam on a carbon target for a quarter of an hour. Then they activated the counter. "There it was: click, click, click, click, click, click," Livingston recalled. They had transmuted carbon into radioactive nitrogen. "It was there waiting for us. In less than half an hour from the time Lawrence brought the news, we were observing it."

Lawrence camouflaged his frustration by ordering up a lab-wide effort to expand the Joliots' findings beyond aluminum, magnesium, and boron to heavier elements. But he communicated his true feelings to his closest colleagues through a series of confessional letters. "We have had these radio active [*sic*] substances in our midst now for more than half a year," he lamented to Joe Boyce at Princeton. "We have been kicking ourselves that we haven't had the sense to notice."

Given the claims emanating from Berkeley about the superiority of the cyclotron for all varieties of nuclear research, the oversight demanded a deeper explanation than simply a miswired switch. One excuse Lawrence cited frequently was that induced radioactivity was so unexpected a phenomenon that the Rad Lab could not be blamed for missing it—every other physics lab in the world had missed it too, until the Joliots came upon it by accident. But this was not quite true. As early as the 1920s, Rutherford had started searching for induced radioactivity in targets blasted with alpha rays. He failed because his detection apparatus was not capable of sensing neutrons and positrons—*that* was an understandable oversight, for neither particle yet had been discovered. The unfortunate wiring of the cyclotron was hardly an excuse for the lab's peculiar failure to search for evidence of a phenomenon that Rutherford had determined to be within bounds of theoretical possibility some fifteen years earlier.

Lawrence remained defensive about the oversight as late as 1940, when he rationalized it as the consequence of the Rad Lab's singular devotion to improving the cyclotron at the expense of unimportant short-term discoveries. The Joliots' findings, he told the Rockefeller Foundation's Warren Weaver, "would always be a matter of little more than academic interest as long as means were not available to produce these radioactive substances in enormously greater amounts." Luckily for mankind, he suggested, the cyclotron had been painstakingly developed to the point that it could perform that service.

The whole Rad Lab felt the disappointment keenly. "I always felt sorry we didn't find artificial radioactivity for Ernest," Henderson lamented years later. "It was in our hands. All I had to do was put the Geiger counter in and put it on target and I'd have had it." Jack Livingood put Lawrence's lament to Boyce in graphic terms that his mentor never would have used: "We felt like kicking each other's butts." The real culprit was the culture of the lab, as dictated by Lawrence. His preoccupation with improving the cyclotron condoned sloppy and inattentive experimental work.

As it happened, the Joliots' discovery of artificial radioactivity came at the moment when Lawrence was fighting his final rearguard action in defense of his deuteron theory. The dual embarrassments would have the mutually reinforcing effect of reorienting Ernest and his laboratory toward much more careful research. This happened just in time, for frustration over the lab's focus on engineering was again on the rise. Franz Kurie, who had come to Berkeley from Yale as a National Research Council fellow, felt the Rad Lab's experimental procedures were too slipshod to foster serious research on the cyclotron. At Berkeley, he told his Yale labmate Don Cooksey, "the field is getting messy . . . Ernest and Malcolm [Henderson] are too excited to go slowly. Their targets are dirty, and they refuse to take long counts."

An imaginative experimentalist who had been skeptical of Lawrence's deuteron theory, Kurie proposed to Cooksey that they goad Yale into building a cyclotron to rival Berkeley's—and to do better work. "I'm thoroughly sold on the cyclotron as the perfect high-voltage source," he wrote, asserting that building one in New Haven would put Yale "on the map in nuclear physics" as a lab "probably second only to Ernest's." He added, with characteristic Eli hubris, that Lawrence himself would benefit from Yale's implicit endorsement of his machine: at the moment, he advised Cooksey, "no one really believes his cyclotron works" due to the low quality of physics coming out of Berkeley. In the end, Cooksey chose the opposite course. A friend of Ernest's since their days as graduate students together, he left Yale to join the Rad Lab, where he would play out his long career as Lawrence's right-hand man.

Despite Kurie's doubts, signs already had emerged that tinkering was yielding to serious science in the Rad Lab. The deuteron was achieving its moment of glory, for the double-barreled particle proved much more useful than the proton in inducing radioactivity in light targets, as was discovered by lab members rushing to replicate the Joliots' results at Lawrence's command.

The Rad Lab's first report on its findings reached the *Physical Review* dated February 27—a crucial two days earlier than a letter from Lawrence's Caltech rival, C. C. Lauritsen, documenting his own experiments. The communications, which appeared on adjoining pages of the *Review*'s March 15 issue, were notable for their divergent approaches to this important scientific effort. Lawrence's brusque four-paragraph letter reported deuteron-induced radioactivity in fourteen light elements and speculated brashly that "in these nuclear reactions new radioactive isotopes of many of the elements might be formed." By contrast, Lauritsen and his Caltech colleagues soberly devoted two full pages to a meticulous accounting of every step they had taken to validate the French findings on carbon and boron alone. They provided precise ionization data, spelled out their theory on how positrons emitted by the bombarded atoms transmuted into gamma rays, and carefully avoided any speculation about the significance of the new phenomenon. The difference was telling. Scientists seeking a step-by-step guide to the novel science of artificial radioactivity needed to study Lauritsen; those seeking a foreshadowing of the new science's possible harvest consulted Lawrence.

For all its audacity, Lawrence's conjecture about the potential to create new radioisotopes was correct. The cyclotron started turning out these new products with amazing regularity. "To our surprise," Lawrence reported to his old friend Joe Boyce, "we found that everything we bombarded with deuterons (about 12 of the elements) is rendered radio active." The cyclotron was about to secure an international reputation as an indispensable tool of nuclear science. Radioisotopes would be the currency of a new physics, chemistry, and biology, and Lawrence's cyclotron would be the world's preeminent mint. Within months, the deuteron fiasco would be forgotten, and universities across the nation and in Europe and Asia would be clamoring for their own machines.

One other discovery by a European laboratory would help to establish the cyclotron as a must-have apparatus for physics research. In March, Enrico Fermi demonstrated that neutrons were the most effective inducers of radioactivity in elements heavier than phosphorus. The finding enhanced the Italian physicist's fame as that rare scientist equally adept at theory and experimentation, for it validated his own hypothesis that the chargeless neutron could penetrate heavy nuclei that repelled deuterons and other heavy, but charged, projectiles. His discovery also underscored the cyclotron's utility as a prodigious manufacturer of protons, deuterons, and neutrons (which were generated by training the deuteron beam on a neutron emitter such as beryllium).

The old guard's natural radiation sources, the thimblefuls of radium and clumps of radium-beryllium they used to make the discoveries that transformed physics—nuclear disintegration, the neutron, and artificial radioactivity—had had their day. Radium had passed its prime as an experimental source, for in its natural form, it did not produce particles with the energies necessary for probing heavy nuclei. But the cyclotron could. Its day had arrived.

At the Rad Lab, every researcher claimed or was assigned an element from the periodic table as an experimental target, with the most senior staffers getting the most promising substances. Martin Kamen, a postdoctoral chemistry fellow from the University of Chicago who showed up soon after the craze took hold and therefore had no seniority, scrounged among the heavy elements thallium and bismuth, which could not easily be activated even at the energies produced by the twenty-seven-inch cyclotron. But no one worked in solitude. With the team-style research encouraged by Lawrence reaching its full flower, collaboration was the order of the day. Kamen soon was drafted by the physicist Jackson Laslett to help with the chemical separation of sodium isotopes, and the chemist Glenn Seaborg was absorbed into a team working with uranium. Ernest presided happily over all this activity without bothering himself too much about the experimental results, notwithstanding the lesson of the deuteron fiasco. "We are having a merry time bombarding atoms," he wrote Boyce. To Jesse Beams he added, "We are finding so many things happening when we bombard nuclei that we are rather bewildered."

The new paradigm of collaborative research pursued in the ramshackle Rad Lab surprised visitors bred in the insular working style still prevalent in academia. Among them was a graduate student from the University of Chicago named Luis Alvarez, a rail-thin, ruddy-faced young man with rust-colored hair, for his mother's Irish genes had outdueled those of his paternal forebears from northern Spain. Luis's fascination with the Rad Lab had been triggered by a lecture Lawrence had delivered at Chicago, and deepened when he accompanied his father, Walter, a distinguished physiologist at the Mayo Clinic, on a visit to Berkeley during the summer of 1934. Luis was disappointed by his first sight of the old wooden building with its peeling white paint; but what he discovered when he stepped over the threshold was "the most exciting place I had ever seen," he would recall. Alvarez exploited his status as the scion of a prominent scientific family to haunt the place for several days, soaking up its unique atmosphere. Graduate students at Chicago, he observed later, "enjoyed a fine camaraderie in the halls . . . but it was considered a serious breach of etiquette for anyone to suggest how a friend's experiment might be improved. By contrast, everyone at the Radiation Laboratory was encouraged to offer constructive criticism of the experiments his colleagues were performing." At Chicago, the students hoarded their meager supplies of chemical reagents and worked behind closed doors. But there were no doors inside the Rad Lab. "Its central focus was the cyclotron, on which everyone worked and which belonged to everyone equally (though perhaps more than equally to Ernest) . . . Everyone was free to borrow or use everyone else's equipment or, more commonly, to plan a joint experiment." The team approach to physics, Alvarez judged, was "Lawrence's greatest invention." Alvarez was determined to become part of it as soon as he received his degree.

What he had witnessed was the harvesting of the twenty-seven-inch cyclotron's copious output of deuterons and neutrons. The stepped-up bombardments ordered by Lawrence coincided, fortuitously, with an improvement of the twenty-seven-inch vacuum chamber by Don Cooksey, who was filling the vacuum in the engineering staff created by the departure of Stan Livingston for a job at Cornell. (Cooksey would relocate for good the following year.) His redesigned tank nearly doubled the energies produced by the cyclotron to 6 million volts and quadrupled the current.

The lab was swimming in a sea of subatomic projectiles. Fermi's associate Franco Rasetti came away from a brief visit in 1935 astonished at the "enormous superiority" of the twenty-seven-inch cyclotron for the production of neutrons, far beyond anything available in Europe. Fermi had conducted his experiments with a 1-gram vial of radium that produced 630 millicuries of radiation, generating about 630,000 neutrons per second; Rasetti calculated that the cyclotron's deuteron beam was throwing off *10 billion* neutrons a second, or the equivalent of several pounds of radium.* Around that time, Berkeley's publicity department (assisted by the Rad Lab's Paul Aebersold) estimated that the cyclotron, which cost less than $100,000, had produced "radiation equal to $5,000,000 of radium." The equation involved a certain amount of fanciful math, but there was no disputing that the cyclotron was beginning to pay its way as a producer of radioactive isotopes on an industrial scale.

Nor was the fund-raising potential of the cyclotron lost on Lawrence. Its initial radioactive products were suited best for physics research, but

---

* The curie is a widely accepted measure of radioactivity, defined as the activity of 1 gram of the isotope radium-226. A millicurie is one-thousandth of that measure.

that was only a start. Wealthy research foundations were especially interested in isotopes for medical research and cancer treatment. The characteristics of a useful biomedical isotope were well understood: they needed to have half-lives of several hours at least, be nontoxic to humans, and vigorously emit gamma rays to replicate, or preferably improve upon, the physiological effect of radium on cancerous tumors. The pinch of the Depression forced many research philanthropies to reduce their grants for basic science in physics and chemistry, but money still flowed abundantly for projects in biology and medicine.

Lawrence tailored his research program and his fund-raising campaign accordingly. "We are now well on the road to the production of neutron radiation of great intensity and are approaching the biologically interesting domain," he informed Ludwig Kast, president of the Josiah Macy Jr. Foundation, which supported health research exclusively. Artfully, he implied that artificial radioactivity, especially as induced by neutrons, had been a discovery of the Rad Lab: "In my last letter I reported the discovery of radioactivity artificially induced in many common substances by bombardment of high-speed deutons . . . During the past two weeks we have found that an analogous effect is produced by neutron rays." This rather deliberate elision of Fermi's role in neutron research was the prelude to a pitch for $2,250 "to increase the yield of neutron radiation tenfold or more." He got the grant, and another $5,000 from the ever-faithful Research Corporation, to work toward the large-scale production of isotopes.

It was not long before he found the medical radioisotope of his dreams. It was sodium-24, produced by the bombardment of ordinary rock salt by deuterons. Radio-sodium had a gratifyingly lengthy half-life of fifteen and a half hours and gamma ray energies that Ernest calculated at about 5 million electron volts. That made the substance much more potent than radium and therefore useful for physics and medical research alike. It was a solid result, which Lawrence announced promptly with a brief letter to the *Physical Review*, followed by a lengthy report in which he left nothing to chance, describing his methodology in painstaking detail and accounting for the possibility of contamination; he was determined not to repeat the deuton fiasco. "Doubtless radio-sodium will find many uses in the physical and biological sciences," he reported, for once without exaggeration. The new isotope's potency and the efficiency of the cyclotron in producing it surprised even experts accustomed to Lawrence's guileless optimism. When he informed Fermi by letter that his machine had produced a millicurie of radioactive sodium, Fermi scoffed. He assumed that Lawrence, in his slapdash way with numbers, had mislaid a decimal point and meant a microcurie, a thousand times less. When he "tactfully" corrected Lawrence by return mail, Lawrence replied with a letter in which was enclosed, sure enough, a millicurie of Na-24. Only a few months earlier, such an enclosure would have been precious almost beyond measure. Now it could be dispatched by mail to silence a doubter.

The obvious value of radio-sodium in scientific research and commercial applications led to a new dialogue between Lawrence and the patent office, more prolonged and rather less rewarding than the process that had led to the cyclotron patent. This time Lawrence launched the patent frenzy himself. A few days after the first production of radio-sodium, he mentioned the achievement to Arthur Knight, the Research Corporation's patent lawyer, proposing that an application be prepared immediately for patents on both the isotope and its method of production. Time was of the essence, he told Howard Poillon at Research Corporation headquarters, for the application had to be filed "before the accompanying announcement appears in print, October 15, thereby keeping open the possibility of patent applications in foreign countries . . . Radio-sodium is destined to be of practical importance."

The US patent examiners, however, proved distinctly uncooperative. They objected that the bombardment of light metals with deuterons had been reported by Lawrence, Livingston, Lewis, and Henderson so long ago that it was no longer patentable (although the scientists had not observed artificial radioactivity at the time); that the use of deuterons to induce radioactivity already had been reported by the Joliots and Cockcroft and

Walton (although the Europeans had not created radio-sodium); and that radio-sodium itself had been discovered by Fermi (albeit by a process different from Lawrence's).

Knight and Lawrence spent months trying to counter these serial objections. Meanwhile, the patent examiners wrestled with the fundamental tension between the old, solitary way of doing science and the novel collaborative teamwork Lawrence had instituted at the Rad Lab, soon to become the standard of Big Science; the patent office's traditionalists had difficulty reconciling team research with their custom of attributing an invention to one inventor or, at most, two. Hoping to untangle the collaboration, Knight solicited an affidavit from Lawrence establishing "just what parts Drs. Lewis, Livingston, and Henderson took in the several experiments" dating back to the original deuton bombardments of 1933—that is, what roles they played in the discovery of radio-sodium. Lawrence replied defensively that "the radio-sodium experiments were instigated by me alone," adding, "I did suggest looking for artificial radioactivity this way and actively supervised the experiments."

More aggravation was to come. Caltech hinted to Poillon that Lawrence's claims for primacy in the discovery of induced radioactivity slighted results obtained first, or at least virtually concurrently, by Lauritsen. As it happened, Lauritsen had slipped an article into *Science* reporting on his own efforts to validate the Joliots' artificial radiation results, about a week before Caltech's and Berkeley's notices appeared side by side in the *Physical Review*.

Poillon, concerned that Berkeley's broad patent claims might provoke a messy interinstitutional fight, pleaded with Lawrence to narrow his claims to avoid controversy. "I know how repugnant it is for any right-thinking scientist to become embroiled in a discussion concerning priority of discovery," he wrote, "especially if material matters [in other words, money] enter into the picture." He warned that Robert Millikan's proud institution in Pasadena was not to be trifled with. "California Technology is quite a 'powerful Katinka' and is out for both intellectual recognition and financial return wherever proper and possible." He closed with the hope that "any question regarding priority in any patent claims . . . be settled without acrimonious discussion."

Lawrence was predictably incensed by the implication that his patent claims might have infringed Caltech's claim to primacy. "Although prosecuting patent applications where there are questions of priority, and the like, is certainly not a pleasant matter; yet I should not hesitate to 'go to bat' to settle a matter of this kind in case I felt the stake were worthwhile," he told Poillon. He acknowledged that Lauritsen had beaten him to publication by a week, but "we had been informed in conversation with our friends down there that actually we were the first ones to produce radio-nitrogen by deuteron bombardment." Still, he had to admit that the evidence for his claim might be murky, since the margin of priority could not have been more than a day or so. "It would be distinctly unpleasant proving that we had the real priority in the matter," he conceded.

In fact, Ernest was losing his taste for the protracted patent battle. He would have been happy to "look out for the commercial aspects of our work, if this can be done in a dignified and proper way," but that prospect seemed to be slipping away. Pursuing the patent now seemed to him "hardly worthwhile." He asked Poillon, Did it really matter? Since the Research Corporation already held his patent on the cyclotron, which was the only practical apparatus for manufacturing radioactive substances, "therefore it is not necessary to have a patent on the substances themselves."

He left the decision on whether to push ahead with the patent on radioactive substances to Poillon, who was not so willing to give up. Knight tried to extract one more affidavit from Lawrence to complete the application, but Ernest feared that the undignified process of trying to claim ownership of natural substances would give the Rad Lab a black eye no matter how it played out. "The more I think about the matter, the less enthusiasm I have," he told Knight. "Even if a patent were obtained, I think there would very likely be a good deal of general criticism." At

Poillon's initiative, the patent case would drag on for another four years. It finally ended in defeat in April 1939, when the patent office rejected the claims finally and decisively based on the prior publications by Cockcroft and Walton and by Lewis, Lawrence, Livingston, and Henderson.

Ernest recognized that the rapidly rising stature of the Rad Lab in worldwide physics was based partially on its generosity to other universities seeking advice, manpower, and radioactive products. For its part, the Research Corporation enforced the cyclotron patent only against commercial and industrial entities; academic institutions got the blueprints for free. Nothing made Lawrence happier than to oversee the expansion of the Cyclotron Republic (as it was labeled by Fermi's friend and assistant Emilio Segrè). This one battle to patent radioactive salt threatened to extinguish all this goodwill, slowing down the propagation of the cyclotron if not halting it altogether.

The cost of cyclotrons already was raising the hackles of academic deans and presidents. The trend line pointed to a future of intense and expensive competition among academic institutions to be first with high-priced research apparatus. Only a few years hence, MIT president Karl Compton, himself a physicist with a cyclotron at his command, would issue his lament about the introduction of "an abnormal competitive element" into academic science. "This situation is one of the growing pains of a new art, and we college presidents are as much responsible for it as anyone else," he told M. C. Winternitz, the retired dean of Yale Medical School. As Compton observed, the demands of Big Science, including the staggering cost of the equipment, were already working changes in the academic world. Lawrence, whose machine and research style gave birth to the new paradigm, hoped to stave off its negative effects as long as he could.

In the meantime, the Cyclotron Republic continued to expand its boundaries. This became evident to Ernest during an extended visit back east in May 1935. The trip had three functions: as a lecture tour, an opportunity for more fund-raising, and an introduction of the Lawrences' first child, seven-month-old John Eric, to his Blumer grandparents in New Haven.

Ernest's lectures, starting at the Carnegie Institution under Merle Tuve's sponsorship, "seem to be rousing 'hits,' " he reported to Edwin McMillan, one of his new postgrads at the Rad Lab (and later, with McMillan's marriage to Molly's sister Elsie, Ernest's brother-in-law). The key was a bit of business he called "the vaudeville": he demonstrated the potency of radio-sodium by drinking a tumblerful of spiked salt water and holding a Geiger counter to his arm to show how rapidly it circulated through the body to the extremities. He was even more thrilled with the reception the cyclotron was receiving on the skeptical and supercilious East Coast. "It has been extraordinarily gratifying (indeed amazing) to find the very high regard for our work by everyone in the east," he wrote McMillan. "Almost any lab of any consequence has consulted me regarding getting started on a cyclotron development. Even Tuve is making plans in this direction!"

The immediate harvest of the lab's rising reputation came in the form of cash from foundation grants; as Lawrence walked the streets of Manhattan, dropping in on the offices of the Research Corporation, the Chemical Foundation, and the Josiah Macy Jr. Foundation, he was tapping new founts of professional respect. The research philanthropies that had not yet made grants, such as the Rockefeller Foundation, signaled that they would look very favorably upon future applications. "I have already enough funds to assure us essential support, but I am trying to get enough to enable us to go forward with full steam ahead," Ernest informed McMillan. "I am vigorously carrying on the money-raising campaign, and things look very promising . . . I'll keep at it until somehow we get the money."

He also was working to secure employment for his trained cyclotroneers. Berkeley had few faculty positions to offer members of the Rad Lab, and, in any case, sending his people out into the world was a crucial factor in spreading the cyclotron gospel. Another lab director might have been inclined to fend off poachers from rival universities; Lawrence welcomed them, or at least refused to stand in their way. "I shall not in my contacts in the east tend to prevent offers to you all," he wrote. "On the contrary, I shall get as many and good offers as possible and allow the decision for next year to be made by the individual concerned in the light of the possibilities." Uttered by another laboratory director, these words might sound like lip service, but the record shows that Lawrence vigorously sought opportunities even for his best people. His technique for retaining those he genuinely thought the Rad Lab could not do without was to obtain foundation grants for them or to pressure the Berkeley Physics Department to place them on the faculty in the rare instances where that was possible. Such was the case with McMillan, who was offered a physics instructorship that summer, allowing him to turn down a competing offer from Princeton. The difficulty that even Lawrence faced in obtaining such appointments for his most valued associates while Depression-era budgeting still prevailed can be seen from the fact that McMillan was the first permanent addition to the physics faculty in five years. It would be another three years before the Rad Lab could secure the next appointment, which would go to Luis Alvarez.

Lawrence expected his cyclotron operators to carry their knowledge out into the world. "We were all supposed to get familiar with his technique," recalled Jackson Laslett, who earned his doctorate at the Rad Lab. "He would say, 'If you go away now and become a member of a physics department in some other school, you may have to do some of these things.' " The required knowledge for a doctoral candidate at the Rad Lab included the fundamentals of metal casting, plumbing, and electrical engineering.

Stan Livingston, who decamped for Cornell in 1934 (and later moved on to MIT), was first among the exodus of cyclotroneers carrying the machine's DNA. The following year, Milton White and Malcolm Henderson moved to Princeton. By 1939, nearly a score of physicists who had been trained on Lawrence's machines were building cyclotrons at a dozen American universities. Others were ensconced abroad in Cambridge, Liverpool, Manchester, and Paris. In Copenhagen, engineers at Niels Bohr's physics institute, overly trusting to the infallibility of its founder and namesake, unwisely launched a cyclotron project in 1935 without the benefit of a Rad Lab consultant or even a tolerably close examination of Berkeley's blueprints. In 1937 they sent up an emergency flare for help; this brought them Laslett, one of Lawrence's most valued assistants, who advised them that their magnet had been ill designed and misengineered. The unit had to be sent away for rebuilding, which meant taking down a wall of the building erected around it and returning it to the manufacturer by ferry.

Most labs planning to build cyclotrons solicited the help of the Rad Lab to avoid having to learn things the hard way. Lawrence's managerial system, which relied on a steady influx of graduate students and postdocs willing to learn how to operate the cyclotron at no pay, meant that the Rad Lab never risked running out of personnel even as experienced hands got snapped up elsewhere. The lab's output now encompassed not only scientific papers and radioactive isotopes but also at least a half dozen postdocs per year, "all of whom know the game from A to Z," Cooksey informed a friend back east.

As 1935 drew to a close, the Radiation Laboratory was poised for a huge leap forward in stature and fame. Ernest Lawrence was being mentioned in quarters and in contexts that only a year or two earlier would have been unimaginable. At the Nobel Prize ceremony that December, honoring the Joliots for their discovery of artificial radioactivity, the presenter took a brief detour to mention Lawrence's preparation of radio-sodium and expressing the hope that "it can be used in the same way as radium salts in medical applications." Somebody in Sweden already had an eye on him. It was still premature for the Rad Lab to dream of a Nobel of its own, but the reference to radio-sodium and its potential applications hinted at grand things in store.

Chapter Eight

John Lawrence's Mice

Dr. John Hundale Lawrence arrived in Berkeley by train for the first time in 1935. Nearly four years younger than his brother, Ernest, at the age of thirty-one he looked rather the elder: his hairline receding, his face already bearing the dour expression of a country doctor. John did not inspire the outpouring of loyalty and devotion that his brother enjoyed, was never quite as adept at communicating optimism and enthusiasm for his inquiries into the unknown, never was hailed for charm or ease of manner, or lionized as the leader of an inspired team of scientists. But he did already rank as a successful researcher in a very promising new field.

Ernest and John were too far apart in age to have shared their high school or college years, but personally they were close, writing each other at least once every couple of weeks. On only a few occasions, John remembered, had Ernest pulled rank as the wiser elder brother. During John's first year at the University of South Dakota, when he had discovered basketball and girls and slacked off in his studies, he received a stern upbraiding. "You've really got to start hitting the ball now," Ernest told him, "because if you're going to get into a good medical school, you really better settle down." Recalled John, "I rose right to the top of my class." His newfound diligence would win him admission to Harvard Medical School.

In professional terms, the brothers' relationship was symbiotic. John brought to their partnership the initiative to expand nuclear research into biology and physiology, and Ernest contributed the instrument that could make that happen. On that afternoon in 1935, John stepped off the train at Oakland with a peculiar cargo: mice, scores of them, caged to travel with him in his third-class compartment all the way west from Boston, fated to be irradiated with neutrons in Berkeley.

John Lawrence became interested in radiation medicine through his work with the pioneering neurosurgeon Harvey Williams Cushing at Harvard. Twenty years earlier, Cushing, a small, trim man with a magnetic personality and a meticulous clinical style, had made his seminal discovery of the syndrome that became known as Cushing's disease, a rapid weight gain around the trunk and face he traced to tumors of the brain's pituitary gland. Cushing may have been an even more important influence on John's life than his brother Ernest; in John's fourth year at the medical school, Cushing had picked him out from the student body and appointed him his clinical assistant.

"What am I going to do about my degree?" John asked.

"I'll take care of it," Cushing replied. "You don't have to finish your fourth year."

Cushing introduced John to the systematic discipline required in medical research and, perhaps more crucially, to the idea that the X-rays produced in physics labs could be deployed to treat tumors. As John recalled, Cushing thought this development could be "as important as, if not more important than, Pasteur and bacteriology." Under Cushing, John studied pituitary syndromes in dogs and mice and contemplated X-ray irradiations of humans. The work pointed him to his brother's lab at Berkeley and David Sloan's X-ray tube, the most powerful in the world. During the summer of 1935, John cadged a small grant from the Macy Foundation to travel third class to Berkeley, and presented himself and his mice at the Rad Lab.

Both brothers were gratified by this convergence in their interests. The divisiveness among scientific disciplines, especially between the basic sciences and the applied science of medicine, was then fostered deliberately by medical schools. "Medical students were advised to stay away from mathematics and physics and chemistry, and take mostly biology," John recalled. Harvey Cushing was one medical expert who understood the value for physicians of the discoveries of radiation physicists, but in this, as in so many other ways, his foresight was uncommon.

Ernest shared Cushing's faith in the value of interdisciplinary research, but his views were not widely shared on campus. The University of California's medical school faculty, housed across the bay in San Francisco, was especially dismissive of Ernest's overtures. Not even the efforts of the medical school dean could quell the interacademic hostility: at one dinner the dean hosted for Ernest, a professor of medicine "got up and made a speech indicating that the cyclotron is of no use in medicine at all, and they were wasting their time on it," Raymond Birge recalled. But it was not just the med school. John was surprised at the meager demand in Berkeley for the isotopes being discovered and manufactured by the cyclotron with the efficiency of an assembly line. "They were being made available to anybody that wanted to use them," John related, "but there wasn't the excitement that we saw back east." More isotopes were being shipped by the Rad Lab across the country, to Chicago, Boston, and New York, than across campus. Evidently the task of forging a brotherhood of scientists would fall to the two brothers themselves.

John's first impression upon stepping into the Rad Lab was the staff's shockingly casual attitude about radiation. He knew that Ernest himself was aware of the potency—and therefore the hazards—of neutrons, for the capacity of the chargeless particles to penetrate human tissue had been a frequent topic of their correspondence. Ernest had been writing of the strength of the neutron beam since 1933, when he advised Poillon that "the radiation is so intense and so powerful and penetrating that we are already worried about the physiological effects on us." Neutrons from the Rad Lab could be detected even in Gilman Hall, the headquarters of Gilbert Lewis's chemistry department, which was separated from the lab by a broad alley. At first Lewis's chemists were mystified by the sudden degradation in their experimental results, but they eventually traced the trouble to neutron interference. Lewis "told me facetiously that he is going to have the Radiation Laboratory declared a public nuisance," Lawrence wrote Poillon, not without a certain perverse pride.

John outlined a safety regime for the lab, including the relocation of the cyclotron controls from their place next to the machine and into a separate room. He ordered a protective shielding of metal canisters filled with water erected around the machine to sap the neutrons of their destructive energy. Yet John's abstract warnings of the perils of the unshielded cyclotron had less visceral impact on the staff than the fate of the very first mouse he irradiated with neutrons. The animal was confined in a small brass cylinder with a single airhole. The cylinder was placed between the magnet poles and nestled right up against the beryllium target that produced neutrons when struck by deuterons. After a one-minute irradiation at low power, the machine was stopped and the cage opened. A stunned silence fell on the room at the sight of a dead mouse.

As it turned out, the mouse had died not from radiation but suffocation, for someone had forgotten to activate its air supply. That was not learned for several days, however, leaving enough time for the putative lessons of the irradiated mouse to sink in deeply. "No one ever got close to this beam after this," John recalled.

Ernest dismissed the episode as an "amusing" distraction. But more seriously, he informed Milton White, then at Princeton: "All of us working with the cyclotron got the jitters and decided that something had to be done about providing protection from the neutrons . . . We all became really scared and decided that it was time to call a halt." John's proposed alterations soon were installed.

After returning back east, John continued to fret over the neutron flux bathing Ernest and his assistants. The Rad Lab, he advised his brother, should perform "complete blood studies on all of the men—and repeat them every so often. I believe we should respect them more and take less chances with them."

It is doubtful that Ernest needed a reminder of the neutron effects on the body; more likely he needed constant goading to take the appropriate precautions. The brothers reported on both the dangers and the therapeutic potential of the particles in a paper for the *Proceedings of the National Academy of Sciences*—their first collaboration—sent off in December and published the following February. The article described how neutrons come by their superior ability to penetrate heavy elements such as lead: the neutron is so light in relation to lead's nucleus that in even a head-on collision, the neutron rebounds with very little loss of energy, "not unlike a billiard ball colliding with a cannon ball." Therefore, it can ricochet through even a thick layer of dense material without losing much energy. But when colliding with a nucleus close to its own weight, like a hydrogen nucleus (that is, a proton), the neutron imparts more of its energy to the target proton. This property allows the neutron to be absorbed more easily in substances with a high hydrogen content, such as biological tissues. Ernest characteristically drew an audacious conclusion from this effect, crowing to Poillon that "preliminary results" on mouse tumors bathed in X-rays from the Sloan tube hinted at "a development which is of far more importance than anything that has been done thus far in the Radiation Laboratory, because it means we have a cure of cancer."

Ernest and John calculated that the biological action of neutrons was as much as one hundred times more powerful than X-rays. They proposed that the allowable daily dosage for a human being should be one-hundredth of a roentgen, about a tenth of the standard applied to X-rays. "This should constitute a warning inasmuch as many laboratories will soon be using neutron generators of such power that individuals in the vicinity of the apparatus will be exposed to many times this allowable dosage in the course of a few minutes unless adequate protective screening is provided," they wrote in their paper for the National Academy.

In the Rad Lab, these words were typically honored in the breach. At one point, Kamen and Jack Livingood became so preoccupied with an urgent order for radio-sodium that they ignored their pocket exposure meters. After spending twenty minutes in the neutron-rich environment next to the cyclotron, they discovered that they had each absorbed several hundred daily doses. The experience "led to much coarse humor about production of monsters among our progeny in years to come," Kamen recalled. They took their comfort from assuming "from what little we knew of elementary genetics that no great consequences would ensue [as long as] our children did not intermarry."

Ernest was now engaged in an unceasing circuit of meetings with foundation boards and speaking invitations all over the country. One tour brought him to Harvard for a series of six lectures in the first week of January 1936. The event would provoke the worst crisis in the long partnership between Ernest and the University of California.

It was evident from the moment Lawrence stepped off the train in Boston that Harvard president James B. Conant had more on his mind than hearing his lecture. At dinner the first night, Conant asked Lawrence what he thought it might cost to replicate the Rad Lab at Harvard. The next question was obvious: Would Lawrence take on the task? Ernest left his host with a list of general specifications that included faculty positions for his top associates and money for the next-generation cyclotron taking shape in his mind: a sixty-inch behemoth to produce medical isotopes and treat cancer.

The discussion soon yielded a formal offer from Conant of a $12,000 salary to cover a full professorship, along with the deanship of a new graduate school of engineering and applied science. The construction of a new cyclotron also was part of the deal.

Conant had been pondering for some time how to move Harvard into the forefront of experimental and theoretical physics. The university was then almost as much of a physics backwater as Berkeley had been when Lawrence first set foot on its campus, especially in comparison to its Cambridge neighbor, MIT. Conant figured that recruiting Lawrence would eliminate with one stroke the two universities' disparity in experimental physics. It would be even better if Lawrence could help Harvard move into the vanguard of theoretical physics by bringing along his friend Robert Oppenheimer, who had been fending off a standing offer from Harvard for years. Graduate Dean George Birkhoff assured Ernest of Harvard's high esteem for Oppenheimer "as a creative theorist" and confided that the university was willing to appoint him as an associate professor at $6,000. "Inasmuch as he is thirty-two years of age, this would give him here a high rank in comparison with his age group," Birkhoff wrote, apparently under the impression that Oppie served as a junior member of Lawrence's team.

As chairman of the Berkeley Physics Department, Raymond Birge was deeply sensitive to the threat Harvard's offer posed to the University of California. He feared that with Lawrence gone, the staff of the Rad Lab and the Physics Department would drain away like a reservoir behind a breached dam. If both Lawrence and Oppenheimer resigned, there would be little to keep McMillan, Alvarez, and other promising young physicists in Berkeley. By his reckoning, the department would drop from first in the nation to twelfth. Not only would brainpower disappear but also funding. The Research Corporation "will give a *little* here," Birge reflected, "but will anyone be left?"

But the truth was that Oppenheimer had no desire to return to Harvard, which as a student he had found intellectually stimulating but socially isolating. He also considered Harvard the wrong choice for Lawrence. Oppie judged that Ernest would never have the freedom there that he had at Berkeley, and he guessed that his friend would find the duties of a dean burdensome. He understood the superficial allure of a Harvard faculty appointment, but he advised Birge, "It's our duty to save Ernest from himself."

Indeed, at first Lawrence seemed to let Harvard's flattery go to his head. Birge knew there was even more to it than that: Ernest had been irritated by the University of California medical school's dismissive treatment of his brother, a physician with a solid background in research who had become an associate in the Rad Lab because the medical school refused to award him a faculty post. The work that John and Ernest were doing at the Rad Lab, Birge recalled, "was pure research, and the medical school had a low opinion of it . . . They thought that anyone that hadn't had extensive clinical experience was just nobody they wanted to do much about. So there was practically no chance for John . . . to get promoted over there." UC Berkeley's medical school would not get over its disdain for John's studies for years, until the situation was finessed by the creation of a separate laboratory for radiation science in 1942. The Donner Lab was built with a donation from William H. Donner, a retired steel executive whose son had succumbed to cancer, and who had been in the audience when Ernest and John presented their first paper on the biomedical uses of radioisotopes. John became its director in 1948.

Bob Sproul and Ernest Lawrence were something of a matched pair: "big, outgoing, hearty men," as Molly Lawrence described them. Their relationship was based partially on personality and partially on trust. Ernest had demanded much from Sproul in funding, space, and institutional support, and promised much in return, and to their mutual benefit he had invariably delivered on his promises. On one previous occasion, Sproul had countered an offer from a rival university—Northwestern, in 1930—by maneuvering around faculty objections to promote Lawrence to a full professor. Perceiving that this was a more serious affair, he invited Lawrence to his office to discuss it.

A few days in advance of the meeting, Birge sat down with a stack of index cards to construct his case for giving Lawrence whatever was necessary to keep him at Berkeley. "To have a quarrel [with Lawrence] would be fatal," he intended to warn the university president, "because [Lawrence's] decision would then rest on emotions and not on sane thinking."

Yet Birge underestimated Robert Sproul, one of the few people who could outmatch Ernest Lawrence in cajolery. Only a consummate politician could have balanced the interests of regents, legislators, professors, and philanthropists to build the University of California into the first-class academic institution it had become. Sproul's back-slapping manner and booming voice were legendary on campus. He took pride in never having lost a faculty member he wished to keep—and among that group, Lawrence stood out. "Hell, he made me," Sproul would say later of the scientist whose tenure at Berkeley coincided almost exactly with his own.

Their years of partnership had given Sproul plenty of opportunities to size up Ernest as someone for whom the latitude to conduct research with appropriate resources and administrative encouragement vastly outweighed mundane matters such as salary and institutional prestige; Ernest's interest was in making science, not necessarily himself, bigger. Sproul perceived that while Harvard's status was alluring, it was so far behind Berkeley in physics that it might need two years or more to catch up—time that would represent reverse progress for someone with Lawrence's determination to remain at the forefront of research. The task was to assure Lawrence that Berkeley would continue to support the Radiation Laboratory at its accustomed level and open the doors for the expansion he was already contemplating.

The only record of their conversation comes from Lawrence, who shared his impressions with Howard Poillon in a letter that documents Sproul's sublime skill at intellectual and emotional combat:

> *Right away, President Sproul made it clear that he was extremely anxious to keep me . . . He immediately assured me that as long as he was president of this university and unless I went crazy, he would back me and our work. He said that the Harvard offer was not only a very great honor, but also offered many opportunities and that he quite understood my inclination to accept. On the other hand, he said that he felt that it was not impossible that he would be able to offer comparable inducements here and asked me to outline what I would like to have . . . I said that if I could have my heart's desire, I would ask that the work in nuclear physics in the Radiation Laboratory receive continued support at about the present level and that in addition a new laboratory*

> *be built with a cyclotron equipment designed especially for medical research and therapy. Also a continuing budget to support a small staff to carry on the medical research work.*

Sproul plainly knew his man. He made the required optimistic noises and asked Lawrence to prepare a preliminary budget. Ernest submitted the plan three days later, proposing that the Rad Lab's existing budget of $15,500 be augmented by $8,100 in salaries for an assistant lab director (this was to be Don Cooksey), two research associates, and a research assistant.

Lawrence cautioned Sproul that personnel expenses at the Rad Lab were destined to rise sharply, for the lab was becoming the victim of its own success. Thus far, its unique standing in the physics community had allowed it to attract a full complement of visiting physicists willing to participate in its research for free or to secure grants or fellowships on their own; he counted ten researchers whose contributions to the lab then came at no expense to the university. But this gravy train was nearing the end of the line. "Many of the leading institutions of this country and abroad are now engaged in building laboratories similarly equipped," Lawrence noted. Consequently, it would no longer be possible to compel "all those interested in nuclear physics and penetrating radiations to come to Berkeley." Soon the Rad Lab would have to pay its scientists a living wage. Lawrence avoided pointing out to Sproul that he had planted the seeds of this competition himself, but plainly the harvest was already being reaped at Berkeley's expense.

The heart of Lawrence's pitch, however, involved two other demands. One was the designation of the Radiation Laboratory as an independent unit of the university, to ensure the "continuity and stability of the work." That was the easy one. The other was for a "new and larger cyclotron" for medical research, including the full-scale production of synthetic radioisotopes. It would have the capability to produce neutrons ten times more energetic than those emitted by the twenty-seven-inch machine, an intensity that "would make it possible to carry on actual clinical therapy of human cancer." The projected cost was $25,000, plus as much as $22,000 a year in staff (two MDs, two physicists, and two technicians) and other operating expenses. Lawrence acknowledged that the cost of the new cyclotron "could hardly be absorbed in the general university budget" and promised to help raise the necessary funds. But the implication was clear: if Berkeley could not commit to the medical cyclotron, Harvard was waiting in the wings.

Lawrence's budget submission launched a game of serve-and-volley with Sproul, who labored to line up voting support and financial pledges from his wealthy regents while Lawrence wondered aloud how soon he could catch a train to Cambridge to continue talking with Conant. The afternoon after Lawrence delivered his budget, Sproul had called him in for another meeting. He ran his finger down the column of figures—$25,000 for the new cyclotron, $40,000 in combined annual operational expenses—and remarked, "It's a pretty large undertaking."

Yes, it's ambitious, Lawrence conceded. "It probably could be done at Harvard." But he agreed to put off his trip back east to give Sproul a chance to corner his main quarry: Regent William H. Crocker, the railroad heir and banker who earlier had put up the money for Sloan's X-ray tube at the medical school. In the meantime, Sproul was able to push the regents to commit themselves as a board to more funding from university resources. The regents' breath was taken away by Lawrence's audacious request for funds at a time when the stock market crash still weighed on the university endowment. But they bowed to Sproul's judgment that losing Lawrence would be even costlier, and voted to cover all the annual operating and maintenance costs that Lawrence had budgeted for the Rad Lab. Additionally, the lab would be recognized as a division of the university, independent of the Physics Department, with Lawrence as director. Birge, who already regarded the lab's relationship with his department as "the tail wagging the dog," voiced no objection; as far as he was concerned, Lawrence was an adornment to Berkeley physics whether he was formally inside the department or out.

As for the new machine, Sproul wrote Lawrence: "I can report only that the Regents have expressed great interest in the possibilities along this line and have offered me their assistance in securing funds for a new cyclotron, together with the annual operating expenses." He assured Lawrence that he had his eye on a deep pocket and had "reasons to be hopeful that our quest in this direction may meet with success."

Sproul's well-honed intuition may have told him that the thrill of the Harvard offer was beginning to fade for Lawrence. The difficulty of shifting so much established work clear across the country could not have been lost on someone whose habitual preference was to stay put. Sproul may even have learned that another consideration had started to dawn on Ernest: Molly was against the move.

One factor in Ernest's receptiveness to Harvard's offer had been the thought of how excited his wife would be about returning to her native New England. In truth, she had long since transferred her devotion to her adopted Northern California and found Ernest's interest in Harvard distressing. Her traditional family upbringing discouraged her from interfering in decisions about where her husband decided to live and work: "It was a question of what he wanted and what was good for his career," she explained later. "I wouldn't have thought of even expressing an opinion." But she was "just in fear and trembling that he was going to pack up and move back to Cambridge. I'd been there, and I didn't want to live there, not with a family." Ernest's disengagement from domestic matters was a fact of life in the Lawrence household, but as the time for a decision grew nearer, he began to sense the chill in Molly's voice when the subject of Harvard came up.

Not only were Molly's feelings now evident, but Oppenheimer's relentless campaigning against Harvard also had found its mark. Ernest finally yielded. Even before all Berkeley's commitments and promises were formally ironed out, he tendered his regrets to Conant. "The most attractive factor in your offer is the assurance of continued support in my research work, with the prospect of being able to do things at Harvard presumably out of the question here," he wrote. "But it now develops that the university administration here is glad to establish the Radiation Laboratory as a permanent university activity . . . From the standpoint of furthering our research program, therefore, it is clear that I should not make a change at this time, as it would involve serious delays in rebuilding the laboratory."

The Harvard affair permanently altered Lawrence's relationship with the University of California. There would be other offers, including a lavish proposition from the University of Texas. But Berkeley would never again face the risk of losing Ernest Lawrence. Their two names would remain entwined for the rest of Lawrence's life, with more atom smashers—and many more achievements—still to come.

The expense of fending off Harvard awakened the University of California regents to the presence of a prodigy on their payroll. No member was more intrigued than John Francis Neylan, a San Francisco attorney whose career in public service would trace a forty-year arc from liberal to reactionary: starting as an advisor to California's progressive governor Hiram Johnson, continuing as chief counsel to the newspaper magnate William Randolph Hearst, and concluding as fire-breathing anticommunist determined to root out "Reds" from the Berkeley faculty. Neylan had been a regent for nearly a decade when the Harvard offer piqued his curiosity and brought him to the door of the Rad Lab.

"It was like a secondhand tin shop, a dinky little place over there on the campus," he recalled years later. He entered, met the implausibly youthful Professor Lawrence, "then I was introduced to Dr. So-and-So from Cornell and Dr. So-and-So from this place and Dr. So-and-So from that place . . . I was flabbergasted. There weren't two in the place that had to shave twice a week. They were just a bunch of children." Ernest, guilelessly playing host, sat Jack Neylan in front of a blackboard and attempted to explain the cyclotron principle to him. "Of course," Neylan recounted, "after the first minute and a half, he was so far out beyond me that I didn't know where he was going."

Ernest spun a grandiose vision of how science being done right there on the Berkeley campus would change human lives, not just through advances in physics but in the fields of health and medicine. Jack Neylan felt a bond being forged. A few days later, he dragged the most eminent regent, a seventy-one-year-old San Francisco lawyer named Garrett McEnerney, over to the Rad Lab. Neylan stood by as Lawrence weaved his spell over this sophisticated gentleman, who had socialized with governors and presidents. "McEnerney had a wonderful talk with Ernest," Neylan recalled. "When we walked out, he said, 'How much of that did you follow?' I said, 'He lost me first time around the track.' "

For the next three decades, Jack Neylan would serve as Ernest's patron, mentor, and guide in the ways of the high and mighty. "Neylan kind of considered Ernest as a sort of protégé," Molly would recall. "He was going to look after him, see that he got taken care of, and got what he needed for his work . . . He was going to teach him how to do the best for himself." In the process, Neylan's political coloration rubbed off on Ernest, too. Years later, when Neylan provoked a confrontation with the Berkeley faculty by implementing an anticommunist loyalty oath, Ernest Lawrence—then at the peak of his influence on campus—was one of the few professors who refused to speak out against it. And Neylan's visceral disdain for Robert Oppenheimer—"so conceited that he just shoved God over"—may well have contributed to the bitter break between Lawrence and Oppenheimer that soured their last years.

The lanky Luis Alvarez returned to the Rad Lab full-time as a postdoctoral fellow in May 1936, finding to his satisfaction that the informality that so impressed him during his visit the previous year still reigned. Lawrence and Cooksey were both out of town, so the graduate student answering Alvarez's ring at the front door of the old wooden hulk brought him over to Jack Livingood. "When can you begin work?" Livingood asked. Alvarez replied: "As soon as I get my coat off."

He detected a few changes in the floor plan since his last visit. The control console had been relocated in accordance with John Lawrence's instructions, though it was now jammed in a crowded room between Cooksey's workbench and a drafting table. The cyclotron room was still dominated by the big yoke-shaped magnet, but the old vacuum chamber with its encrustations of red sealing wax had been replaced by an improved vacuum tank known, in deference to its designer, as "Cooksey's can."

Around the back was another laboratory space, which was to be Alvarez's research home. Upon entering, he was physically staggered by a powerful stench emanating from John Lawrence's mouse cages. In the room, a female graduate student evidently immune to the smell sat working on a spectrograph, an instrument for measuring electromagnetic radiation. When Alvarez asked her how she could stand the fumes, she assured him cheerfully that one grew used to them, advice he accepted dubiously at the moment but soon discovered to be true. More persistent was the sharp smell of the circulating oil that cooled the big magnet and its electrical transformers. One of the transformers was in need of repair, which Livingood assigned Alvarez as his inaugural task. Returning home at lunchtime that day, his clothing soaked in warm, dripping oil, Alvarez presented his wife, Gerry, with her first eye-watering dose of the fumes "that would signal my whereabouts for years to come," he recalled.

By the time Lawrence returned from an East Coast fund-raising trip, Alvarez had become tolerably acclimated to the smells and to the lab's other quirky phenomena, including its miasma of radio-frequency interference—so potent that one could touch the metal base of a light bulb to any exposed electric conduit in the lab and make it glow (a parlor trick frequently employed by staff members to impress their visitors). The boss greeted his new recruit with the news that he had secured $70,000 in funding for the new sixty-inch cyclotron, and that Alvarez's role would be to design its magnet. "I pleaded ignorance of all things magnetic," Alvarez recalled, "to which Ernest characteristically replied, 'Don't worry, you'll learn.' "

Alvarez was fortunate to have joined the Rad Lab just as the cyclotron was evolving into a machine reliable enough to meet Ernest's expectations consistently. Through 1935, the twenty-seven-inch had exhibited a persistent tetchiness that kept grad students and technicians busy troubleshooting in its innards for hours on end. A string of especially perplexing breakdowns had bedeviled the lab that year, at a time when Lawrence was struggling to fill a mounting stack of orders for radioisotopes from all over the country. To Merle Tuve he had lamented an "epidemic of trouble, connected with raising the power output." He allowed his innate optimism to shine through, however, promising Tuve that "it will not be long now before we will have this trouble licked and will be able to go ahead satisfactorily." He was correct. Less than three weeks later, the electrical flaw causing the breakdowns was identified and fixed.

But other problems persisted. In mid-October Lawrence guaranteed Poillon that there would be a steady supply of radio-sodium coming out of the lab—eventually. At the moment, however, "our apparatus runs only spasmodically, and a good share of the time we have it dismantled for repairs and alterations." He balanced his "too dark" assessment with a typical assurance that "there can no longer be any doubt but that ultimately the apparatus will produce really enormous amounts of radioactive substances." He concluded the letter with the audacious observation that "the apparatus is now almost to the engineering stage of development: that is, we are about at the point now where further desired improvements are of a primarily engineering character, which will make the apparatus thoroughly reliable and practically effective." The words must have elicited from Poillon a knowing, if wry, smile; in the midst of his travails, Lawrence was looking ahead to the time when the cyclotron's inconstancy would be merely a quaint memory. Poillon likely understood that the moment was not nearly as close at hand as Lawrence claimed, but was not necessarily too far off, either. Lawrence was given to extravagant predictions of this sort; if he had not consistently delivered on them, no one would believe him.

Indeed, the year 1936 introduced a string of technical improvements yielding higher energies, higher currents, and unprecedented reliability. This started with the replacement of the machine's timeworn old vacuum tank with Cooksey's new version. The Cooksey can, which was the key to a near doubling of energy to 6 million volts, worked almost flawlessly from the moment of its installation. The new tank changed the cyclotron's reputation from a device whose efficient management was something of a black art, toward one that could be operated predictably under almost any condition.

Following the installation of Cooksey's chamber, the next project was to bring the beam out of the vacuum tank and into the open air. The goal was to liberate the beam from the interference of the machine's powerful magnetic and electrical fields—"one of the formerly objectionable features of the cyclotron," as two visiting researchers from the University of Illinois put it. The Rad Lab called the process "snouting," after the pig-snout shape of the tube that was to carry the beam from the vacuum chamber and out into the air. The design involved applying an electrical force to the ions on their final circuit around the vacuum tank to bend their path toward the tank wall, which they would penetrate via a platinum "window" one ten-thousandth of an inch thick. The resulting effect was striking: a bright streak ten inches long, glowing lavender from its interaction with nitrogen ions in the atmosphere. The beam not only facilitated more sophisticated experiments but also provided a new "vaudeville" for Lawrence to exploit, for it never failed to elicit gasps of astonishment from his visitors.

The most important upgrade took place during the summer of 1937, when Lawrence achieved a four-year-old goal of expanding the magnet poles to thirty-seven inches, accommodating a larger vacuum tank designed by Cooksey and allowing a step up in maximum energies to 10 million volts. The old tank was shipped to Yale, where it would become the core of the new cyclotron. On July 8 Cooksey's new "can" was maneuvered into the lab by block and tackle anchored by his treasured new Packard, a yellow sedan he dubbed the "Creamliner." Three weeks later, the now thirty-seven-inch cyclotron gave birth to a powerful beam.

Cooksey's new tank compiled everything the lab had learned about how to build a cyclotron into a rugged, gleaming, machined tank with glass insulators and arrangements for air and water cooling. In Ed McMillan's judgment, it was the first version of the cyclotron to show "signs of professionalism, [with] nicely machined surfaces and things welded together, bolted together, and gasketed together." For the first time, the unit also incorporated enhancements developed elsewhere than the Rad Lab. Just as Lawrence had warned Sproul that the cyclotron diaspora was bound to end Berkeley's monopoly on cheap grad-student labor, it was now ending Berkeley's monopoly on cyclotron technology. By mid-1937, a dozen cyclotrons were under construction or operating around the world; the flow of information and innovation had become a two-way street, with advances in ion sources, radio-frequency systems, and magnetic controls filtering back to the Berkeley cyclotron from its offspring elsewhere. The Rad Lab was not too proud to build these enhancements into its machine, where warranted. The inflow of new ideas was especially helpful because one purpose of the thirty-seven-inch was to pretest design details for the sixty-inch, which would represent such a tremendous leap in specifications that as little as possible could be left to chance.

Few of these improvements would have worked as well as they did without the ministrations of the lab's newest employee. He was Bill Brobeck, a rusty-haired twenty-nine-year-old who wandered into the lab one day in the summer of 1937 out of sheer curiosity. Raised in Berkeley and possessed of engineering degrees from Stanford and MIT, Brobeck had thrown over a job at a local power company out of boredom and was searching for something new to occupy himself, preferably a position that would not turn him into a drone who "pushes a slide rule all day or punches a calculator." Fortunately, he could pursue this quest at his leisure thanks to his late father, a San Francisco corporate attorney who had left his family with enough money to ride out the Crash and the Depression in comfort.

Brobeck kept up with the engineering profession through frequent visits to the Berkeley campus library. There one day he happened upon an article by Franz Kurie describing the cyclotron. Surprised to learn that the machine was located on campus a few paces from where he sat, Brobeck strolled over. In the Rad Lab, he encountered Don Cooksey hunched over the initial plans for the sixty-inch. Brobeck applied for a job; informed that the lab had no money to pay him, he explained that he was willing to work for nothing. This was enough to get him an appointment with Ernest Lawrence, who, he recalled, "was very pleased to have another person with engineering interests, because there were lots of engineering problems." Brobeck, for his part, was taken with Ernest's democratic approach to lab management. "The janitor was just as important a person as the Nobel scientist who came through, because he had his job to do," he reflected. Having been educated at two of the most elite schools in the country, Brobeck found the Rad Lab's lack of intellectual snobbery refreshing—more so because as an engineer ignorant of nuclear physics, he was bound to be the odd man out in the Radiation Laboratory.

It was obvious even to a nonphysicist such as Brobeck that "scientific knowledge was pouring in." On one wall of the Rad Lab hung an isotope chart devised by Alvarez, with brass hooks to hold cards identifying every new isotope and listing its properties—most of them discovered in that very building. But Brobeck could not avoid casting a disparaging eye on the quality of the lab's engineering; too many important things fell through the cracks as a result of the sharing of responsibilities for physics and machine tending, he judged. The lab really had an engineering staff of one: Don Cooksey, who was unquestionably skilled but who found it a challenge to maintain the technical standards of cyclotron design all by himself.

With his engineer's soul, Brobeck could not help but be offended by the Rad Lab's slapdash operation and maintenance standards. That included the building itself. The chief virtue of the old wooden shack, he judged, was that "no one objected if you drove a nail into it anywhere." The high-voltage transformers in the courtyard also were showing their age; Brobeck appraised them as "equipment Marconi would have recognized." As for the lab's operational technique, he was "amazed at how haywire things were, how sloppy the work was, and how many things could be improved." The cyclotron ran, "but it was held together with string and sealing wax, literally."

Brobeck recognized the underlying problem as one that afflicted any factory facing relentless pressure for production: "There was a strong tendency to keep the machine running at all costs by patching and improvising to get back 'on the air' after a failure." Time could not be set aside for comprehensive maintenance and repair. "They were doing physics. If they got a neutron, they were happy. It didn't make any difference if it broke down half an hour later if they had their results. The next guy could fix it." This approach led to frequent failures and lengthy outages, during which the staff finally would strip away the accretion of Rube Goldberg–style patches applied since the last major repair and restore the machine to full working order. As for the staff's amusing game of illuminating light bulbs from the metal surfaces energized by the force field in the cyclotron room, to Brobeck's horrified eye, this posed a fire risk, especially with oil flowing everywhere—often onto the floor. Such indifference to electrical hazards, he knew from experience, would not have been tolerated in any professional setting.

Brobeck assumed personal responsibility for imposing order on this freewheeling place, which had grown up with the spirit of small science and had not yet developed the professional maturity required of big-dollar research on an institutional scale. His background allowed him to see the cyclotron not as an experimental apparatus but as a machine requiring regular upkeep. Brobeck introduced such rudimentary industrial practices as preventive maintenance. His checklist for the thirty-seven-inch cyclotron would eventually encompass two dozen weekly tasks, including cleaning water filters, checking oil levels, blowing dust out of electrical equipment, and checking flywheel belts—procedures very similar, he observed, to "those used in automobile service stations," and no less crucial. What was most important was that Brobeck's standards went into place in time for the construction and launch of the new machine: the "Crocker Cracker," a cyclotron so immense that, as the cyclotroneer Stan Van Voorhis recalled, the first photographs showing its scale had distant cyclotroneers "sitting open mouthed, almost believing that the camera must lie."

Resolving the operational problems of late 1935 introduced a new occupational hazard for the Rad Lab staff: tedium. By early 1937, Cooksey reported to a friend, "The boys are all complaining because the cyclotron has become so dull."

This complaint was an outgrowth of the lab's chronic shortcoming: the lack of time for basic scientific experimentation.

Its main cause was the increasing pressure to produce radioisotopes in bulk. By early 1937, the lab was regularly producing material for "two dozen physicists, half a dozen biologists, and several chemists," as an early chronicle reported. Ed McMillan expressed a widely shared Rad Lab sentiment when he tweaked Ernest, tactfully: "We hope very soon to be able to satisfy the hoards [*sic*] of biologists that are swarming around looking for radioactive samples, and perhaps to get in a little bombardment for ourselves." Lawrence told Cockcroft in late 1937 that in nuclear physics the lab had found "nothing particularly exciting at this moment to report"; but in fact McMillan and Alvarez, among other scientists, were convinced there would be no dearth of exciting discoveries to report if only they had time to look. Yet Ernest considered the willing distribution of isotopes to all applicants to be a crucial component of the lab's development. He turned away few requests and received fulsome gratitude in return. His response to the mounting demands was not to reorder priorities to give his staff and graduate students some breathing space, but to add a late-night shift to keep the cyclotron running around the clock.

Still, as grinding as the routine became, the allure of the lab remained strong. Even Martin Kamen, then on a temporary fellowship and consistently overtaxed by his duties supervising the production of isotopes, found the prospect of returning to his permanent post at the University of Chicago "too painful to contemplate." His hope was somehow "to earn the right to stay on at Berkeley working with E.O.L. into the indefinite future." He would earn the right with a spectacular feat of research, but to his misfortune it would not, alas, be indefinite.

Foreign visitors were especially perplexed by the apparent absence of scientific ambition at the lab. Maurice Nahmias, who the Joliots had dispatched to Berkeley to pick up pointers for their Paris cyclotron, was dismissive of the lab's preoccupation with the easy work of identifying new radioisotopes, as opposed to the hard labor of cutting-edge physics. He ridiculed their devotion to the machine as "a mania for gadgets or a post-infantile fascination for scientific meccano games."

Alvarez seconded Nahmias's perception that the cyclotron was used "as a radioactivity factory first of all because great numbers of new radioisotopes could be discovered that way with very little effort." He felt some empathy with Lawrence over what he called the "gold rush of isotopes," because he recognized the importance of public relations and "missionary work" for the lab. But he bristled at the countless hours lost to hard science from the dreary routine of "finding leaks, adjusting equipment, repairing oscillators, and developing cyclotron technology . . . After spending days in cyclotron repairs we grumbled when a physiologist or a biologist turned up to claim the fruits of the first bombardment. We grumbled among ourselves, that is; we knew the strength of Ernest's convictions and were much too loyal to allow outsiders to discover our ambivalence." This was an early hint of one of the emerging drawbacks of science funded on a large scale: the bigger the financial contribution, the more its donors wished to see tangible results from their money. But that desire was fundamentally at odds with the incremental pace and the serendipity of basic science.

Alvarez may have felt the constraints more sharply than most. He was admired by his colleagues for what Kamen described as his "knack for ingenious experimentation," though perhaps less so for his unabashed ambition in exploiting it. At one point, Kamen recalled, Alvarez tried to cadge more experimentation time from Lawrence by displaying a chart showing a decline in publications by members of the lab. Lawrence, unmoved, pointed out that Alvarez himself had produced an impressive number of research papers during the very period he claimed that scientific research had been hobbled. Alvarez's own record, he argued, proved that the lab had reached the optimum balance between work *on* the cyclotron and work *with* the cyclotron.

Yet Ernest was not oblivious to the tensions in the lab. "We are trying to preserve a reasonable balance between using the cyclotron and improving it," he wrote Malcolm Henderson, now at Princeton, in January 1937, "and accordingly about half our time is being used for physics and the other half of the time we are making improvements on it." Perhaps not all his younger colleagues would have concurred with his calculation of the balance, but at least he was trying to strike one. And with few exceptions, those who worked in the lab recalled their time there as "magical years" pervaded with "enthusiasm and zeal for accomplishment," in Kamen's words. Those who found the team research Lawrence pioneered uncongenial simply moved on, but they were a distinct minority.

During 1937, another constraint on experimentation appeared: the rising demand for isotopes from Ernest's brother, John. This was endured silently not only because of its family pedigree but also because Ernest so heartily endorsed the science underlying John's experiments. Meeting the biochemical needs of John's group alone was "a full-time occupation," grumbled Kamen, whose duties included the production of the radioactive phosphorus John needed for experiments in the treatment of leukemia and other blood diseases. John's medical hypothesis was based on the observation that phosphorus naturally concentrates in bone marrow. This suggested that phosphorus would be a better carrier of therapeutic radiation than radio-sodium, since the latter distributes itself all through the body as salt, attenuating its effect. Starting around Christmas 1937, when he administered his first dose of a phosphorus radioisotope to a patient at the university medical center, John's need for the substance became "insatiable," Kamen recalled. This imposed a burden on the Rad Lab not only because the necessary bombardments took hours but also because the isotope had to be painstakingly cleansed of contaminating radioactivities and other impurities to render it safe to administer to a patient.

The following year, doctors working with John Lawrence conceived a new use for the cyclotron: the direct irradiation of patients. The legend took hold in Berkeley that this fancy was born in the Lawrence brothers' neutron treatment of their own mother. As related by Raymond Birge in 1960, after both Ernest and Gunda were dead, the story was that Gunda had been diagnosed with terminal cancer and been saved by the cyclotron. "She was the first person to get this treatment, and the neutrons cured the cancer completely," he reported. Birge, who used the story to show that "there's no question about the medical value of the instrument," was retailing hallway gossip that may have spread in part because the idea of a scientist curing his own mother of a dreaded disease packed an irresistible dramatic punch. But his version was garbled and exaggerated. The facts were these: Gunda's chronic abdominal pain and swelling had been diagnosed by the Mayo Clinic as a cancerous tumor in November 1937. Carl reported to his sons that the doctors declared it inoperable and gave their sixty-eight-year-old patient three months to live, at which point John brought her to California to receive X-rays from David Sloan's powerful tube. John took credit for ordering especially aggressive treatment under the medical center's chief radiologist, Robert Stone—"I'd stand by and encourage Dr. Stone to give as big a dose as he could"—and under this onslaught, the tumor shrank and, over a period of ten years, disappeared.

"She was cured, no question about it," John recalled. It is impossible today to reconstruct what actually occurred, because the exact nature of Gunda's tumor is unrecorded and whether other factors contributed to her recovery is undocumented. But she had not been treated with neutrons, and obviously was not the first patient subjected to neutron therapy. X-ray treatment, which she did receive, was by no means novel at the time, though the intensity of her treatment might have been unusual.

In fact, neutron irradiation of live tumors was more Ernest's enthusiasm than John's. Although the younger brother had pioneered the technique, his targets were tissues that had been surgically removed from diseased mice and placed in the beam, which appeared to destroy the tumors at lower exposures than those that killed living mice. That was an indication that patients might survive bombardments that killed their tumors. But John quickly cooled on the concept, partially out of doubt that his experimental conditions fairly replicated the effects on tumors in the human body. Meanwhile, his brother pushed ahead enthusiastically.

Ernest invited Stone to bring cancer patients from San Francisco for treatments at the thirty-seven-inch one or two days a week. Paul Aebersold constructed a removable wooden chamber as a "treatment room" next to the imposing machine. ("[T]he patients will hardly know they are next to such a monster," Ernest proudly informed one cancer specialist.) But the new mandate interfered further with the work of the Rad Lab; Kamen observed that the hassle of constantly erecting and dismantling the wooden box while crossing his fingers that the machine would operate as needed during each session "made a pill-popper out of Aebersold."

The neutron therapy experiments would continue through 1943 at Berkeley's new sixty-inch cyclotron with Stone's guidance, Ernest's enthusiastic support, and John's increasing disaffection. "I could see that nothing really great was happening," John recalled. On the contrary, some patients developed severe skin reactions that lasted for years. Stone himself eventually disavowed the technique as a useful therapy. In 1947, delivering the annual Janeway Lecture of the American Radium Society, he reported that of his original 252 patients, only 18 were still alive less than ten years after the experiment. Even considering that all had been considered terminally ill when treated, Stone thought this a dismal outcome and recommended that the therapy be discarded.

So it was for more than two decades. In 1970 the concept was revived after new findings showed that neutron therapy was most effective at dosages a fraction of those Stone had used. To this day, neutron therapy as pioneered by the cyclotroneers in 1938 remains an important part of the arsenal against certain cancers, including those of the prostate and salivary glands. Lawrence and Stone may have been too aggressive, but they were on the right track.

Chapter Nine

Laureate

As Lawrence had pointed out to Alvarez, outstanding results in nuclear physics were now emerging steadily from the Rad Lab. One reason was the increasing power and efficiency of the Berkeley cyclotron, which then outstripped those of every other cyclotron in existence. Outsiders might disdain the lab's focus on radioisotope hunting, but filling in the roster of isotopes throughout the periodic table was important work.

By the end of 1939, the Radiation Laboratory was leading the world in its discoveries of nuclear transformations. In 1935 the dominant lab in this category had been the Cavendish, which specialized in reactions induced by alpha rays and protons; Berkeley could only compete in research on deuterons. But by mid-1937, Berkeley was accounting not only for more than half of all discoveries of deuteron reactions but a significant share of neutron and proton reactions; and by December 1939, it comfortably dominated discoveries using alpha rays, deuterons, and neutrons, and held its own in the use of protons. This was an extraordinary range of experimentation unmatched by its rivals: in 1939 Birge calculated that the Rad Lab had discovered more than half of all isotopes identified by cyclotrons worldwide. Big Science had plainly demonstrated its value. As cyclotrons around the world contributed to the list of known isotopes, rumors and expectations began to surface that the accelerator's inventor might be in line for the Nobel Prize.

There was more to Rad Lab science than the identification of new radioisotopes, due in part to the recruitment of world-class physicists such as McMillan and Alvarez. Soon after returning to Berkeley, Alvarez decided to focus his research efforts on the lab's most intriguing experimental project: the search for an elusive decay process known as K-capture. This is different from beta decay, in which a neutron is transformed into a proton by the emission of an electron. In K-capture, the nucleus absorbs one of the two electrons from its innermost electron "shell"— the "K" shell—thereby transforming a nuclear proton into a neutron. When that happens, an orbital electron drops down from a higher shell to fill the gap in the K shell. That action emits a recognizable X-ray signature.

Despite Ernest's personal interest in the search for these X-rays, the Rad Lab had no more success finding them than other labs. The problem, as it turned out, was that everyone was looking in the wrong place. Alvarez determined that K-capture occurs only in heavy atoms with many protons—that is, those with high atomic numbers—and in isotopes with long half-lives. By resourceful engineering, he fashioned an experimental apparatus to find the reaction in elements of atomic number 23 (vanadium) and above.

Alvarez would say later that he was inspired by the pleasing thought of undermining "the infallibility of Bethe's Bible." The "bible," a heroic synthesis of everything then known and accepted in nuclear physics, had been published in 1936 and 1937 by Cornell's Hans Bethe, one of the preeminent theorists in the field, with the assistance of Robert Bacher and Stan Livingston. In its first volume, Bethe asserted that K-capture was "practically unobservable." It was typical of Alvarez's competitive personality that he would describe the work that first made his reputation as an effort to take one of his respected elders down a peg. But he could not have been unaware that solving the riddle of K-capture would be an important scientific achievement in its own right. Alvarez's experiment, which involved his spending long hours in the LeConte basement counting clicks on his handmade X-ray detector, set a new standard for rigorous physics at the Rad Lab. With it he finally identified the telltale X-ray emissions in gallium (atomic number 31).

Even pieces of the cyclotron diverted from the refuse bin were advancing science. When the old vacuum tank was replaced in 1936, Lawrence handed over strips of molybdenum from the interior of the discarded tank to the visiting Emilio Segrè. The parts had spent their working lives under bombardment by deuterons. For Segrè, a friend and collaborator of Enrico Fermi's who was heading home to a professorship at Italy's poverty-stricken University of Palermo, these pieces of radioactive shrapnel were priceless. At his Palermo laboratory, he worked the Berkeley molybdenum assiduously. Finding it "a fertile mine of radioactivity," Segrè extracted phosphorus, cobalt, and zirconium isotopes, and then, in early 1937, a real prize: element 43, which had never been seen in nature and was suspected by some scientists not to exist at all, leaving an irksome gap in the periodic table between elements 42 and 44 (molybdenum and ruthenium). "The cyclotron evidently proves to be a sort of hen laying golden eggs," Segrè wrote Lawrence fulsomely. Ernest might have preferred that the discovery of the first artificially produced element, presently dubbed technetium, be made in his own lab, but at least the Rad Lab had played an important role in the achievement. He continued to provide Segrè with metal scraps and even acceded to Segrè's request to irradiate a quantity of uranium oxide mailed from Italy so that he could continue searching for nuclear reactions. "I would beg you to put them somewhere near the cyclotron with a paraffin block on the back so that they may be strongly irradiated by neutrons," Segrè wrote, adding presciently: "U[ranium] seems to me to be rather promising."

Nineteen thirty-seven was shaping up as the year of the cyclotron. Lawrence basked in its glory and made sure the entire lab was warmed in the glow. "Of course all of us here are pleased with the world-wide epidemic of cyclotron construction," he assured William W. Buffum, general manager of the Chemical Foundation, one of his major patrons. The lab was hosting a steady stream of eminent international visitors, he reported, adding that "two of our young men have been invited to go abroad next year" to assist Bohr in Copenhagen and the Joliot-Curies in Paris with their cyclotron projects. Later that year came the award ceremony for the Comstock Prize, given every five years by the National Academy of Sciences for outstanding work in physics. In his keynote speech, General Electric research chief W. D. Coolidge lauded Ernest's "boldness and faith and persistence to a degree rarely matched." In his acceptance remarks, the honoree alluded to his treasured principle of interdisciplinary collaboration. The "essential unity of science," as he described it, means that "an advance of the horizon of knowledge in any direction uncovers territory of all the sciences."

One week after that ceremony in Rochester, New York, *Time* appeared on the stands with Ernest's bright blue eyes staring out from the front cover over the caption "He creates and destroys." To the unabashedly chauvinistic *Time*, the youthful Professor Lawrence's career symbolized the emergence of American science as the lodestar of international research. Its judgment was underscored by the passing of Lord Rutherford only eleven days earlier: "Ernest Rutherford was one of the old pioneers in atomic physics and Ernest Orlando Lawrence is one of the new," the magazine declared, marking the handoff of science's torch to a new generation and from Old World to New—and from small science to Big.

Ernest had sought to get *Time* to hew to his practice of spreading credit broadly, but was only partially successful. The names of Edlefsen, Livingston, and Sloan appeared in the article, but Ernest's acknowledgments of the assistance of the Research Corporation and Chemical Foundation were lopped from the piece as a result of *Time*'s decision to squeeze a Rutherford obituary into the available space. Also omitted were his mentions of the biomedical team working with John Lawrence; consequently, the article gave readers the impression that the radioisotope and neutron work was solely a two-brother effort. Robert Stone of the medical school, who was intimately involved in the research, "blew his top" at the omission, which he found especially irksome given that Berkeley's nepotism rules had been waived to facilitate John's recent appointment to the Berkeley faculty. Ernest mollified Stone by blaming *Time*'s editors for the slights.

The *Time* cover story, an all-purpose certification of celebrity in that era, enhanced Ernest's fame among the lay public, which was happy to offer unsolicited advice and criticism to the brilliant young American scientist. Wrote one Alan Wells of Cambridge, Massachusetts: "I realize it is none of my business, but is there nothing you can do to the atom's nucleus but smash it? Divinity seems to think it wise to uncover the earth's secrets a little at a time with sort of a loving unfoldment instead of smashing it all of a sudden just for the hell of it." This and similar correspondence went into Ernest's file cabinet, in a new folder labeled "Crank."

Only a few weeks after the *Time* cover story, however, the locomotive of adoration was sidelined by a more substantial critique. Its source was none other than Hans Bethe of the Bethe Bible. He had been studying the electrical and magnetic fields of Cornell's little sixteen-inch cyclotron, which had been built by Livingston in 1935 as the very first non-Berkeley machine. In a letter published in the *Physical Review* on December 15, Bethe and his associate M. E. Rose declared that the Einsteinian law of relativity placed a hard cap on the cyclotron's energy at about 11 million volts, which was roughly the rated energy of Berkeley's thirty-seven-inch accelerator.

Bethe's reasoning was that as a particle's mass increased with its velocity, it would reach a point where it would become so massive that it must either become immune to the focusing effects of the magnetic and electrical fields, or its capacity for resonance—that is, reaching the gap between the dees at the exact moment necessary to receive the necessary electrical jolt—must be destroyed. A cyclotron designer could sacrifice either resonance or focusing, but not both. That meant "a very serious difficulty will arise when the attempt is made to accelerate ions . . . to higher energies than obtained thus far."

Bethe's stark conclusion that "it seems useless to build cyclotrons of larger proportions than the existing ones" appeared at a delicate and potentially mortifying moment for the Rad Lab. GE's Coolidge, in his Comstock Prize presentation, had vouched for the vast potential of new and bigger cyclotrons, asserting that "the limit to the particle energies which can be generated in this way is not yet in sight." Lawrence was beating the bushes for money to build his new 100-volt, sixty-inch cyclotron, which by Bethe's reckoning was ten times beyond useless. In fact, Bethe declared, thirty-four-inch pole faces were "ample" to reach the relativistic energy limit. That suggested that Berkeley's thirty-seven-inch machine was already overbuilt.

Yet Lawrence greeted Bethe's attack with surprising complacency—even condescension. "I am awfully glad that Bethe and Rose are working on the theory of the cyclotron, as it is not a simple problem and the more people think about it the better," he wrote Lee DuBridge, a Caltech-trained physicist who had become dean of arts and sciences at the University of Rochester. "However, I think it would be well for Bethe not to draw too general conclusions as to what can be done, as there are many ways of skinning a cat."

Lawrence had several feline pelts already at hand. For one thing, the Rad Lab's Robert R. Wilson had been studying the shape of the magnetic and electrical focusing fields in the thirty-seven-inch cyclotron, and had found that shaping the dees so that they narrowed toward their outer edges helped focus the beam. Using Wilson's findings, Ed McMillan calculated that particles could produce a serviceable beam even when they were significantly unfocused and out of resonance. In fact, the cyclotron had been working perfectly well with beams far more defocused by imperfections in the magnetic field—those bothersome irregularities combated by shimming—than they would be by the relativistic effects that Bethe posited. Over the course of an edgy exchange in the pages of the *Physical Review*, Bethe scoffed at McMillan; McMillan demonstrated the truth of his calculations; and Bethe backed down. Publishing an addendum to his initial paper, he conceded that the true relativistic limit might be double what he originally proposed. Privately, he explained to McMillan that "we considered the existence of a relativistic limit so important that we thought we should communicate it to cyclotronists as quickly as possible, without endeavoring to give accurate figures."

In truth, there was no doubt that a relativistic limit to cyclotron energies existed; the question was where it was. Stan Livingston had feared that he had hit it as early as 1931. But he had been wrong, and the record of the subsequent seven years suggested that the limit still remained well over the horizon. Hands-on cyclotron operators, a species of which Bethe was not a member, understood that there were so many imponderables about how the machine worked that empirical experience still trumped even the most carefully polished theory. "Although the principle of the cyclotron is simple," Cockcroft wrote after building a thirty-six-inch accelerator at the Cavendish, "the fact that it works is rather surprising."

The constraints on the size and power of the cyclotron came from something other than physics. The real limitation, Lawrence advised Mark Oliphant, then planning a sixty-inch machine, was "funds available." That imposed a constraint Lawrence had become a master at overcoming.

The aplomb with which Lawrence responded to Bethe's challenge reflected the Rad Lab's increasing confidence in the quality of its work. Sometimes this was manifested as a new willingness to admit error, as when Niels Bohr visited Berkeley in April 1937 and casually demolished an Oppenheimer theory based on results from the cyclotron.

At issue was an experiment that Lawrence had conducted with James Cork of the University of Michigan on the disintegration of platinum under bombardment. They expected to find that the disintegration yield rose in a curve consonant with increases in the energy of the bombarding deuterons. Instead, they discovered several points at which the yield abruptly jumped out of the curve. Oppenheimer duly contrived "an elegant theory to rationalize these results," and when Bohr came for his visit, the work was trotted out for him as a showpiece of Rad Lab science.

The presentation took place before a standing-room-only crowd in the LeConte Hall auditorium. Lawrence presented the data, and Oppenheimer followed with a "typically stupefyingly brilliant exposition of its theoretical consequences," which his hundreds of spectators strained to make out through his "nim-nim-nim" mumbling. Then came Bohr's turn. His words too were almost inaudible in the cavernous hall, but Lawrence and Oppenheimer were near enough to hear him declare flatly that the results were incompatible with his liquid-drop model of the nucleus, and therefore that Lawrence's data were invalid and Oppie's theory nonsensical.

This was a frightful blow, especially as Lawrence's name was starting to be heard in connection with the Nobel Prize, for which Bohr's endorsement was indispensable. A few years earlier, Ernest might have dug in his heels as he had at the Solvay Conference, perhaps even suggesting that Bohr's theory might have to bow to the higher energies achieved by the cyclotron. This time he displayed the maturity to take another look, and had the experimental talent to help him do so. He appointed McMillan and Kamen to ferret out any flaw in the work. They soon discovered that the problem was that old nemesis, contamination. Lawrence and Cork had gone to great lengths, all meticulously documented, to cleanse their platinum targets of all impurities. But their procedure, as it turned out, actually had baked laboratory dust onto the targets, which accounted for the irregular findings.

The next step was to repeat the experiment. This involved months of difficult chemistry to distinguish the various radioactivities from bombarded foils of platinum, iridium, and gold. After all that, McMillan and Kamen still had anomalous results, but of a very different kind. What they discovered were new examples of nuclear isomerism, in which isotopes of identical mass numbers and changes show dramatically different radioactive characteristics. This had been thought to occur only in limited cases, but they showed that the phenomenon was far more widespread than had been known; indeed, they posited a "fantastic number of isomeric nuclei" in their subsequent report to *Physical Review*. This work held up over time, and the new discovery in his lab provoked by Bohr's skepticism ultimately enhanced, not diminished, Lawrence's claim to the Nobel.

Lawrence could also take pride in the cyclotron's improved reputation as a reliable laboratory instrument. Cockcroft's judgment notwithstanding, the operational principles of the machine were rapidly becoming standardized. Cooksey witnessed firsthand the conversion of a skeptic into a believer in April 1938, while he was traveling back east on what amounted to a sales mission. Accompanied by Kenneth Bainbridge, a member of the physics faculty at Harvard, he stopped at the Bartol Research Institute, which had completed its cyclotron three months earlier. Its builder, Alex Allen, flipped a few switches and produced a beam "at once," Cooksey reported back to Lawrence. Never having seen a cyclotron in action and having heard tales of its difficult temperament, Bainbridge's eyes widened. "Why, he just turned it on!" he exclaimed.

Cooksey carried the gospel up the Eastern Seaboard. At Bell Laboratories in New Jersey, he described the machine's capabilities for Karl Darrow, the lab's eminent physicist and a writer of popular science articles. Responded Darrow, "I now realize that this country may need a thousand cyclotrons." That comported nicely with a standard Lawrence proposed to the Cavendish-trained Arthur L. Hughes, who headed the Physics Department at Washington University in St. Louis—that "there should be a cyclotron laboratory in every university center" devoted to nuclear physics, biology, and clinical medicine.

Lawrence could afford to scoff at suggestions that the cyclotron was still in its shakedown stage. During Cooksey's East Coast tour, physics professor Robley Evans of MIT suggested that his university viewed its cyclotron project, then just getting under way with Stan Livingston in charge, as a sort of probationary test. It was still up to the cyclotron's inventors, he wrote Cooksey superciliously, to show "whether or not the cyclotron is now a piece of standard equipment, a laboratory tool, which can be put up by your experts in a short time, at reasonable expense, and without undue local development work . . . I rely on receiving from you the necessary personnel, plans, advice etc." Ernest called MIT's bluff, informing Evans that he was welcome to hire experienced cyclotroneers from Berkeley to do the job, but warning that the good ones would demand at least an assistant professorship to move.

Evans was not out of line in questioning whether the cyclotron yet ranked as a simple-to-use laboratory apparatus. Following his own ten-day tour of East Coast cyclotrons, Livingston wrote Cooksey fretfully, "Don't let this get out, but I didn't find a single cyclotron operating! They were all making changes or in the throes of some development." Cooksey seconded Livingston's alarm by scribbling on the letter, "Please do not show this around."

As the apostle of building the biggest cyclotrons that technology and funding could produce, Ernest Lawrence was not shy about envisioning the new Berkeley cyclotron being funded by William Crocker's donation on a grand scale. "For medical purposes alone, such a large installation was hardly justified," he confided to Arthur Hughes. To Tameichi Yasaki of Tokyo's Institute for Physical and Chemical Research, which audaciously was building a direct copy of the machine (with Lawrence's permission) even as the original was rising in Berkeley, he explained that the Crocker was so big "because we can get the money." In effect, he was saying that there was no virtue in allowing research funding to lie fallow; the cyclotron might be overbuilt for medical research alone, but who could know what wonders that excess power might uncover?

Indeed, the Crocker Cracker was big. Following its magnet's installation in the new Crocker Laboratory building on campus, Cooksey assembled the entire staff for a photograph to show how big. Thirty-seven men were arranged standing or seated inside the eleven-foot-tall structure. The towering magnet weighed 220 tons, its sixty-inch pole faces nearly twice the diameter of its predecessor's at the old Rad Lab. One cyclotroneer joked that "its neutrons would reach Chicago"; another wondered, less facetiously, if its beam would be so potent that the targets would be "too hot to handle."

Ernest now discovered one hazard of overambitious planning: the cost overrun. In the glow of the Harvard offer, he had outlined for Sproul a machine budgeted at $25,000 for construction and $22,000 a year to staff; but the magnet alone turned out to cost $30,000, and the projected staff expenses ballooned in parallel with the expanding ambitions of the research program. By late 1937, the project's budget had swelled to $75,000 for the building (the sum donated by William Crocker before his death that September) and $68,600 for the cyclotron itself, to be put up by the Chemical Foundation.

And even that was not enough. Ernest shifted his fund-raising efforts into high gear. Fortunately, cyclotrons were still all the rage, especially among biomedical foundations anxious to keep the supply of medical isotopes flowing. A new funding source had entered the field: the National Cancer Institute, established by Congress in August and seeded with $400,000 in annual grant-making authority. Lawrence promptly made an "immediate and urgent" appeal to the director of the National Advisory Cancer Council, the institute's parent, for $30,000 to equip the Crocker "properly as it should be for clinical and experimental use." The approval came through in less than two weeks. To Lawrence's gratification, the council also voted to spend up to $100,000 a year for the next two years to "stimulate" cyclotron-based research into cancer treatment around the country. This task was entrusted to a two-person committee: Arthur Holly Compton of the University of Chicago—a Nobel laureate who was building his own cyclotron with the help of two former Rad Lab men—and Ernest Lawrence, with Compton serving nominally as chair. ("It is not necessary that anyone be named on this committee in addition to you and me," he assured Lawrence.) Ernest drew up a preliminary list of first-year grants covering almost every cyclotron in the country save his own, apportioning $10,000 each for Chicago, Columbia, Harvard, Michigan, and Princeton.

As it happened, however, the well was about to run dry. The National Advisory Cancer Council reconsidered its decision to scatter its funds in small increments, choosing instead to concentrate its financial firepower on a few larger projects. Meanwhile, the Research Corporation, still smarting from the Depression-related shrinkage of its endowment, began to withdraw funding from programs it considered marginal; under this pressure, Harvard and MIT, neighbors on the banks of the Charles River, took steps to join forces rather than build individual machines. Lawrence's vision of a cyclotron at every major university, like Darrow's vision of one thousand machines, looked increasingly fanciful as academic institutions and their patrons began to wonder whether it was healthy for science to be based so heavily on the sheer sums available to support it.

Among the cyclotron labs seeking sustenance, Berkeley remained the most voracious and still the most successful. But Lawrence had to beat every bush. He now made his first appeal to a source he had been carefully nurturing for more than a year. This was the Rockefeller Foundation, which soon would supplant every other source as Ernest Lawrence's preeminent philanthropic sponsor.

Ernest Lawrence and Warren Weaver had first met in 1933. Weaver, the director of the Rockefeller Foundation's Natural Sciences Division, had mapped out an ambitious funding program focusing on experimental biology. Although biomedical research was not yet on the Rad Lab's agenda, both men filed away the encounter as something to be followed up in the future, in part because they had a common interest in encouraging interdisciplinary research: Weaver, a mathematician who had taught briefly at the small Throop Polytechnic Institute in Pasadena before its transformation into the California Institute of Technology, had as his "special concern" the development of connections "between the biological and the physical sciences (biochemistry, biophysics, chemical genetics, molecular biology, etc.)," as he wrote in a later report to the Rockefeller trustees.

In January 1937, with the Crocker Cracker's blueprints on the drafting table, Ernest invited Weaver to visit Berkeley. On a bracing afternoon, he escorted Weaver into the Radiation Laboratory and showed him the twenty-seven-inch cyclotron, not skimping on the vaudeville of the lavender deuteron beam shooting into the air. Oddly for someone who had been rubbing shoulders with physicists for nearly twenty years, Weaver mistakenly described the beam in his diary as comprising 5-million-volt "electrons"—particles that were not in the cyclotron's inventory. (The beam was composed of deuterons, which comprise a proton and a neutron.)

But Weaver did appreciate what was surely the main point urged upon him by his host: "Biology and medicine will have first call on the machine." Lawrence plied him with therapeutic results from his brother's biomedical research showing that neutrons "are some 5.5 times as effective as X-rays on cancerous tissues, while only about 4.3 times as effective as X-rays in their effect on ordinary tissue," as Weaver jotted down. "This differential . . . may prove extremely important." He also recorded Lawrence's assurance that a recent visitor, "a very distinguished biologist," had declared "that this technique could well be as important to biology and medicine as the discovery of the microscope."

Weaver came away dazzled by the cyclotron's potential as a biomedical tool. Lawrence had laid the groundwork for a request that he finally submitted toward the end of the year. By then, the funding for the sixty-inch cyclotron was running perilously short. The National Cancer Institute's $30,000 grant had turned out to be insufficient to cover new demands for shielding and other safety provisions. These demands arose from the growing awareness that researchers had become entirely too complacent about the potent electrical and nuclear forces with which they were working. In March a physicist named Wesley Coates—a former doctoral student of Ernest's—brushed against a high-tension cable in Columbia University's X-ray lab and received a fatal 5,000-volt shock. The tragedy underscored the necessity of shielding the high voltages at play in the lab.

Even more alarming was what the Lawrence brothers witnessed that September at the Fifth International Congress of Radiology in Chicago.

The meeting's topic was the efficacy of radiation for medical treatments of all kinds, but the impression made on Ernest and John was very different. They could not stop thinking about the men they had met "who had obvious scars on their hands where they'd had skin grafts," John recalled decades later. "You would shake hands with some man, he'd only have two fingers or something like that, and he might've been a famous radiologist." Ernest's reaction was uncharacteristically blunt. In a message to Weaver, he called the event a "congress of cripples."

Weaver informed Rockefeller Foundation president Raymond B. Fosdick in November that "an unexpected emergency" had developed at the Rad Lab. Three emergencies, actually: first, Lawrence recognized that he must spend more "in order that this giant new machine may really be safe." Second, the price of steel and other materials had been on the rise. Finally, the Rad Lab's usual sponsors were tapped out. Weaver told Fosdick wryly that most of them "warmly express the hope that the Rockefeller Foundation will be able to come in at this critical juncture."

Lawrence's request was for $30,000. His application emphasized that the new lab would be equipped not merely as a cyclotron installation but "also as a biological laboratory"; in a separate letter, Bob Sproul appealed to Fosdick's idealism by underlining the "breath-taking possibilities" presented by the cyclotron. "There is a spirit about it that should be fostered," he wrote, portraying the entire project as "an investment for the future of mankind." The three-way appeal from Weaver, Lawrence, and Sproul convinced Fosdick and the foundation trustees: the grant, to be disbursed over two years, came through before the end of January.

It still was not enough. Lawrence filled out the rest of his budget with contributions from the John R. and Mary Markle Foundation and the Macy Foundation. He also accepted a grant from the federal government's Works Progress Administration for the hiring of otherwise unemployed shop craftsmen—ten positions each in 1937 and 1938—even though this grant raised objections from the other donors after the WPA demanded an acknowledgment of its contribution in any papers the Rad Lab published. Some of the private foundations' publicity-shy directors quailed at becoming named as partners with a New Deal program detested by their corporate patrons. The strongest complaint came from the Macy Foundation's Ludwig Kast, who fretted to Howard Poillon that if the WPA were overturned by the Supreme Court as "subversive to the spirit of the Constitution," public disclosure of the WPA's support of a Macy grantee might suggest "that we have something to do with sinister communistic tendencies." At Poillon's suggestion, Lawrence deleted the WPA acknowledgment from papers that involved medical work financed by the Macy.

Two days before Christmas 1937, a load of 196 tons of steel was delivered to the Moore Dry Dock Company on the Oakland waterfront, to be machined into the Crocker Cracker's huge magnet. Wound with twenty-five tons of copper, this behemoth was transferred to the cyclotron's unfinished new home before the end of March. The work was tracked assiduously by a parade of fascinated foundation executives; for ten days in April, it was the turn of Rockefeller executive Frank B. Hanson. In the shadow of the hulking magnet, which now stood inside a three-story wing of the new Crocker building, Ernest showed Hanson where a medical treatment room would be built "in such a way that the patient will not see the machine, which will be noiseless," and how ample clearance was being left around the unit for the radiation shielding on which so much of the Rockefeller money was being spent. There also would be quarters for John Lawrence's "rat colony," Hanson reported. After inspecting the unfinished Crocker Cracker, Hanson was escorted across the street to the Rad Lab for the obligatory demonstration of the lavender beam and was sent home suitably awestruck.

In January 1939, the cyclotron's six-ton vacuum tank, engineered by Bill Brobeck, was slid in between the magnet poles. The sixty-inch machine laid easy claim to being the last word in cyclotron design. Everything was custom engineered—"no hand-me-down magnets, no industrial discards . . . 'no patch work,' " as the Joliots' emissary Maurice Nahmias had derided the old standards. Every part had been modeled and tested, even against the seismic shocks of shaky Northern California. Reflecting Lawrence's lack of conceit when it came to engineering, the machine incorporated the latest innovations of cyclotroneers across the country in designs for transformers, power transmission, the oscillator, and the ion source.

Finally came the moment for "beam hunting," or the tuning of the system. Here the reality that cyclotroneering was still very much a black art reasserted itself. For four months, the beam remained elusive as Lawrence, Alvarez, McMillan, Cooksey, and Brobeck tinkered with shims, relaid transmission cables, rewired the oscillator, and attempted myriad other tweaks and trims to coax a detectable stream of protons from the machine. The lab's initial confidence in its handiwork had prompted Lawrence to agree to a live radio broadcast of the start-up over the Columbia Broadcasting System on April 15; when no beam had emerged by April 4, the broadcast was canceled.

Nature finally capitulated on April 17, when resonance was finally detected. A stream of protons reached the collector exactly one month later. On June 7 came another milestone, a 17-million-volt deuteron beam slicing through the air for a distance of five feet. Lawrence calculated the cyclotron's yield as the equivalent of more than one ton of radium—more radium than was known to exist on earth. Triumphantly, he reported the Crocker Cracker's performance to the *Physical Review* in a letter cosigned by Alvarez, Brobeck, Cooksey, McMillan, and three other physicists; in the new paradigm of Big Science, success belonged to an expansive family. One aspect of the report, however, was uniquely Lawrencian. "We are convinced," he wrote, "that much higher energies could be obtained from a cyclotron of larger dimensions." The sixty-inch, a huge leap forward in cyclotron technology, had just begun to work, but Ernest Lawrence was already thinking ahead to the next step.

The year 1938 had closed in a whirl of expectation and hope at the Rad Lab. The sixty-inch was taking shape. Its cousins around the nation had

1

Lawrence family portrait, probably circa 1910.
Clockwise, from left: Carl, Ernest, Gunda, John.

2

A teenage Ernest (*center*) joyriding with unidentified friends.

Niels Edlefsen's first cyclotrons, basically glass flasks slathered over with sealing wax. Livingston doubted that these produced the resonant acceleration Lawrence claimed.

M. Stanley Livingston's 4-inch cyclotron, uncovered to show the interior, including the lone semicircular "dee" and a strip of copper serving as the target. Lawrence and Livingston finally achieved resonant acceleration with this device.

5

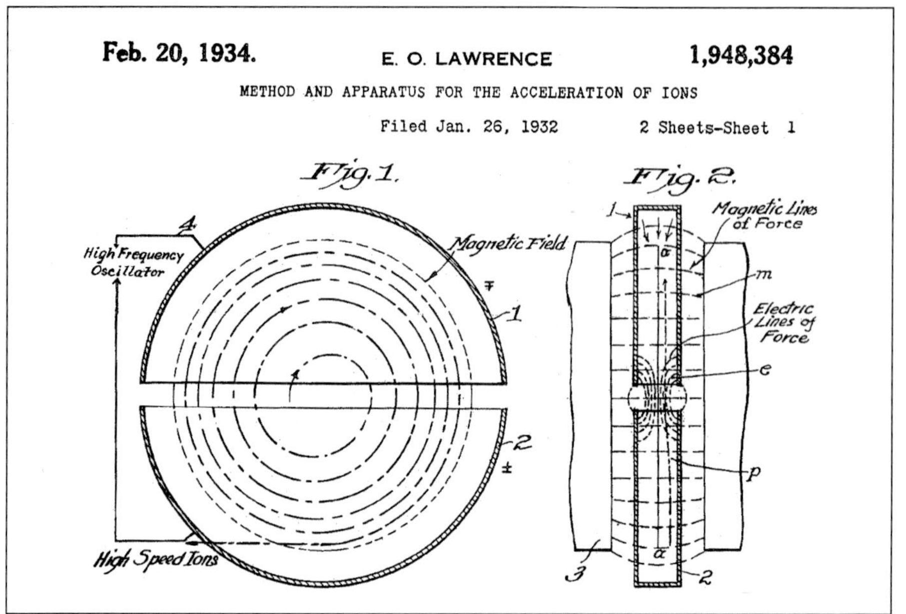

A page from Lawrence's 1932 cyclotron patent application showing the spiral paths of energized ions (*Fig. 1*) and the action of the magnetic field to keep them corralled within the dees (*Fig. 2*).

6

J. Robert Oppenheimer (*left*) and Ernest Lawrence in the first, warmest stage of their friendship, at Oppenheimer's Perro Caliente ranch in New Mexico, probably around 1931.

7

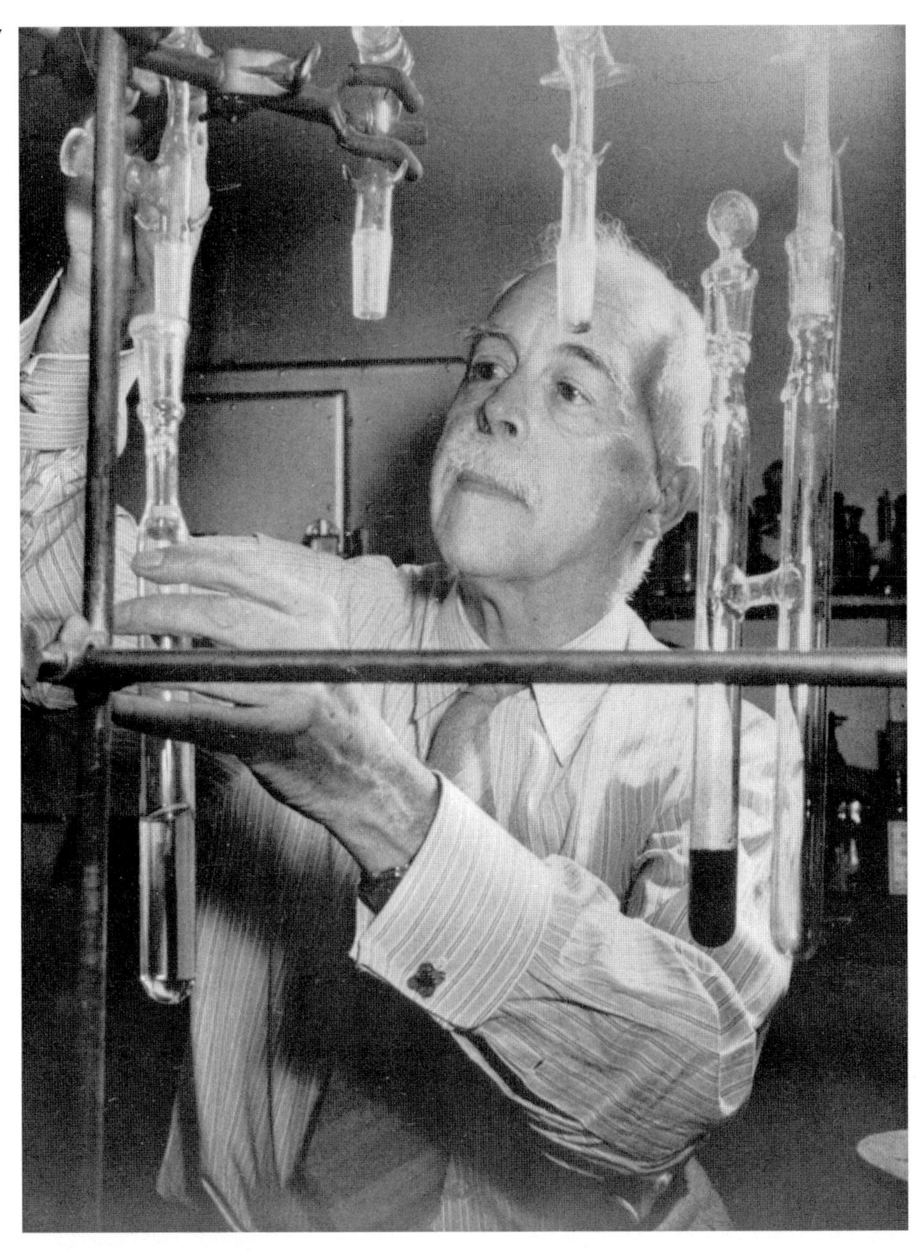

Chemistry dean Gilbert Lewis was the undisputed head of the Berkeley science faculty when Lawrence arrived.

8

The first "Rad Lab." In 1931 Berkeley president Robert Sproul gave Lawrence this ramshackle campus building, slated for demolition, to house the new 27-inch cyclotron. Bill Brobeck later judged its chief virtue to be that "no one objected if you drove a nail into it anywhere."

9

The 27-inch, with its Livingston-designed vacuum chamber nestled between the pole faces of its magnet, and its distinctive horseshoe-shaped support yoke. Livingston (*at left*) is wearing his characteristically somber expression.

10

The 1933 Solvay Conference on Physics brought together the science's aristocracy, including a score of current or future Nobel laureates. Lawrence, the sole American invited, is standing, second from right. Among the other notables were (*circled, left to right*) Erwin Schrodinger, Frederic Joliot, Irene Joliot-Curie, Werner Heisenberg, Niels Bohr, Enrico Fermi, E.T.S. Walton, Marie Curie, George Gamow, Ernest Rutherford, Wolfgang Pauli, John Cockcroft, Rudolf Peierls, Lise Meitner, and James Chadwick. Attending but not shown: Albert Einstein.

11

First-generation Rad Lab scientists and staff, arranged before the 27-inch in 1933: (*from left*) Jack Livingood, Frank Exner, Stan Livingston, David Sloan, Lawrence, Milton White, Wesley Coates, Jackson Laslett, and Telesio Lucci.

12

Cooksey's 37-inch vacuum "can" in 1937 brought the machine's engineering to a new level of precision.

13

The first external beam was produced on March 26, 1936, by "snouting"—so named because of the pig's-snout shape of the portal carrying ions out of the vacuum tank. By November 1937, when this photo was taken, the beam was refined to the point that it allowed experimentation far from the interfering field of the cyclotron and treated visitors to thrilling displays of its lavender glow.

14

The second Rad Lab. Built in 1938 with money from Berkeley regent William Crocker, the Crocker lab housed the new 60-inch atom smasher—the "Crocker Cracker"—in its two-story rear wing.

15

Rad Lab staff and associates seated in the lap of the magnet yoke of the 60-inch, under construction in the still-skeletal Crocker Lab. Physicist Maurice Goldhaber compared this Big Science photograph to his memories of Ernest Rutherford holding his experimental apparatus in his own lap. *Front row, left to right:* John Lawrence, Robert Serber, Franz Kurie, Raymond T. Birge, Ernest Lawrence, Don Cooksey, Arthur Snell, Luis Alvarez, Philip Abelson. *Second row:* John Backus, Wilfred Mann, Pail Aebersold, Edwin McMillan, Ernest Lyman, Martin Kamen, David Kalbfell, Winfield Salisbury. *Back row:* Alex Langsdorf, Sam Simmons, Joseph Hamilton, David Sloan, J. Robert Oppenheimer, William Brobeck, Robert Cornog, Robert R. Wilson, Eugene Viez, Jack Livingood.

Bill Brobeck in 1938, next to discards from the lab's neutron-shielding equipment: cans filled with water.

17

The 60-inch enhanced the lab's usefulness for the biomedical research that paired Ernest (*front*) with his brother, John, seen here together at the machine's control console.

18

Molly and Ernest on the Crocker Lab stoop with children Margaret and Eric, 1939.

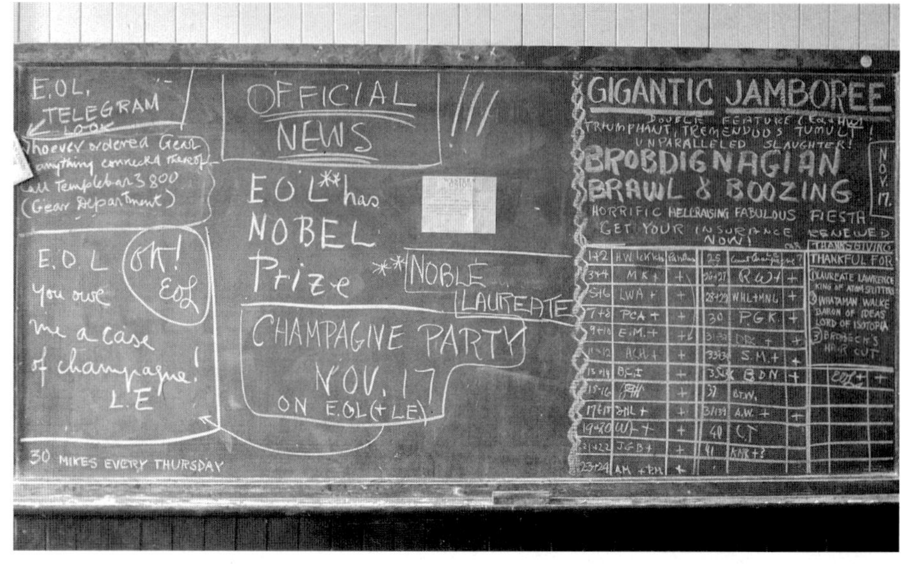

19

Nobel Laureate, November 9, 1939. The Rad Lab blackboard bears the news, along with the announcement of the celebration to follow. The "L.E." who wrote the message in the lower left-hand corner was Lorenzo Emo, an Italian count who was among the paid staff.

20

As the European war made travel to Sweden for the Nobel ceremony impossible, Molly, Ernest, Gunda, John, and Carl celebrated the Nobel the following February at an event in Berkeley.

21

Iconic photo of America's scientific elite: Five eminent scientists and administrators meeting with Lawrence to review the grant application for the 184-inch cyclotron on behalf of the Rockefeller Foundation, at the Rad Lab, during the last week of March 1940. Less than two years later, they would come together again, this time as the civilian leadership of the atomic bomb program. *From left:* Lawrence, Arthur Compton, Vannevar Bush, James B. Conant, Karl Compton, Alfred Loomis.

22

Dinners at DiBiasi's Restaurant in Albany, near the Berkeley campus, were an annual event for Rad Lab members and their guests. This photograph by Donald Cooksey dates from the early 1940s. Helen Griggs, seen at the far right, was Ernest's secretary and the future Mrs. Glenn Seaborg.

23

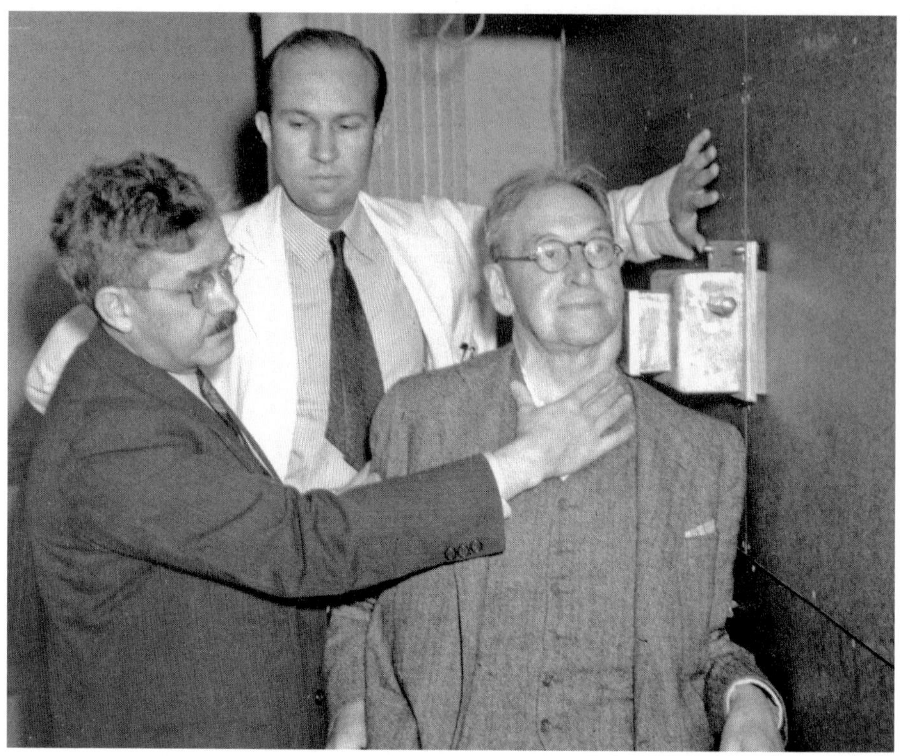

First direct irradiation of a patient, November 20, 1939. Robert Stone of the University of California medical school (*left*) positions patient Robert Penny at a portal of the 60-inch cyclotron in a room specially constructed for medical treatment, as John Lawrence looks on.

24

Plotting the campaign for the 184-inch grant? The Rockefeller Foundation's Warren Weaver (*left*) confers with Alfred Loomis and Ernest Lawrence in a Washington hotel room in 1940. Loomis's influence helped Weaver and Lawrence secure the $1.15 million grant, although its size strained the Rockefeller Foundation's resources.

25

Crucial meeting of the atomic age, September 22, 1941. Australian physicist Mark Oliphant (*left*) confers with Lawrence in the shadow of the massive 184-inch cyclotron magnet, awaiting construction of its home in the hillside above the Berkeley campus. Oliphant informed Lawrence that British scientists had determined that an atomic bomb was feasible; three days later, in Chicago, Ernest passed the information to James Conant and Arthur Compton and committed himself personally to the bomb project.

26

After the reconstituted S-1 committee finally got atomic bomb planning under way, its members met at the exclusive Bohemian Grove in California in September 1942, with Lawrence as host. *From left:* Harold Urey, Lawrence, James Conant, Lyman Briggs, Eger Murphree, Arthur Compton.

27

Lawrence donated the unfinished 184-inch to the bomb effort by converting it to a mass spectrograph for the separation of uranium isotopes. *From left*: J. Robert Oppenheimer, newly appointed to head the Manhattan Project's bomb design lab at Los Alamos; future Nobelist Glenn Seaborg, developing plutonium for the bomb program; and Lawrence inspect the machine's control console. The separation technique, implemented at Oak Ridge, would produce the uranium core for the Hiroshima bomb.

The "alpha racetracks" like this one at Oak Ridge separated fissionable U-235 from natural uranium at a high volume, though not enough to build more than a single bomb.

29

Glenn Seaborg (*left*), working with émigré physicist Emilio Segrè, isolated the "fiendishly toxic" element 94 at the Rad Lab and dubbed it plutonium. In 1966 the discoverers prepared to turn over to the Smithsonian Institution the cigar box in which they deposited their very first sample of the elusive element. The box had been donated by the inveterate cigar smoker Gilbert Lewis.

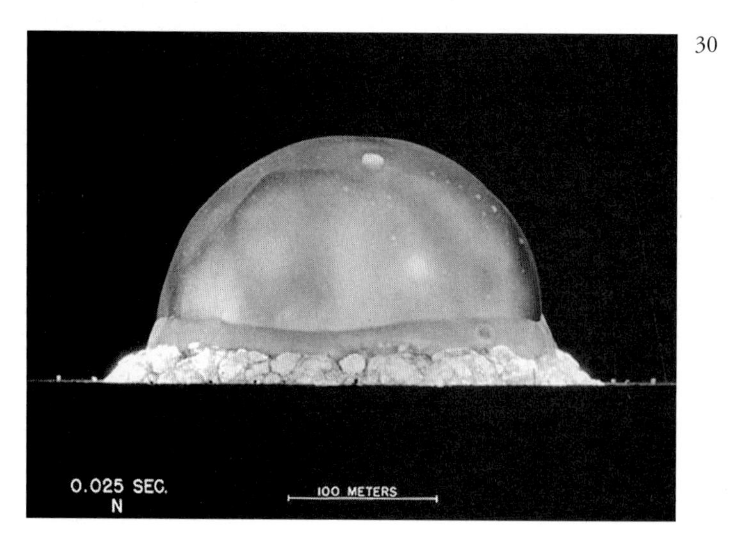

30

The Trinity fireball at 25 thousandths of a second after detonation, witnessed by Lawrence and others from a watching post twenty miles from the test site.

31

Oppenheimer and Gen. Leslie Groves, head of the Manhattan Project, inspect the test site after the blast.

Headquarters
Atomic bomb Command
August 9, 1945.

To: Prof. R. Sagane.
From: Three of your former scientific colleagues during your stay in the United States.

We are sending this as a personal message to urge that you use your influence as a reputable nuclear physicist, to convince the Japanese General Staff of the terrible consequences which will be suffered by your people if you continue in this war.

You have known for several years that an atomic bomb could be built if a nation were willing to pay the enormous cost of preparing the necessary material. Now that you have seen that we have constructed the production plants, there can be no doubt in your mind, that all the output of these factories, working 24 hours a day, will be exploded on your homeland.

Within the space of three weeks, we have proof-fired one bomb in the American desert, exploded one in Hiroshima, and fired the third this morning.

We implore you to confirm these facts to your leaders, and to do your utmost to stop the destruction and waste of life which can only result in the total annihilation of all your cities if continued. As scientists, we deplore the use to which a beautiful discovery has been put, but we can assure you that unless Japan surrenders at once, this rain of atomic bombs will increase manyfold in fury.

To my friend Sagane
with best regards from
Luis W. Alvarez

Formally signed
Dec 22, 1949

Letter handwritten by Luis Alvarez on behalf of himself, Robert Serber, and Phil Morrison of the Rad Lab to their former colleague Ryokichi Sagane, warning of continued nuclear destruction if Japan did not surrender "at once." The letter was dropped on Nagasaki with the plutonium bomb on August 9, 1945. It failed to reach Sagane until after the war and was returned years later to Alvarez, who put his signature to it and delivered it again to Sagane as a mordant keepsake.

33

Two patrons meet: As Berkeley president Robert Sproul (*right*) looks on, General Groves presents Lawrence in March 1946 with the Medal for Merit, then the U.S. government's highest civilian honor for service to the country. Among other participants in the bomb program to receive the award were Bush, Conant, Fermi, Loomis, and Oppenheimer.

34

Leaders of Lawrence's Berkeley and Livermore empire during the postwar bonanza, which enriched the university and secured the Rad Lab's position as the preeminent center of high-energy physics in the nation, perhaps the world. *From left:* Glenn Seaborg, Edwin McMillan, Lawrence, Donald Cooksey, Edward Teller, Herbert York, Luis Alvarez.

Finally installed and operating as an accelerator after the war, the 184-inch was big enough to accommodate the burgeoning Rad Lab staff in 1946, when it was converted into a synchrocyclotron to exploit the "phase stability" discovery by Edwin McMillan (*standing eleventh from right*). Lawrence can be seen seated with his hands clasped on the lower pole face, near the center of the photo, with the omnipresent Cooksey at his left.

Ernest Orlando Lawrence, overlooking the 184-inch cyclotron at what is now Lawrence Berkeley National Laboratory.

shaken off the torpor observed by Livingston in midsummer and were all operating again. From overseas came a cable signed by Bohr—"all institute wishes [to] express thanks and admiration"—for the Copenhagen machine finally had been finished and made operational with the help of Rad Lab veteran Jackson Laslett. From Soviet Russia there was word of a project to build a cyclotron—known there as a "Lawrence apparatus"—under the supervision of physicist Igor Kurchatov, later to become famous as the father of the Soviet atomic bomb. That machine was developed without advice or assistance from Berkeley but with funding cadged from the Soviet regime very much by the Berkeley method: by stressing its potential contributions to medical research. (Testing on the Russian cyclotron's vacuum chamber began in Leningrad on June 1, 1941, but the machine never ran, as the German invasion of Russia prompted its physicists to flee the city three weeks later.)

In this triumphal atmosphere, rumors were heard again that Ernest might land the Nobel Prize. The high caliber of the competition, which included Fermi and the team of Cockcroft and Walton, made for long odds, but news reporters and newsreel photographers gathered at the lab and camped at the Lawrence home on hilly Tamalpais Road on the date of the scheduled announcement from Stockholm. Microphone cables crisscrossed the living room, posing an obstacle course for a pregnant Molly as she waited impatiently for Ernest to traipse home from campus. At lunchtime came the word: it was Fermi. (Cockcroft and Walton would have to wait until 1951 to receive recognition for the work they performed in 1932.) The newsmen packed up their gear, relieving Molly, who had cringed at the thought of being photographed in her laden condition and having to brace up for a royal reception in Sweden. Lawrence tendered his congratulations to Fermi via Segrè, who was now in residence at the Rad Lab and who observed that despite Ernest's graciousness, he was "clearly disappointed."

For the rest of that year and into the next, the sixty-inch, despite its painstaking engineering and construction, exasperated its handlers. Brobeck regimented the operating crews into shifts, but the increased specialization in the Rad Lab failed to overcome the machine's vulnerability to bugs that could knock it off-line for more than a week at a time. The burden fell most heavily on Martin Kamen and others tasked with isotope production. "The new cyclotron still continues to be a creature of chance and in its present condition cannot be expected to run continuously for another month," he wrote in late November to Paul F. Hahn, a physiologist at the University of Rochester who was anxiously awaiting a shipment of radioactive iron for his research on the bloodstream. Kamen offered Hahn some "excreta" from the thirty-seven-inch, including iron shavings and the remnants of a recently used probe, "to tide matters over until the advent of the iron utopia I have been so glibly promising for the last three months."

Kamen's chemistry skills had steadily raised his profile in the Rad Lab, to the point where his original panic over being exiled back to the University of Chicago gave way to the fear of collapsing from overwork. He celebrated the "dramatic improvement in my prospects," which included marriage—his second—to a bride who shared his deep love of music. He also carried the evidence of his scientific labors everywhere, for his heavy exposure to cyclotron targets, as he recalled, "kept me in a steady state of radioactive contamination, which rendered me *persona non grata* around assay equipment." During one collaboration, he and Philip Abelson were bedeviled by the erratic performance of an ionization chamber until Abelson noticed that its behavior corresponded to Kamen's wanderings around the room. He ordered Kamen to disrobe, garment by garment. Finally, they got to the last piece of clothing: Kamen's trousers. They draped these over the balky apparatus and localized the radiation source to Kamen's fly, which evidently had collected a heavy burden of waste radio-phosphorus from Kamen's work for John Lawrence.

Yet once the sixty-inch shook off its quirks in the spring of 1939, it began to perform as spectacularly as its designers expected. Lawrence touted its achievements proudly; to Gerald Kruger of the University of Illinois—to whom he had once confessed being "almost driven to distraction" by the eccentricities of the twenty-seven-inch—he proclaimed the new machine's "amazing smoothness and stability." To J. Stuart Foster, who was planning a cyclotron at McGill University in Montreal, where he was an associate professor of physics, Ernest boasted of its 33-million-volt alpha particles and an output of neutrons and radioisotopes "no less than prodigious."

Now it was about to bulk especially large in Kamen's career and establish a new milestone in the Rad Lab's reputation. The episode began with a prickly remark by Columbia's Harold Urey, a chemist who maintained a lively skepticism about the cyclotron's usefulness. Urey questioned publicly whether radioactive isotopes would ever play an important role in biological research, for the simple reason that no long-lived radioactive isotopes had been found for hydrogen, carbon, nitrogen, or oxygen, the essential building blocks of life. This was true at the time of Urey's writing: of the known radioisotopes, carbon-11 had a half-life of only twenty-one minutes; nitrogen-13, only ten; and oxygen-15, only two. Hydrogen's naturally occurring radioactive isotope, tritium, which has a half-life of more than twelve years, would be isolated late in 1939 by Luis Alvarez and Robert Cornog of the Rad Lab, using the sixty-inch cyclotron; but they published the bulk of their findings only in 1940. All four elements had stable isotopes, however, which Urey was assiduously producing in his own lab for a wide range of studies. He had also begun negotiating a contract with Eastman Kodak for large-scale industrial production of these nonradioactive substances.

Lawrence was deeply irritated by Urey's doubts. It was not merely that he took Urey's words as a personal affront, but he understood that if Urey's prediction proved correct that radioactive tracers would not be found for these essential biological materials, an important pillar of the Rad Lab's fund-raising edifice would be undermined. So one day in September he issued an urgent summons to Kamen. The young chemist sprinted up three flights to Lawrence's office in LeConte Hall to find him in a state of agitation over Urey's needling. "What can we do about this?" Lawrence asked.

Kamen thought he might have an answer, but it would not be easy. In collaboration with a biology grad student named Sam Ruben, he had been probing the mysteries of photosynthesis using carbon dioxide tagged with carbon-11, but the isotope's short half-life had brought them to a dead end. The one prospect of success lay with carbon-14, which was expected to be radioactive. The problem was that although its existence had been conjectured for years, no one had found it. Ed McMillan had staged the Rad Lab's most determined search for the isotope by propping a bottle of granular ammonium nitrate in the path of the thirty-seven-inch machine's neutron beam for several months. That effort ended when the bottle was accidentally knocked from its perch and smashed on the floor. No one was even sure if carbon-14 was long- or short-lived—its elusiveness might be caused by its having such a short half-life that its radioactivity dissipated before it could be measured; or such a long life that the telltale emissions were too rare to be spotted. Kamen told Lawrence that he and Ruben would have been happy to continue their search, but the oversubscribed cyclotron was unavailable for the lengthy bombardments the program required.

"I would need it full-time," Kamen said.

"You have it," Lawrence replied instantly.

Lawrence granted him priority access to both the thirty-seven-inch and sixty-inch cyclotrons for a "systematic and energetic campaign" to find long-lived radioactive isotopes of carbon, nitrogen, or oxygen, with carbon-14 the main quarry. Kamen staggered down the stairway in a daze and made a beeline for Ruben's lab, located in a derelict old annex to the chemistry building known as the Rat House. The two researchers were a Mutt-and-Jeff pair—Ruben, who had boxed for a boys' club coached by former heavyweight champion Jack Dempsey and starred on the Berkeley High School basketball team, towered over his squat, unathletic partner.

That said, Kamen was very much Ruben's superior in the academic pecking order, for the latter was a freshly minted PhD still seeking a faculty position in the Chemistry Department.

The search for carbon-14 was a messy, exhausting job. Their carbon preparation was a paint-like graphite suspension known as Aquadag, which they smeared onto probes for insertion into the cyclotron tank to be bathed in the deuteron beam for days at a time. Periodically Kamen would withdraw a probe and chip off the irradiated Aquadag, trying not to think about the exposure he might be absorbing. The target material went to Ruben for analysis while a resmeared probe got plugged back into the tank.

The launch of the carbon-14 project coincided with another round of anxious expectation about a Nobel Prize for Ernest Lawrence. The few doubts being heard stemmed from rumors that the Nobel committee might suspend the prizes for the duration of the European war, which had begun with Germany's invasion of Poland on September 1. Instead, the committee made one last round of awards in 1939, and then no more until 1943.

On November 9 the announcement came, first by telegram and then by phone from the Swedish consulate in San Francisco. Ernest had been working out his nervous anticipation on the tennis court. The news reached him there in mid-set.

At the Rad Lab, the telegram was tacked on the blackboard next to the scrawled words "EOL has NOBEL PRIZE" and the announcement of a congratulatory "BROBDIGNAGIAN BRAWL & BOOZING Horrific Hellraising Fabulous Fiesta," scheduled for DiBiasi's on November 17. The occasion would live up to its billing—a raucous gathering at which the attendees tried to top one another with randy limericks and songs. Aebersold's contribution was set to the tune of "The Ramblin' Wreck from Georgia Tech":

> And then he bombed some common lead and turned it into gold
> The prexy jumped around with joy and loudly shouted "Hold

> I am convinced the thing is good—no more I'll have to go
> To the solons up in Sacrament' to ask them for some dough."

There was a cake in the shape of the sixty-inch, marked with the words of a telegram from Art Snell, a cyclotroneer at the University of Chicago. "Dear Ernest, Congratulations," he wrote. "Your career is showing promise."

Lawrence was the University of California's first Nobel laureate—indeed, the first from any public university in the United States. But the prize carried a greater significance. By honoring both the invention of an essential instrument for large-scale research and the creation of a laboratory model to put it to use, the Nobel committee validated a sea change in science. In the presentation speech, Professor K. M. G. Siegbahn of the Royal Swedish Academy of Sciences praised the cyclotron as "without comparison, the most extensive and complicated apparatus construction carried out so far." The debate behind the scenes had focused on Lawrence's role as the quintessential pioneer of Big Science. "Well, what has Lawrence done?" Bohr waspishly lectured G. P. Thomson of the Cavendish, who had been stumping for Cockcroft and Walton. "Invented an instrument which would have been more or less obvious to anybody unfamiliar with the difficulties of experimental technique, made it to work, and done nothing with it except to incite a large number of very able experimental physicists all over the world, unsuccessfully, to emulate his methods." Cockcroft and Walton had built a big machine, yes, but as Bohr observed, it was not as big as Lawrence's, nor had they carried it through to the invention of a new paradigm of research experimentation. (George Paget "G. P." Thomson was the son of J. J. Thomson and a Nobel Laureate in his own right—in 1937, for research into the properties of the electron.)

The last word came from Mark Oliphant, one of the few Cavendish alumni to support Lawrence's candidacy over his former colleagues, and one of those who noticed the shift in the Nobel committee's interest from pure theory to the hard toil of experimentation: "It is extremely encouraging to find that the Nobel Prize Committee . . . is now recognizing the tremendous importance of technique in scientific investigations," he wrote the new laureate. "The technical side of the subject is now recognized as equally important with advances that follow from the use of these techniques, and more important, I hope, than the theories that endeavor to explain them." An assiduous fund-raiser himself and the recipient of Lawrence's confidences about a planned new cyclotron to dwarf the Crocker Cracker, Oliphant was fully alive to the real value of the prize beyond the $40,000 check that came with it: "It is certain that you will have no difficulty now in raising funds for your 'father of all cyclotrons.' "

The formal presentation of the Nobel Prize was held on February 29, 1940, in Berkeley. Kamen and Ruben were bringing their work to a close while the frenzied planning for the ceremony proceeded around them. In the second week of February they staged a marathon seventy-two-hour bombardment by the Crocker Cracker. Kamen was alone in the lab when the run ended late on the third night. Thunder and heavy rain filled the air, punctuated by inhuman shrieks emanating from a recording of a grisly French melodrama being played for a French class on a nearby balcony. Before dawn, he withdrew the last probe from the vacuum tank, scraped the flecks of Aquadag into a bottle, and deposited it at the Rat House for Ruben. On his way home, disheveled and hunched over in the pouring rain, Kamen was picked up by a Berkeley police cruiser and slotted into a lineup for the surviving witness of a multiple murder committed that night. He passed the test and was released to go home, where he collapsed into bed and slept for twelve hours. After awakening, he watched Ruben complete the assay—from a distance, for he was radioactive. Their counter clicked; there were emissions just above the background level of radioactivity, but that was enough. It was February 27, and the search was over. They had carbon-14, and by their calculations, its half-life was at least 1,000 years. (In fact, it is about 5,730 years.) The material they had isolated would be the most important biological isotope of all, and the key to a wide range of research requiring precise biological tracing and dating.

Quivering with the uncertainty of pioneers, dreading that they might have been led astray by some minuscule miscalculation, they wrote a short note for publication and hastened to Lawrence's house for his approval. They found him in bed with one of his chronic sinus infections, resting up for the Nobel ceremony two days hence. Overjoyed at having bested the disdainful Urey, "he jumped out of bed, heedless of his cold, danced around the room, and gleefully congratulated us," Kamen recalled. What they did not anticipate, however, was Lawrence's reaction to the publication the following March of their note in the *Physical Review*. Ruben's name was listed first, before Kamen's.

The explanation was simple; the consequences for Kamen's career rather more complex. Ruben had requested the lead position. This struck Kamen as presumptuous, for he was the senior member of the team and the one whose work had been the more important. But Ruben pleaded that his quest for tenure in the Chemistry Department rested on a knife edge, not least because of the acrid hint of anti-Semitism in the Berkeley administration. Lead authorship of a landmark paper could tip the balance in his favor. Kamen, with a surfeit of empathy and a lack of foresight, assented. But when Lawrence saw the publication and heard the lame explanation, he turned his back on Kamen and strode away in wordless fury at the Rad Lab's subordinate position. The discovery of carbon-14 was the high water mark of Kamen's career in the Rad Lab; its publication was the beginning of his painful journey out of Lawrence's good graces. Even worse, his act of solicitude for a good friend resulted in his forever being known as the junior partner in a momentous discovery.

But that came later. At the Nobel award ceremony in Berkeley, Kamen was still basking in the glory of having demonstrated, once again, the spectacular capabilities of the cyclotron. Birge delivered the news to the audience in Wheeler Hall, accurately describing carbon-14 as "certainly much the most important radioactive substance that has yet been created." At his suggestion, the entire audience turned to Kamen and applauded.

But the day belonged to Ernest Lawrence. He took full advantage of the occasion, reminding his listeners that the cyclotron was the source of discoveries of "immediate practical significance" and pointing to the next goalpost: "The domain of energies above a hundred million volts . . . a territory with treasures transcending anything thus far unearthed.

"To penetrate this new frontier," he said, "will require the building of a giant cyclotron, perhaps weighing more than four thousand tons"—twenty times larger than the Crocker. "Of course, such a great instrument would involve large expenditures," Lawrence said. "Perhaps I might say that the difficulties in the way of crossing the next frontier in the atom are no longer in our laboratory. They constitute a very considerable financial problem, which we have handed over to President Sproul!"

If Sproul was startled at those words, he kept his reaction under wraps. But he could not be unaware that Lawrence's program would once again strain the university's resources beyond the breaking point. More sources of funding plainly would be needed. Fortunately, the new laureate would find them waiting in the wings, starting with an individual who would do more to shape the course of scientific research, with less public renown, than almost any other figure in the late twentieth century.

Chapter Ten

Mr. Loomis

All during the 1930s the legend had grown, percolating through the physics fraternity by word of mouth: tales of a fabulous hilltop laboratory, exquisitely equipped, owned by a mysterious millionaire who moved through the highest circles of business and politics as quietly as a shadow. The lab was located in a secluded community of country mansions an hour or two north of New York City, and not just anyone could gain admission. It was not enough to know somebody. If you were an unremarkable talent, it was almost impossible to obtain an invitation; but if you were an accomplished scientist or one of manifest promise, somehow the invitation would come to you.

The millionaire's name was Alfred Lee Loomis, and his domain was known as Tower House. The sprawling Tudor monstrosity in the plutocrats' retreat of Tuxedo Park, New York, had been built at the turn of the century by a wealthy banker and abandoned amid the grief of a family tragedy. Alfred Loomis spent part of the fortune he had made on Wall Street to buy the derelict house in 1926, and then fitted it out with the latest in scientific equipment and embarked there on a new career, his third. He had already made his mark as a successful lawyer and as an investment banker. Now he became a physicist. Measured by his influence on the lives of his fellow Americans, that would be his most important role, though his least known.

The "peculiar accomplishment" of Alfred Loomis, it would later be written, was "being a public figure without letting the public in on it." Within a select circle, however, he was an eminent and respected figure. The visitors' roster at Tower House listed Albert Einstein and Werner Heisenberg. Niels Bohr, during a crucial visit to the United States just before the onset of war, stopped there to see Loomis before continuing to Washington; it was said that he was dining at Tower House when he received a telegram from Europe relaying the earthshaking news of the fissioning of uranium.

Ernest Lawrence first came to Tower House in 1936. The deep friendship between him and Loomis that began during that visit would change both men's lives.

Loomis was a throwback. Luis Alvarez, who was privileged to serve as a protégé of both Loomis and Lawrence, called him "the last of the great amateurs of science." Alvarez was using the word *amateur* in its traditional sense: someone who engaged in a pursuit not for money or fame but purely for the love of knowledge and the abstract quest for truth. This was a nineteenth-century paradigm of science, the milieu of Charles Darwin, Lord Henry Cavendish (the gentleman scholar in whose honor Cambridge's physics lab would be endowed and named), and the distinguished Egyptologist Lord Carnarvon. Come the turn of the century, professionalism transformed careers in the laboratory and the university classroom into middle-class pursuits, which is why Loomis's conversion of Tower House into a private scientific preserve hinted at a kind of antique eccentricity. Yet that was the wrong impression to take from Alfred Loomis's devotion to science. His outlook was very modern indeed.

Loomis was born on November 4, 1887, to a family of eminent Yankee physicians. They were not wholly of the American aristocracy but tolerably close, with connections and affluence that endowed Alfred with upper-class tastes and bearing. His mother was a Stimson, a grand old family in banking and law. One of Alfred's cousins and contemporaries, Henry Stimson, would distinguish himself in public service with one stint as secretary of state and two as secretary of war.

Alfred followed a conventional educational path for his class: prep school at Phillips Academy in Andover followed by Yale. As a youth, he displayed prodigious talents for magic and for chess, at which he could play two blindfold games simultaneously. Yet in later life, his attitude toward these skills reflected a curious facet of his personality: having mastered them, he abandoned them almost entirely to move on to new challenges. Luis Alvarez would observe that in his thirty-five-year association with Loomis, he never saw him perform a magic trick for an adult audience (though occasionally for children) or even mention the game of chess, noting, "his homes contained not a single visible board or set." It was as though each stage of his life required perfect concentration, undistracted by old pastimes. The pattern would recur over and over, but there was an important exception. Throughout his life, Loomis remained fascinated with technology—"gadgeteering," as he described it. In his youth, there were model planes and radio-controlled automobiles; and in maturity, he would play a key role in the development of artillery technologies, radar, and ultimately the most important "gadget" of all: the atomic bomb.

After Yale and Harvard Law School, Loomis took a job at his Stimson cousins' Wall Street law firm. Winthrop & Stimson was a familial partnership that prided itself on its integrity and the rectitude of its clientele. The firm shunned those who came seeking a defense of indefensible behavior, which was common enough in the unprincipled corridors of the financial district. If that meant that Stimson partners earned rather less money than lawyers at other firms, so be it; they made enough, and kept their scruples besides.

That stage of Loomis's life ended with World War I. At the age of twenty-nine, he enlisted in the officer's corps, where his mathematical skills, honed at Yale, were promptly put to the test. At Aberdeen Proving Grounds in Maryland, he helped invent a new device for measuring the velocity of a cannon-fired shell, essential for accurate range finding, winning a patent for the work. Loomis obtained other patents, including one for a mechanical horse-race toy, but he was proudest of the Aberdeen

Chronograph, which became standard field equipment for the army and the navy. It was the only invention he cited in his entry in *Who's Who.*

After his sojourn among the scientists and engineers of Aberdeen, Loomis returned home to find the practice of law at the Stimson firm unutterably dull. An escape hatch was provided by his brother-in-law Landon Thorne, a Wall Street wunderkind who proposed that they form an investment banking partnership. They made a perfectly complementary pair, Thorne a skillful salesman who hawked the securities transactions contrived by the introverted Alfred. "Loomis had ninety ideas a minute," one of their financial backers observed. "Thorne knew how to pick the good ones and put them to work."

The partners rode the growth of the electric utility industry to wealth and influence in the postwar economic boom. By the 1930s, however, utility holding companies came to symbolize the corporate corruption and greed that the average American saw as the root cause of the Great Depression. Loomis's experience as a utility financier would come to color his opinion of Franklin Roosevelt, who was determined to break up the very firms that Loomis had organized—among them the Commonwealth & Southern Corporation, which faced a new rivalry from the Tennessee Valley Authority, one of the earliest New Deal initiatives. "He thought [TVA] would destroy the business world," recalled Loomis's daughter-in-law Paulie.

By the time of FDR's accession, Loomis already had achieved his goal of becoming a millionaire—$50 million was the figure later cited by Henry Luce's *Fortune* magazine—and managed as well to protect his wealth from the crash of 1929. Like other financiers, he was deeply demoralized by the post-crash hostility in Washington and the nation toward Wall Street and the banks, so in the first years of the New Deal, Loomis resigned from most of his board seats and sold almost all his corporate stock. "Without so much as a backward look," his biographer observed, "Loomis quit Wall Street for good."

The foundation for the next dramatic change of his life's course had been laid years before. The catalyst was his friendship with Robert W. Wood, a Johns Hopkins physicist whom Loomis had met at Aberdeen. After World War I, when Wood volunteered to tutor Loomis in physics, Loomis responded with an uncommon offer: "He suggested that if I contemplated any research we might do together which required more money than the budget of the Physics Department [at Johns Hopkins] could supply, he would like to underwrite it."

That was in 1924. Wood took up Loomis's proposal to work in the developing field of ultrasonics, which involved sound waves with frequencies inaudible to humans. Ultrasonics had promising potential for physics, chemistry, and biology, ranging from the detection of underwater objects to the removal of diseased tissue, but the research required hugely expensive equipment. Without hesitation, Loomis escorted Wood to General Electric's research headquarters in Schenectady, New York, where they ordered an immense high-powered generator to be shipped to Loomis's residence in Tuxedo Park and installed in the garage.

Within two years, the burgeoning lab had outgrown its drafty quarters. So Loomis purchased the decrepit Tower House, its stained-glass windows long since shattered by vandals, its vast salons draped with spider webs and infested by field mice (also, according to local legend, by ghosts), and proceeded to turn it into a scientific palace.

As other wealthy men collected art, Alfred Loomis collected scientists. "Queer things" went on at the old house on the hill, the locals later told *Fortune*. "Strange outlanders with flowing hair and baggy trousers were settling down there for weeks and months on end. They were performing all kinds of crazy experiments—cooking eggs and killing frogs with sounds that nobody could hear, making turtles' hearts beat in a dish, etc., etc."

Yet the lab was no dilettante's plaything. In 1926 and 1928, Loomis and Wood sailed to Europe to tour its great laboratories. They met with Ernest Rutherford himself—"very abrupt in conversation," Loomis would recall. In the midst of their exchange, the great man suddenly burst out: "You damned American millionaires! Why can't you give me a million volts, and I will split the atom." The nonplussed Loomis replied, "Well, we don't know how to make a million volts that can be useful to you. We can only make sparks jump."

Tuxedo Park became an obligatory stopover for eminent scientists visiting the United States. The place teemed with first-class scientists (and during the summer months, with their families). Loomis's guests arrived by private railcar and were housed in luxury, but they came to do serious work. In the decade after 1927, sixty-six papers were published in scientific journals based on research carried out at Tower House. Regular convocations there attracted participants from the highest echelons of international research; a conference in January 1928 honoring the German Nobel laureate James Franck featured presentations by Franck (his first lecture in the United States), Robert Wood, Karl Compton of Princeton, and W. F. G. Swann, delivered to an audience bathed in the light of the restored stained-glass window of the Tower House library. All through the Depression, the *Physical Review*'s customary publication invoice to researchers for their submitted articles came with a note stating that if they or their university could not pay the bill, it would be covered by an "anonymous friend" of the American Physical Society. The anonymous friend was Alfred Loomis.

The circumstances of Ernest Lawrence's first visit to Tuxedo Park are murky. Loomis recalled merely that Ernest came out for a long weekend during 1936 and that an invitation to the world-famous inventor of the cyclotron would have been unremarkable. "Every famous scientist had been out there as a routine thing," he recalled.

What he did remember is that he and Lawrence immediately "hit it off." Their personalities and backgrounds were as divergent as Ernest's and Robert Oppenheimer's, but like that other relationship, theirs thrived on their complementarity: the country lad from a state university and the Yankee scion from Andover and Yale; the extroverted fund-raiser and the backstage financier; the professionally accomplished scientific genius and the eager amateur. But there also was that indefinable something that arises between two men destined to be lifelong friends. Separations lasting months seemed to be no more than "gaps in time that weren't any bigger than if I should go upstairs, change my suit, and come down. We would go right on where we left off," Loomis recalled. Their friendship "had all the earmarks of a 'perfect marriage,' " Luis Alvarez asserted: Ernest's "ebullient nature plus his scientific insight and his charisma . . . attracted Alfred to him, and Alfred in turn introduced Ernest to a world he had never known before, and found equally fascinating."

When Lawrence visited New York, which happened with increasing regularity as his fund-raising demands expanded, he invariably stayed at Loomis's Manhattan town house. At dinner parties there and at Tuxedo Park, Alfred and his extroverted mistress, Manette (who would become his second wife in 1945), would introduce him to their society friends, who later would discover to their astonishment that the big-boned, easygoing fellow with the sparkling blue eyes who had entertained them all night with vitality and ease was the brilliant master of an impenetrable science. None could equate this engaging young man with the stereotypical aged, graybearded laboratory scientist of their fancies. "He was just a handsome, big fellow," Manette would recall, "full of loving and full of fun, and very easy to make friends with."

Loomis's assistance to Ernest went beyond the social niceties. In 1936 he established a private fund for Ernest to spend on unscheduled and unrecorded needs of the Rad Lab, including travel and equipment. For several years, the size of the fund remained concealed from university officials. Loomis's donation checks were endorsed to "Ernest O. Lawrence, personal," with the money to be spent at Lawrence's sole discretion. "He probably didn't even have to account to Mr. Loomis," Raymond Birge reflected.

Despite Lawrence's reputation for personal integrity, an unaudited private fund controlled by a faculty member eventually made university accountants uneasy. They soon asked Loomis to make his contributions through conventional channels. In November 1940 he donated $30,000 in stock to the university through President Sproul's office. Loomis instructed Sproul that the university was free to use the funds "in any way that seems fit," but expressed the strong desire that the money be spent "to further scientific research in connection with the various scientific undertakings in which Professor Ernest O. Lawrence is now engaged . . . I hope, therefore, that you will permit Professor Lawrence to interpret broadly the uses to which these funds are to be put." Eventually Berkeley's auditors would bring the fund fully into the formal regime established for all such trust funds, requiring that it be deposited in an interest-bearing bank account and designated as a university asset. Even then, however, they assured Lawrence that the funds could be disbursed "subject to your discretion without the usual University restrictions." In practice, the fund, which Loomis periodically topped off, remained entirely subject to Ernest's control until nearly the end of his life, its outflow documented indifferently but, by all evidence, devoted properly to supplementing the work of the Rad Lab.

Strangely, it was not until 1939 that Alfred Loomis first visited Lawrence's home turf in Berkeley—but then he became nearly a permanent member of the Rad Lab. That initial visit lasted six months, during which Loomis ensconced himself at the elegant Claremont Hotel, located on a verdant hillside not far from campus. Every day he made his way to the Rad Lab in a seven-passenger limousine, which would remain parked next to the wooden building, attended by its chauffeur, while Loomis toiled inside. This was the only sign of plutocracy Loomis displayed at Berkeley; inside the Rad Lab, he spent his days perched on a laboratory stool on the second floor, immersing himself in the minutiae of cyclotron engineering and the physics of the atomic nucleus. He would seek out the younger members of the staff "to learn about us and from us," recalled Alvarez, who was Loomis's junior by a quarter century. "I had never before discussed physics seriously with anyone as old as Alfred." From this experience, Alvarez drew the lifelong lesson that "a scientist can stay active as he grows older only by staying in touch with the youngest generation."

• • •

Soon after Loomis's first visit to the Rad Lab, Lawrence enlisted him in a campaign to satisfy his latest obsession: to build the biggest cyclotron ever conceived, its magnet outweighing the Crocker's by a factor of twenty, and its cost by a factor of ten. In the hallways of the Radiation Laboratory, it was known as the "he-man." Nor was the resonant swagger of its nickname accidental. Lawrence had brought in the sixty-inch despite the warnings of naysayers such as Bethe. That machine, now turning out radioisotopes and bombarding tumors with neutrons in the Crocker Lab, was the most exquisitely engineered cyclotron in the world, thanks to Brobeck's painstaking oversight. After its uncertain launch, its stolid reliability had become almost boring; the Crocker Lab seemed more like an industrial factory every day, so why not push past the boundaries of the predictable once again?

Another factor driving Lawrence's ambition was the speed at which the rest of the cyclotron world was catching up to Berkeley. Thirteen machines of thirty-five inches or more had been commissioned or were already operating in the United States by late 1939. Two sixty-inch giants were on their way, one under construction by Merle Tuve at the Carnegie Institution; another by Mark Oliphant at England's University of Birmingham. Ernest's generosity with plans and personnel had fulfilled his goal of making the cyclotron the indispensable centerpiece of any self-respecting university physics department, but the princelings were beginning to make the king uneasy on the throne.

As so often happens in science, their discoveries only whet the appetite for more research, which in turn created a demand for bigger, more expensive accelerators. Lawrence had the best shot at acquiring the funding to lead the way. "I hope your new apparatus is really big," Chadwick wrote him in April 1938, when the dream of the he-man was just taking shape. "I feel that one ought to make a serious attempt to get up to 60 or 70 million volts . . . With such particles, we should begin to learn the true mechanics of the nucleus." This was not a stretch of the imagination; as Chadwick observed, studies of cosmic rays, which provided even higher energies but were hard to control, already hinted at new particles and new forms of energy swirling within the still-mysterious nucleus. "I think the phenomena in the cosmic rays point the way to us," he wrote.

But Lawrence was thinking about 100 million or even 200 million volts, not Chadwick's mere 60 million or 70 million. He was not complacent about the challenges, though the obstacles he foresaw were geographical and financial, not technical. Aware that there was no room on campus for a machine with a magnet weighing two thousand tons and that its likely profusion of high-energy particles rendered unsafe any site in a populated area, he set his eyes on Strawberry Canyon, a bucolic ravine east of campus and high in the hills. Initially, he penciled out a construction budget of $500,000; by the end of the year, the budget for construction and ten years of operation looked more like $2 million. That fall, he acknowledged to the Rockefeller Foundation's Warren Weaver that "in some quarters it might be considered no less than shocking that we should be looking towards a larger cyclotron almost before the 60-inch is in operation," but no one who knew him could be surprised that he was thinking ahead; the essence of Big Science lay in pushing the limits of research and its tools ever outward. Yet Ernest seemed unwittingly to be nearing his own relativistic limit. War had come to Europe and lurked over the horizon for the United States; the uncertainty of the international situation made talk of an expensive project in basic science seem premature, not to say foolhardy and insensitive.

But two developments that fall of 1939 placed it back in the realm of reality. The first was an offer from the University of Texas to install Lawrence as a vice president at a salary of $14,000 a year, with command over a lavish research budget accommodating the largest cyclotron he could imagine. The second was the Nobel Prize. The first restored the imperative that the University of California take Lawrence's demands seriously. The second, which recognized the very process by which Big Science grew bigger, made Lawrence even more the darling of every scientific research foundation in the land. His vision still seemed outsized, but it could not be dismissed easily.

The fund-raising campaign for the he-man cyclotron kicked off on January 7, 1940, when Warren Weaver arrived in Berkeley to visit with Lawrence and Sproul. Ernest had spent weeks stoking himself up to a high level of optimism. The proposal was "going ahead splendidly," he wrote Loomis just after Christmas. "I can hardly wait for the day when these plans will be realized." He now envisioned a cyclotron with a 4,500-ton magnet and pole faces of 184 inches, a scale that took away the breath of even Rad Lab veterans. "If it looks anything like the artist's conception," physicist Robert Cornog told a friend, "it will be the eighth, ninth, tenth, and eleventh wonder of the world." After Sproul countered the Texas offer by conditionally approving the Rad Lab's annexation of Strawberry Canyon, Ernest requested preliminary plans from the university's supervising architect, Arthur Brown Jr. This was a major step, for Brown was among the most distinguished architects in the Bay Area, with San Francisco City Hall, the iconic Coit Tower, and numerous Berkeley campus buildings to his credit.

The new proposal staggered Weaver, who had been bracing for a grant application for a machine less than half that size. Only a month earlier, Lawrence had given him preliminary estimates of $750,000 for construction and another $250,000 for operating costs over a ten-year period. The cautious Weaver, adding an expansion factor for Ernest's ambitions but still thinking too small, had privately recalculated the budget at $1 million for construction and another $500,000 for ten years of operation. That was a goal that could be reached, barely, if the Rockefeller Foundation, the university, and private industry all contributed; at least, he remarked later, it would be "not an entirely forlorn hope." But as he later informed Sproul, the new plans, which would require $2.65 million for construction and operation, "carried me . . . far beyond any figures which I had ever discussed" with the Rockefeller Foundation's president, Raymond Fosdick. At the minimum, it meant a grant of $1.5 million from the foundation alone.

Weaver sat through his lunch and dinner meetings with Sproul and Lawrence in a daze, trying not to make promises he could not keep. He was painfully aware that Sproul had recently been named to the foundation's board, though the appointment had not yet been announced. Weaver advised his hosts carefully that the foundation might be able to contribute $1 million as an absolute limit, but that was only a guess. Sproul mentioned that he was willing to ask the regents to provide $85,000 a year for ten years, but observed that this would be an unprecedented commitment by the university to a single project—equal to the research funds of all the other departments combined. The rest, he made clear, would have to come from the Rockefeller Foundation and private industry.

Lawrence, viewing the meeting through the scrim of his boundless optimism, overlooked these off-key notes, detecting only a resounding validation of his vision. "Dr. Weaver has come and gone, and his visit was a great success all around," he wrote Loomis. "From the moment of his arrival it was clear that he was very keen for the project; and, as his visit here progressed, he became even more enthusiastic . . . Weaver and Sproul agreed that the project was of such impelling importance that it simply had to be carried out without delay. It was sweet music to my ears to hear them say that they agreed that the project was going to be carried out, and it was only a question to find ways and means."

Back home in New York and safely out of range of Ernest's ebullience, Weaver outlined his doubts in letters to both Sproul and Lawrence. He acknowledged, for the record, that "there seems every disposition here [in other words, at the foundation] to agree that this project is, from the scientific point of view, of the greatest possible interest and potential importance." Then came the dash of cold water: "Were financial and world conditions more favorable, there would be a strong probability that the Rockefeller Foundation would . . . view it as an opportunity which justified support even up to the whole of the capital costs.

"But the joker—and a very serious joker it is, if you will permit the

Hibernicism—lies in the phrase 'were financial and world conditions more favorable.' "

A contribution of $1.5 million was out of the question. The foundation, straitened by the long economic slump and beleaguered by scores of desperate applicants, had just refused a request for the same sum from a medical school (unnamed by Weaver) that faced the collapse of its endowment without it. The board would fear a backlash should it grant the same sum for a project "which an unsympathetic and inaccurate critic might describe as a single instrument for one man." Would it not make more sense, he suggested, to put off the entire project for a few months, even a year? "Professor Lawrence is fortunately still young, there is a great deal of rich experience which can be gained with the 60-inch cyclotron, and there is negligible danger that anyone else will run away with the ball," Weaver told Sproul. This might just be the right moment to take "a somewhat more leisurely examination of all possibilities."

Weaver's words could not have sounded a more discordant note to Lawrence's ears. His letters arrived in Berkeley on Friday, January 26; at dawn the following Monday, Ernest was on the long-distance line, making an emotional plea for the Rockefeller Foundation to fund the cyclotron without delay. "I may be sort of panicky," he told Weaver, "but I am afraid of the general international situation." If things got worse in Europe, philanthropic spending would grind to a halt in the United States; even if the war clouds dispelled, millions of dollars would shift to Europe for economic reconstruction. "That is the reason I am sort of in a hurry . . . It means so much to me—it is almost a matter of life and death."

He offered to shrink the cyclotron to 150 inches, which would reduce the cost to $750,000. He would strip every conceivable frill out of Brown's architecture to create "more of a factorylike building." He would cut his grant request to $500,000 and try to make up the balance from other sources.

To an exasperated Weaver, these ideas were all beside the point. He warned Lawrence that Fosdick was of a mood to ratchet back all foundation spending: the endowment was in such dire condition that less than $1.5 million would be available for *all* grant making for the coming year—a sum that would be consumed entirely by Lawrence's vision. Moreover, Weaver counseled, there would be no point in building a smaller cyclotron now, if everything might be won simply by sitting out a prudent delay. He reminded his impatient supplicant that the full-sized cyclotron of his dreams had important supporters whose backing would yield dividends, if he would simply wait. Only a few weeks previously, Weaver confided, Fosdick had been buttonholed at a benefit dinner by Dave Hennen Morris, a wealthy former ambassador to Belgium and a board member of the Research Corporation. Morris was a Lawrence fan who labored ceaselessly to solicit donations for the Rad Lab from his circle of distinguished friends. In a typical appeal a few days after the Nobel Prize announcement, he had tried to get Edsel Ford—son of Henry, and the president of Ford Motor Company—to subscribe to a $650,000 construction fund by describing the cyclotron as an "epoch-making" device that would "link the names of those connected with it alongside of Newton and Einstein." Ford declined to contribute, but during his encounter with Fosdick, Morris alluded to the breadth of support for Lawrence among other influential donors. "You've just got to play ball with us on the cyclotron," he said.

Weaver assured Lawrence that he himself was "perfectly willing to bleed for this thing," but "we have simply got to take more time." Lawrence asked, "Is there any point in my coming east at the moment?" Blanching at the thought of the indefatigable Lawrence pestering Fosdick in person, Weaver responded sharply, "I don't see it."

But Ernest had one last ace in the hole: Alfred Loomis. Though not a Rockefeller Foundation trustee, Loomis was an intimate of several board members. With the cyclotron grant hanging by a thread, he undertook to hustle them for votes. His principal quarry was Karl Compton, the president of MIT and Arthur Compton's brother, who just had been named a trustee. (His appointment, like Sproul's, had not yet been announced.)

Aware that Compton's background as a physicist would give his opinion special weight, Loomis invited him to spend a week at his private retreat on Hilton Head Island, South Carolina. The goal was not necessarily to change Compton's mind, for he was likely to hold a positive view of Lawrence's project anyway, but to encourage him to express his opinion in especially forceful terms. Having been swaddled in the moneyed elegance that was Loomis's stock-in-trade, Compton duly sent Weaver a written recommendation describing the project as "one of the most interesting, the most potentially important, and the most promising projects in the whole present field of natural science . . . I should definitely place it in the number one position by a large margin among the various scientific projects of which I have knowledge at the present time." He concluded with the declaration that "no one could possibly question the selection of the University of California and Ernest Lawrence as the institution and the scientist to whom the project should be entrusted."

At Loomis's urging, Weaver also asked several prominent physicists for "a considered statement of your opinion and advice" on whether to press for immediate funding for the 184-inch. The return mail brought fulsome endorsements from Niels Bohr, Mark Oliphant, and Frederic Joliot, among other names likely to impress the trustees.

That was enough to persuade Weaver to abandon his strategy of delay. He presented Lawrence's application at the board's upcoming meeting on April 3. But now time was short. In mid-February he had asked Lawrence for ammunition against the most likely objections to the new cyclotron. The first question was whether the proposed machine would be powerful enough to produce mesotrons. These particles (known today as mesons) were thought to be the carriers of the strong force, which binds positively charged protons in the nucleus, counteracting the electromagnetic repulsion that would otherwise drive them apart. But thus far they had been detected only in cosmic rays; the possibility of a laboratory demonstration of their existence might well justify construction of the cyclotron that could achieve it. A machine that fell short of the necessary energies, on the other hand, would look like an enormous waste of money. "Is someone going to come along," Weaver asked, "and say that now a new instrument obviously must be built . . . to produce mesotrons?"

Weaver warned that some might question whether it was necessary to build a new cyclotron at all, since cosmic rays carried energies equivalent to those to be expected from the new cyclotron. "Some of the cosmic ray enthusiasts are very likely to urge that nature has furnished us with particles of extremely high energies," he observed. Why not spend a decade or so exploiting nature's gifts before building an expensive machine to generate the same energies? This query bore the unmistakable fingerprints of Arthur Compton, who reigned as the world's leading expert on cosmic rays. As a Rockefeller Foundation grantee in cosmic ray research, Compton was not shy about reminding Weaver of the potential of such studies to yield the discovery of the mesotron and other fundamental particles; what's more, he stressed, unlike the beams of multimillion-dollar cyclotrons, cosmic rays were gifts from nature and came free of charge.

Weaver saved the thorniest issue for last. This was the painful question of how the Rad Lab had overlooked so many milestones in nuclear physics during the previous decade: "I suppose that the outstanding developments in the investigation of the nucleus during the last few years would include the discovery of the positron by Anderson in 1932; the discovery of the neutron by . . . Chadwick in 1932; the discovery of the phenomenon of induced radioactivity by Curie and Joliot in 1934; the identification of the mesotron from cosmic ray studies; and the discovery of the phenomenon of nuclear fission by Hahn and others in 1939 . . . Is it not a fact that the five entries I have mentioned are outstanding ones, and that no one [*sic*] of these discoveries was made using a cyclotron?"

Unsurprisingly, Lawrence replied in high dudgeon. He disposed of the mesotron question by assuring Weaver that he, Oppenheimer, and Fermi were all agreed that the likely energy of the mesotron was 80 million volts, so that even a 150-inch machine, which he expected to produce projectiles of 100 million volts, should be potent enough to yield the elusive particles.

As for whether it might be best to let cosmic ray studies play out before building the cyclotron, Lawrence observed that the goal of physics was not merely to "make discoveries of natural phenomena" but to do something with them. This goal could not be met by relying purely on nature's bounty: "The discovery of the mesotron in cosmic rays will be of little value in the course of time unless there is developed a means of . . . controlling them, and learning of their manifold properties . . . It means a great deal more to civilization, let us say, to find a new radiation or a new substance that will cure disease than it would to discover a super nova."

Finally, he turned his attention to Weaver's irksome list of missed milestones. The simple reason that Berkeley's cyclotroneers had been beaten to the punch, Lawrence asserted, was that they had remained focused on cyclotron development, in effect, as a service to the future:

> *Every one of these discoveries were "in the air" and would have inevitably been made within cyclotron laboratories within certainly a few months. As we were building up the cyclotron beam, we could not possibly have escaped the discovery of artificial radioactivity, for exampel* [sic], *for more than a month or so after the announcement by Joliot. The development of the cyclotron was begun several years before these discoveries on faith that the availability of controlled atomic projectiles within the laboratory would lead to important scientific progress. If the cyclotron development had been, let us say, a year earlier, I repeat, there is every reason to believe that some of these discoveries referred to would have been made using cyclotrons.*

Lawrence mustered his best arguments for this elaborate defense of his work. But he dodged the truth about several of those discoveries, especially artificial radioactivity and fission: the cyclotroneers had been capable of making them first, had they only tried. It was not the inadequacy of the machine but the Rad Lab's inattention and its narrow focus that prevented the lab from winning credit for them. Weaver had put his finger on the real shortcoming of the cyclotron lab, which was its scientific judgment, not its technical expertise. That fault could properly be laid to Ernest Lawrence, who was still learning how to balance engineering and hard science.

Meanwhile, Loomis continued his campaign to secure the grant. He launched a major initiative during the last week of March, when he sponsored a tour of the Rad Lab for a group that functioned as an ad hoc scientific advisory committee for the foundation: both Compton brothers, Karl and Arthur; Harvard president James B. Conant; and Vannevar Bush, president of the Carnegie Institution of Washington.

Bush was the most important figure in the group for reasons that went beyond his influence with the Rockefeller trustees. A tall, wiry New Englander of fifty, the grandson of Yankee sea captains and the son of a nonconformist Universalist minister, Bush's upbringing had bequeathed him both a salty independence of mind and a respect for formality and traditional values. Trained as an electrical engineer, in the 1920s he had invented a machine known as a differential analyzer: an analog computer whose digital offspring would dominate the information age. Subsequently he became a vice president at MIT (under Karl Compton) before taking up the Carnegie presidency, a position that placed him at the crossroads of government policy and academic research. He was already thinking about the role America's scientists might play in the world war lurking on the horizon. For a year, Bush had been meeting regularly with Conant, Karl Compton, and other leading science administrators to show concern about the country's listless response to a crisis that might easily spread beyond European borders and to ponder the imperative of technological preparedness. "We were agreed," he wrote later, "that the war was bound to break out into an intense struggle, that America was sure to get into it one way or another sooner or later, that it would be a highly technical struggle, that we were by no means prepared in this regard." He was determined to have a hand in moving the country's technical establishment into the forefront of wartime planning. Now, meeting Lawrence in person for the first time, he found himself agreeing with Loomis that the Berkeley physicist should be part of the effort.

Guided along the Rad Lab's corridors by Loomis and Lawrence, the visitors stopped briefly in Don Cooksey's second-floor office. Seated at ease in three-piece suits under a blackboard displaying the dees of a simple cyclotron, they allowed Cooksey to snap a photograph of them grinning companionably at one another as though in shared appreciation of an inside joke. The snapshot was fated to become a historical artifact, for within a year, these same men would come together again—as leaders of America's effort to create the atomic bomb.

The tour of Berkeley would play a critical role in allowing men acquainted chiefly as professional colleagues to take one another's measure as individuals. The process was helped along by a party Loomis threw that weekend at the Del Monte Lodge in Monterey, a few hours down the coast from Berkeley. His goal was to expose them at close quarters to the charismatic energy of his protégé Ernest Lawrence, and he was surely successful at finalizing their approval for the cyclotron grant. "You can't get a group together for a long weekend without Ernest's effect on them," he reflected later. By the end of the weekend, "there was no opposition." The advisory committee transmitted its unanimous endorsement of the 184-inch machine to the foundation before leaving the West Coast. Weaver and Loomis put the finishing touches on the lobbying effort by persuading Fosdick that the cyclotron would stand as a bookend with the foundation's other great scientific investment: the 200-inch Hale telescope planned for Mount Palomar in Southern California, cementing the Rockefeller Foundation's stature as the world's preeminent supporter of Big Science.

On the morning of April 3, Lawrence picked up the phone at the Rad Lab to hear Weaver's voice on the other end. "Our trustees voted $1,150,000," he said. With the $250,000 in operating costs that Sproul had managed to pry from the Berkeley regents, that essentially provided everything Lawrence had requested.

"The full original budget," Ernest marveled over the long-distance line. "It's hard to tell you how I feel."

To Poillon he showed less reticence, declaring himself to be "walking on air." The money was a landmark, for no single research laboratory had ever received a grant of such magnitude; none even had shown the audacity to ask. But it was not only the money, it was the public expression of esteem from the leading figures in science and business, delivered via a unanimous vote by nineteen distinguished representatives of industry and academia sitting as the Rockefeller Foundation board. The era of Big Science was now launched by a partnership of foundation, university, and industry. And it had all happened without a single dissenting vote. "Great and small, they all backed the *plan* and *you*," Dave Morris wrote Lawrence the next day. "Do get full emotional satisfaction from such rare unanimity: You deserve it."

In the weeks following the board's approval, Loomis continued to place himself at the service of the 184-inch cyclotron. Bringing Lawrence back to New York, he exploited his personal business contacts to secure tons of copper and iron for the giant machine. War preparations were already tightening the supplies of both commodities, but Loomis pulled strings, even yanked them, to get Lawrence what he needed at a preferential price. As Ernest related to Alvarez, "After spending some time with the Guggenheims, during which a favorable price for copper was negotiated, Alfred said, 'Well, now we have to go after the iron. I think Ed Stettinius is the right man.' " A call was duly placed to the chairman of the United States Steel Corporation: "Hello, Ed, this is Alfred. I have someone with me I think you'd like to meet. When can we come over?"

But there were some things that Loomis could not control. The Rockefeller Foundation grant required the cyclotron to be completed and placed in operation by June 30, 1944. For understandable reasons, it would fail to meet that deadline.

Part Three

THE BOMBS

Chapter Eleven

"Ernest, Are You Ready?"

"It was a cool September evening." Arthur Holly Compton serenely begins his account of his meeting with Ernest Lawrence and James B. Conant, the president of Harvard University, in Chicago on September 25, 1941. "My wife greeted Conant and Lawrence as they came into our home and gave each of us a cup of coffee as we gathered around the fireplace. Then she busied herself upstairs so the three of us might talk freely."

Compton's guests had come to the city to receive honorary doctorates from the University of Chicago. But that only provided the opportunity for this more momentous encounter, which had much to do with Conant's role as an important science advisor to the Roosevelt administration and Compton's as head of a blue-ribbon committee charged with appraising the military usefulness of atomic energy. Lawrence, who had demanded the urgent face-to-face meeting, had come bearing news of an extraordinary breakthrough in that field. Their conversation took scarcely more than an hour. But when it was over, America's wartime planning, and the lives of all three men, had been set on a new course. The country was on its way to building the atomic bomb.

The roots of the meeting had been planted more than two years earlier by the discovery of nuclear fission. That news, which broke in January 1939, launched physicists upon flights of learned speculation about the enormous energy released when the uranium nucleus split following its absorption of a stray neutron. Most intriguing was the possibility of a chain reaction: if neutrons emitted in fission struck neighboring nuclei and caused *them* to split, they might in turn emit even more neutrons, producing more fissions. If enough neutrons boiled out of each shattered nucleus with just the right energy, the process might continue on its own until no more uranium nuclei remained to shatter.

Whether this process would produce explosions or merely heat became the focus of a rather abstract debate, for the practicalities of harnessing the energy were elusive. As befit the man who had challenged Ernest Rutherford's disparagement of atomic power as "moonshine" back in 1933, Ernest Lawrence's first instinct was to take the news of fission as vindication. "It may be that the day of useful nuclear energy is not so far distant after all," he wrote to his fellow cyclotron builder Alexander Allen.

Among those who let their imaginations roam was Robert Oppenheimer. When Luis Alvarez burst in on his seminar with news of the first report of fission's discovery by Otto Hahn and Fritz Strassmann, Oppenheimer had responded instantly, "That's impossible." But within hours, he had withdrawn his snap judgment. And within a week, recalled one of his students, he had filled his office blackboard in LeConte Hall with "a drawing—very bad, an execrable drawing—of a bomb."

Oppie shared his conjectures widely. "In how many ways does the U come apart?" he wrote a fellow physicist. "At random, as one might guess, or only in certain ways? And most of all, are there many neutrons that come off during the splitting, or from the excited pieces? . . . [S]hould be quite something." To another, he expanded on the theme with a sort of minatory thrill: "I think it really not too improbable that a ten cm cube of uranium deuteride . . . might very well blow itself to hell."

One man who took such apocalyptic speculation seriously was Leo Szilard, the energetic and resourceful Hungarian physicist who had tried to patent a prototypical cyclotron just before Lawrence's practical invention. Szilard was tormented by a vision of uranium's explosive potential placed in the hands of Adolf Hitler. "You know what that means?" he told Edward Teller, a fellow Hungarian émigré. "Hitler's success could depend on it." Szilard urged his colleagues to verify the explosiveness of the reaction promptly, the better to steal a march on any German research that might soon get under way, and to place their research results voluntarily under wraps. But his promptings about secrecy fell largely on deaf ears, due in part to widespread skepticism that there was anything worth concealing. To Szilard's friend Enrico Fermi, the possibility of an explosive reaction was so remote that Szilard's concerns seemed driven by paranoia, not physics.

Yet Szilard had learned from bitter experience that sometimes paranoia is the prudent approach. In 1933, when Hitler came to power and he was a junior instructor living at the faculty club of the Kaiser Wilhelm Institute of Physics in Berlin, he had kept two suitcases packed in his rooms. When the Reichstag building burned down, which became the pretext for Hitler's suppression of political dissent, Szilard was dismayed by the failure of his German friends to grasp the developing reality. "They all thought that civilized Germans would not stand for anything really rough." The day after the fire, he fled to Vienna on a nearly empty train; one day after that, he would learn, all Austria-bound trains were jammed with would-be refugees, who were stopped and interrogated at the border. "This just goes to show," Szilard wrote, "that if you want to succeed in this world, you don't have to be much cleverer than other people, you just have to be one day earlier than most people."

Szilard was determined that scientists in America should have that one-day advantage. It would be a source of anguished frustration to him that his warnings met with complacency, especially from fellow refugees such as Fermi, who should have been far more sensitive to the peril of being late. "We both wanted to be conservative, but Fermi thought that the conservative thing was to play down the possibility that this [chain reaction] might happen, and I thought the conservative thing was to assume that it would happen and take all the necessary precautions."

Physicists at the Radiation Laboratory and elsewhere picked away at the secrets of uranium fission all that year, trying to answer Oppenheimer's questions as well as others more fundamental. What triggered the reaction? Why was it not seen in uranium in its natural state? The fact that uranium deposits occurred naturally around the world without disintegrating themselves suggested that extraordinary conditions had to be present.

It was Niels Bohr who came up with the crucial insight. Natural uranium's fission cross section—that is, the probability that the nucleus would fission under given circumstances—was highly sensitive to the energy of the impinging neutrons. The explanation, Bohr recognized, lay in the prevalence of distinct uranium isotopes. The most abundant isotope, U-238, was harder to nudge toward fission than to move a recalcitrant donkey, and it responded only to fast, or highly energetic, neutrons. But naturally occurring uranium also contained U-235, which was much more fissile—highly likely to split after absorbing a neutron of almost any speed. U-235, however, was present only in the tiny ratio of 1 atom in 139, or about 0.7 percent.

Bohr's insight motivated physicists to ask whether fissioning concentrated U-235 would produce the abundant neutrons needed to sustain a chain reaction, and if so, how could one separate U-235 from U-238 or increase its prevalence in a sample? Since isotopes are chemically identical, a nonchemical means had to be found to accomplish this. The neutrons produced from the fissioning nucleus, known as secondary neutrons, "became the object of a worldwide search," Luis Alvarez recalled.

Everywhere in the world, curiously, except the Rad Lab. Lawrence judged that the glory that might come from being the first to discover secondary neutrons was not worth delaying the completion of the sixty-inch cyclotron, which was needed to meet the increasing demand for medical isotopes. The task of finding fission neutrons was dumped in the lap of Alvarez, then a junior researcher still searching for a career-making project. He did not see the task as a ticket to fame, so he designed what he called a "quickie experiment" involving a single neutron detector placed in a stairway outside the cyclotron room. He spent all of five minutes bombarding uranium oxide, and, detecting no neutrons in his apparatus, gave up. Only later did he realize that if he had moved his counter a bit nearer the cyclotron, bombarded a bit more uranium, and counted for an hour instead of five minutes, he would have found that very day the secondary neutrons being sought all over the world.

Joliot's team accomplished that task in March, estimating the yield of secondary neutrons from U-235 at about 3.5 neutrons per uranium fission. Szilard and Fermi, working in separate labs at Columbia University, came up with a figure closer to 2.0—still a large quantity under the circumstances. "Chances for reaction now above 50%," Szilard wired a friend. He felt no triumph in the discovery, though, recalling later: "That night, there was very little doubt in my mind that the world was headed for grief."

Szilard was convinced that a bomb was possible and, given Hitler's goal of conquering the world, even probable. In early July he persuaded Albert Einstein to put his name to a letter alerting the president of the United States to the threat. That was the genesis of what would become one of the seminal documents of the atomic age, Albert Einstein's message to Franklin Roosevelt, dated August 2, 1939. Drafted mostly by Szilard with Einstein's input, the two-page, eight-paragraph document buries the urgency of the situation under dry and conditional prose, its momentousness communicated largely by Einstein's signature at the bottom. "Some recent work by E. Fermi and L. Szilard, which has been communicated to me in manuscript, leads me to expect that the element uranium may be turned into a new and important source of energy in the immediate future," it read. "Certain aspects of the situation which has arisen seem to call for watchfulness and, if necessary, quick action on the part of the Administration."

The letter mentioned the possibility of "extremely powerful bombs of a new type." It closed with the ominous observation that German scientists might have already started working on such a project.

The missive was duly placed in the hands of Alexander Sachs, a

Russian-born economist with a scientific background and, more to the point, access to the White House inner circle as an advisor to FDR. Sachs finally gained entrance to the Oval Office on October 11, a few weeks after the Nazi invasion of Poland had instigated the European war. After a brief exchange of jocular anecdotes and the pouring of two glasses of Napoleon brandy, Sachs read FDR a digest of Einstein's letter that he had prepared himself, in the hopes of reducing its abstruse science and circumlocutions into language the president would grasp quickly.

He chose his words well. "Alex," Roosevelt said, "what you are after is to see that the Nazis don't blow us up."

"Precisely," Sachs replied.

Roosevelt summoned his military aide, General Edwin M. "Pa" Watson. Handing over Sachs's documents, he said forcibly, "This requires action."

Even the fretful Szilard could not have been unhappy with the pace of action that followed. Before Sachs left the White House that day, Watson was already jotting down names for a committee to investigate the military applications of fission. No scientific bureaucracy existed in the government to assume the task, so Watson created one on the fly. It would be led by Lyman J. Briggs, a career government scientist then directing the National Bureau of Standards, which by default served as the government's physics laboratory. Lieutenant Colonel Keith Adamson and Commander Gilbert C. Hoover, two members of Watson's staff, were its other members. On October 21 Briggs presided over the so-called Uranium Committee's first meeting, with Szilard, Teller, and Eugene Wigner, another émigré Hungarian physicist, on hand as technical advisors. It was a lightning-fast schedule for government work.

But fault lines between the officers and the scientists opened up immediately, as the talk turned to how much money the scientists considered necessary for their research. When Teller said that Fermi would need tens of thousands of dollars to build a preliminary reactor to test conditions for a chain reaction, Adamson responded scornfully. At the army's Aberdeen Proving Grounds, he sneered, "we have a goat tethered to a stick with a ten-foot rope, and we have promised a big prize to anyone who can kill the goat with a death ray. Nobody has claimed the prize yet." He informed the scientists crisply that men and morale, not fancy weaponry, were what won wars, and continued in that vein until he was interrupted by Wigner, a slight, red-haired man whose diffident personality concealed a razor-sharp mind. "It's very interesting for me to hear this," he said politely. If it is correct, he suggested, perhaps the army's budget for armaments ought to be cut stringently. Taken aback, Adamson snapped, "All right, you'll get your money." The committee voted to send Fermi an initial grant of $6,000.

After that, however, the Uranium Committee's work ground to a halt. Szilard and Wigner, who had left the initial meeting confident that the urgency of research was understood at the highest levels of government, now came face-to-face with the natural sluggishness of the bureaucratic process. The experience was like "swimming in syrup," Wigner would recall. Szilard was mystified. "I had assumed that once we had demonstrated that in the fission of uranium neutrons are emitted, there would be no difficulty in getting people interested, but I was wrong."

Szilard and Wigner were not the only physicists feeling frustrated. So too was Ernest Lawrence.

The war had thrust itself upon Lawrence's consciousness in an especially personal way. His brother, John, who had sailed for Britain a month before the Nazi invasion of Poland, was scheduled to return home at the beginning of September from a nation at war, on an ocean crossing that now seemed immeasurably perilous. Carl and Gunda wired Ernest anxiously from South Dakota for word of John's plans; Ernest wired back the soothing news that John was to sail from Liverpool on the liner *Athenia*—a British ship, but an unarmed passenger vessel immune from attack according to the Hague conventions governing the conduct of war.

No sooner had Ernest's wire been dispatched than horrifying news arrived: the *Athenia* had been sunk by a German U-boat torpedo, in the war's first attack on British shipping. During the next two nights, the family received only sporadic, contradictory reports about the disaster: some said all hands had been lost; others, that hundreds of passengers had been saved. Ernest spent the hours pacing by the radio, silent and withdrawn among friends and colleagues from the lab. Unnerved by the sensation of utter powerlessness, he only summoned his reserves of sunny optimism to field phone calls from his parents. At last, a miraculous cable arrived from John, pronouncing himself "safe and sound" aboard a British destroyer. His story turned out to be one of outstanding heroism: having remained aboard the foundering *Athenia* to minister to injured passengers and crew, he was the very last passenger to board a lifeboat.

The experience instantly altered Ernest's attitude toward politics in the lab. It was no longer possible to view conditions even in distant Europe as irrelevant to the work going on in Berkeley. Physicists visiting the Rad Lab now found themselves being drawn by Ernest into discussions of the military applications of the latest research rather than the progress of the sixty-inch and the achievements of the cyclotron team. Arthur Compton, arriving in Berkeley to report for the National Advisory Cancer Council on the nuclear medicine program funded by its $30,000 grant, was sidetracked into what Lawrence described for Alfred Loomis as a discussion "regarding the war situation." He added that Compton, "like all of us, is very anxious that we scientists do everything we can in the direction of preparedness, and we discussed ways and means." Among the topics they had discussed was how to use the new cyclotron in the war effort, he told Loomis. "We certainly will not be unmindful of the possibilities of discoveries of military value in the energy range above one hundred million volts."

As Lawrence became more interested in the military applications of nuclear research, he grew more dismayed at the lack of progress under the dead hand of the Uranium Committee. Briggs had taken Szilard's concerns about publicity to heart but applied them in a peculiarly counterproductive way. He instituted a regime of "compartmentalization" in which physicists working on one aspect of fission were denied access to research on different aspects, even if they might be relevant to their own. The bottleneck frustrated scientists up and down the line: Merle Tuve, who had the training, equipment, and willingness to advance the field, complained that he was "hard pressed to get any data on uranium fission," including such fundamental information as nuclear cross sections. Even Harold Urey, who was a member of the Briggs committee, had no luck obtaining access to other scientists' work to help him develop an isotope separation process for uranium.

To be fair to Briggs, his office was not the only place developing an obsession with secrecy. In June 1940 Ed McMillan and Philip Abelson of the Rad Lab published an account in the *Physical Review* of their discovery of element 93, a radioactive daughter of uranium. (The element subsequently would be named neptunium.) After the article's appearance, Lawrence received a scolding from none other than James Chadwick, via an envoy dispatched from the British Embassy in Washington, for allowing the publication of research that could aid the Nazi regime. The *Physical Review* soon acceded to a system by which it would accept articles on nuclear reactions but keep them in a vault, to be published after the conclusion of the war.

But there was a difference between suppressing the publication of research findings and restricting the personal give-and-take among scientists that was indispensable for scientific progress. European scientists were sharing more among themselves than were the Americans, albeit through personal contacts, not in the pages of widely published journals. As a result, they were making startling discoveries that only underscored the constraints faced by their American colleagues.

One such pioneer was Otto Frisch, a nephew of Lise Meitner, the gifted Austrian physicist whose work with Otto Hahn had led to the discovery of fission, for which she then proposed a theoretical foundation from her political exile in Sweden. (History judges her to have been unfairly excluded from the Nobel Prize for the discovery, which went to Hahn alone in 1944.) Frisch, who had assisted his aunt with her experiments, had been evicted from the University of Hamburg under the Nazi racial laws, which barred Jews from high academic positions beginning in 1936. Mark Oliphant came to his rescue by inviting him to move to the University of Birmingham. "Just come over," Oliphant told him. "We'll find you something to do."

As a German national, Frisch was barred from Oliphant's main research project, a secret effort to develop radar. Instead, he probed the explosive characteristics of U-235 and soon realized that a bomb could be made from roughly a pound of the separated isotope. "That set me thinking," he recalled later. "I felt one pound is, after all, not such a lot."

Frisch and his fellow refugee Rudolf Peierls, who also was becalmed at Birmingham, calculated how much equipment would be needed to acquire a pound of 235 by thermal diffusion, which employs temperature differentials to separate the 235 and 238 isotopes by weight. (The heavier isotope gravitates toward cooler temperatures, the lighter U-235 toward heat.) Their figure of £1 million was forwarded to Henry Thomas Tizard, the Oxford University chemistry don spearheading his country's scientific effort in the war. Tizard's formation of a committee to study the theory under the leadership of G. P. Thomson, son of the legendary J. J. Thomson, marked the start of organized atom bomb research in Britain.

Filled out with Oliphant, Chadwick, and Cockcroft as members, the group met for the first time in April 1940 as the MAUD Committee. (Although the initials appear to be an acronym, the name actually came from a cable to Cockcroft from Meitner asking him to send a message from Niels Bohr to "Maud Ray Kent"; Cockcroft had interpreted the words as an anagram for "radium taken," which suggested that the Nazis were acquiring radioactive substances for fission experiments. In fact, they referred to Maud Ray, a former governess to Bohr's children, who lived in the county of Kent.)

The MAUD Committee was Britain's counterpart to Briggs's panel, but any resemblance was entirely superficial. The MAUD members were all accomplished nuclear scientists. They were "electrified" by the conclusions of Frisch and Peierls that the critical mass of U-235 might be only a pound and that a chain reaction could build up rapidly to explosive force, Oliphant reported later. It would take the MAUD Committee fifteen months to determine decisively that a bomb was practical and to outline the necessary steps. By the time it had done so, the American effort had been laboring for nearly two years without reaching a conclusion.

The Briggs committee's stranglehold on American fission research showed signs of loosening, if slightly, only in June 1940, two months after the MAUD Committee was formed. That was when Vannevar Bush emerged from a meeting at the White House with the treasured initials "OK-FDR" scrawled on a single sheet of paper. The document's four short paragraphs outlined Bush's proposal to establish a National Defense Research Committee to coordinate all technical research with military applications under his chairmanship. The meeting with President Roosevelt had taken ten minutes. After that, Bush would recall, "All wheels began to turn."

Bush wrote later that many in Washington regarded the establishment of the NDRC as "an end run, a grab by which a small company of scientists and engineers, acting outside established channels, got hold of the authority and money for the program of developing new weapons." To this he had a simple answer: "That, in fact, is exactly what it was . . . The only way in which a broad program could be launched rapidly and on an adequate scale."

Bush's appointment as the nation's science czar would soon give Ernest Lawrence entrée to the highest councils of government. But Bush's first act on nuclear weapons development was not especially encouraging: he slashed Lyman Briggs's request for additional funding for Fermi's atomic-pile research to $40,000 from $140,000; his reasoning was that no evidence existed yet that the pile was practical or that it could lead to a weapon. For Fermi, this was a disappointment, and for Szilard, an all-too-familiar setback. The Briggs committee had finally seemed willing to press ahead, but now it was the NDRC holding back.

Lawrence would soon jolt it in the right direction. Bush had asked Ernest to take on a roving brief for the committee, injecting himself into any area encountering problems as "a sort of fire department." This underscored Bush's esteem for the breadth of Lawrence's scientific knowledge and his emollient managerial technique. To Lawrence, however, the assignment seemed both inchoate and potentially overwhelming. Tactfully, he turned it down. That turned out to be fortunate, because it left him free to take on a new project that soon landed on the NDRC's plate.

The project was an outgrowth of the secret research Oliphant had been conducting at Birmingham. It involved an invention known as the cavity magnetron, a source of high-powered microwaves. The British had sent Cockcroft to the United States to solicit engineering help to turn the magnetron into a serviceable radar apparatus. Alfred Loomis was brought into the talks; after hosting the British delegation for a week at Tower House, he pressed Bush to make radar a priority of the NDRC. This led to the formation of a microwave committee with Loomis as chairman and Lawrence as a member, and in turn to the NDRC's subsequent decision to establish a crash program at MIT. Lawrence agreed to recruit the new lab's staff.

His first call was to Lee DuBridge, a Caltech-trained physicist who had first met Lawrence in 1934 at one of the "vaudeville" demonstrations of radio-sodium. DuBridge went on to build a cyclotron at the University of Rochester, which brought the two physicists closer. Reaching DuBridge by phone in early October, Lawrence informed him curtly that he was needed to lead an important defense initiative. "I can't tell you about it, but I assure you it's very important," he said. Bush could not fail to be impressed by DuBridge's instantaneous assent and what it indicated about Lawrence's standing in the scientific community. "If Lawrence was interested in the program," DuBridge explained later, "that was what I wanted to be in." He boarded a train for New York that same evening.

Lawrence's role did not end there. He and DuBridge worked together to fill out the personnel roster of the project, which was code-named the "Rad Lab" in the perhaps naïve assumption that the name might confuse the enemy into thinking it was merely a spinoff of Lawrence's lab in Berkeley. They started with physicists they knew personally: "We just got our good cyclotron friends together," DuBridge would recall. Lawrence did not spare his own staff: among the first scientists he summoned to MIT were Edwin McMillan, who was then deeply involved in the search for element 93, and Luis Alvarez. Both accepted the call out of a combination of fidelity to Lawrence and duty to country—the latter communicated to them also by Lawrence, who emphasized that their work would be crucial to the war effort.

"It was essentially an order, although he didn't phrase it that way," McMillan recalled. "He told Alvarez and me . . . that this great project was starting and that we must get into it, that Hitler has to be stopped." McMillan thought wistfully of his orphaned research into the transuranics—those elements heavier than uranium, such as element 93—but reckoned that "it would have been very bad grace for us to have said, 'Well, we've got other things to do.' " Ernest tried to soften the blow by assuring McMillan that he would be needed for only a few months, but McMillan doubted him. "I had the strong feeling that I was going to be away a long time. And I was right." Many of the scientists Lawrence and Loomis assembled at the MIT Rad Lab would eventually move on to the team that was to build the bomb.

Having proved his worth by organizing the MIT Rad Lab from scratch, Lawrence thought himself well positioned to press his concerns with the NDRC brass about the slothful pace of the Briggs committee. He was about to learn the hard way about the perils of pressing too hard.

He started his campaign auspiciously enough with an overture to James Conant, who visited Berkeley in May to deliver the keynote address at Charter Day, the university's annual founding celebration. The time had come to "light a fire under the Briggs committee," Lawrence urged Conant. "What if German scientists succeed in making a nuclear bomb before we even investigate possibilities?" He blamed Briggs for placing the NDRC in doubt about the potential of uranium research in war. At a further meeting with Loomis and Arthur Compton at MIT on March 17, Lawrence startled them with the news that he was prepared to manufacture U-235 on his own by converting the 37-inch cyclotron—rendered obsolete by the completion of the 60-inch and the planning for the 184-inch—into a mass spectrograph to separate uranium isotopes electromagnetically.

Compton relayed Lawrence's words to Bush the next day, repeating Ernest's dreary assessment of Briggs as "slow, conservative, methodical, and accustomed to operate at peacetime government bureau tempo." Briggs's management, Lawrence had argued, left American science hamstrung by comparison with the British and, more perilously, with the Germans, even though the United States boasted "the most in number and the best in quality of the nuclear physicists of the world." Thinking that the voluble Lawrence himself stood the best chance of communicating the urgency of the situation to Bush, Compton and Conant decided to send him to New York to make his pitch to the NDRC chairman in person.

This was a blunder, as Conant should have realized. Having worked as Bush's deputy at the NDRC for nearly a year, Conant knew that his boss could be "very vindictive when people went out of channels." Sure enough, Bush interpreted Lawrence's visit as an attempt to short-circuit his own prerogatives of command. He laid into the dumbfounded Lawrence the moment he walked in the door. "I told him flatly that I was running the show, that we had established a procedure for handling it, that he could either conform to that as a member of the NDRC and put in his 'kicks' through the internal mechanism, or he could be utterly on the outside," Bush related to his friend Frank Jewett, the head of Bell Laboratories. He further informed Lawrence that he intended to back up Briggs "unless there was some decidedly strong case" for accelerating uranium research, which he had yet to see.

Ernest was properly apologetic ("He got into line," Bush told Jewett), but realistically, Bush was working at a disadvantage. In the end, science was going to dictate the course of the NDRC's program, and the science of the atomic nucleus was on Lawrence's side. In fact, Bush acknowledged privately that most atomic physics went "over my head."

There were other reasons to keep Ernest Lawrence in the loop. He was not only one of the most accomplished nuclear physicists in the country but also a peerless organizer of research, as he showed in the creation of the Rad Lab at MIT—so much so that Bush would soon ask him to orchestrate the research and development of an underwater communications technology for submarines in San Diego. Lawrence placed that project under McMillan, whom he extracted from the radar program for the purpose. "You did a very fine piece of work for the radiation group [meaning the radar group]," Bush wrote Lawrence in July in gratitude before adding imperiously, "I think with a clear conscience you can now put your prime efforts on the submarine matter. It very much needs your heavy attention." As if attempting to smooth any ruffled feathers, he closed the letter with the reassuring words "I've been putting a lot of thought on the uranium matter."

Bush fulfilled that promise a few weeks later by placing Lawrence on a special committee formed under Compton to review the Briggs committee's work. Within a week, they had canvassed all the members of the Briggs panel and come away even more shocked at their complacency than they expected.

"Two facts were at once apparent," Compton wrote later. "The first was that uranium fission would one day be of very great importance to the world. The second was that not a single member of the Briggs Committee really believed that uranium fission would become of critical importance in the war . . . This committee had been considering these possibilities for a year and a half . . . No one had felt keenly enough its possible contribution to the war effort to move away from another field of study." Perhaps this latter finding should not have been so surprising: after all, Lawrence himself was still thoroughly preoccupied with his other work—namely, the construction of the 184-inch cyclotron on the hillside above the university.

On May 17 Compton sent Bush his panel's unanimous report on the Briggs committee. The report discussed three possible military applications for fissionable uranium: dropping radioactive materials over enemy territory; generating power for submarines and other oceangoing vessels; and development of a bomb based on U-235. The most modest of these options, the radioactive spray, would take at least a year to perfect after the first chain reaction was achieved. That could come in as soon as eighteen months if Fermi's program had full government support. Given the still-unsolved challenge of separating U-235, bombs seemed to be three to five years off.

Compton described these findings later as "on the whole hopeful." But to Bush, they suggested that uranium research could be safely laid aside for the duration of the war. The bottom line, he felt, was that there was "certainly no clear-cut path to defense results of great importance." As a check on his own reaction, he created yet another technical panel, this one headed by Jewett, to review Compton's report. By the time they convened, there was one more piece of technical evidence to weigh, for Lawrence had uncorked another surprise: Glenn Seaborg of the Berkeley Rad Lab had discovered and refined a small quantity of a new radioactive element, a daughter of element 93, which was itself a daughter of neutron-bombarded U-238. Chemically distinct from uranium and therefore theoretically separable by chemical means, element 94 appeared to be about five times more fissionable than uranium. At the moment, 94 had not been named, but in due course it would become known as plutonium.

Lawrence tried to communicate his excitement about Seaborg's discovery in writing, via a memo to Jewett's committee declaring that "an extremely important new possibility has been opened for the exploitation of the chain reaction." A power plant based on element 94 would require perhaps a hundred pounds of the material, rather than the hundred tons thought to be necessary for uranium reactors. Most strikingly, in a chain reaction in 94, "energy would be released at an explosive rate which might be described as a 'super bomb.' "

Had Lawrence addressed the committee personally with his customary verve, he might well have convinced its members of the importance of the new discovery. But he was stuck in Berkeley helping Molly tend to their daughter Margaret, who had undergone an emergency appendectomy. Also absent was Compton, who was away on a summertime jaunt to South America. So the significance of 94 sailed over the heads of Jewett's committee, which endorsed increased funding for chain reaction research but remained cold to the prospects of a bomb. With a second equivocal report in hand, Bush was on the verge of terminating fission research altogether as a war program.

Then Ernest rescued it for good.

The MAUD Committee's final report on July 15, 1941, came to conclusions dramatically different from all Briggs's deliberations and Bush's technical studies. The report treated the inherent uncertainties of research and development not as disqualifying obstacles (the American mind-set) but as challenges well within the ability of British and American scientists to overcome. It projected that an "effective uranium bomb" could be made with 25 pounds of U-235 and that a plant to produce 1 kilogram (2.2 pounds) per day of the separated isotope—nearly enough for three bombs a month—could be built for the equivalent of about $20 million. The committee made short work of US concerns that the bomb might come too late to affect the war. Instead, it concluded that the first bombs could be produced by the end of 1943, and therefore were "likely to lead to decisive results in the war." The members urged that work proceed "on the highest priority" and in close collaboration with the United States.

G. P. Thomson, the MAUD chairman, sent the report and its technical appendices to Briggs for forwarding to Bush and awaited the inevitable invitation to collaborate with the United States. For a month, he heard nothing. Then, learning that Oliphant was heading across the Atlantic for a meeting on radar, Thomson asked him to make "discreet enquiries" about the American reaction to the documents. Immediately upon landing in Washington, Oliphant called on Briggs, only to learn in dismay that "this inarticulate and unimpressive man had put the reports in his safe" without even showing them to his own committee. Oliphant also met with Bush and Conant, who both told him they had heard nothing of the British findings.

Fed up with the same phlegmatic American approach to fission's military potential that had driven Szilard to distraction, Oliphant decided to make his case to the one man he thought would comprehend the MAUD Committee's findings, and who had the standing to beat the American establishment over the head with them. He wired Ernest Lawrence, pleading for an urgent meeting. "I'll even fly from Washington to meet at a convenient time in Berkeley," he wrote. Lawrence, whose friendship with Oliphant grew out of the bonds forged during the deuteron fiasco of 1933, happily tendered the invitation.

At Berkeley, Lawrence drove Oliphant up to the hill above campus for an obligatory visit to the ravine being readied for the 184-inch, whose towering magnet was standing in an open field like a monolith at Stonehenge. Back in his office, Lawrence asked Oppenheimer to join them. Oliphant outlined the MAUD report for both men, relieved at last to be addressing an audience receptive to the potentialities of a fission-induced explosion and to the need for urgency. A few days later, he wrote Lawrence to express his confidence that "in your hands the uranium question will receive proper and complete consideration, and I do hope that you are able to do something in the matter." His confidence was not misplaced. Lawrence had already telephoned Arthur Compton to set up a meeting in Chicago. The stage was set for one of the most important events of the prewar period.

On that "cool September evening," Compton, Conant, and Lawrence arranged themselves around the fireplace in Compton's living room. Lawrence began by repeating what he had learned from Oliphant. Addressing Conant, he underscored the significance of element 94, described how it could be made from a chain reaction in U-238 and chemically separated, and mentioned new strides being made at the Rad Lab in the physical separation of U-235 from U-238. He repeated Oliphant's fears that the Nazis were counting on their own atomic bomb to decide the war. "If they succeeded first," Compton recalled Lawrence's warning, "they would have in their hands the control of the world."

Lawrence spoke forcefully, even passionately. As the practical details piled up in his mind, Conant, who had come to Chicago still imbued with the skepticism about atomic weaponry prevailing in Bush's circle, finally began to reconsider.

Conant had advised Bush to put the uranium project on the shelf for the duration of the war, but that advice had been based on his impression that a bomb was purely conjectural. Now he turned to Lawrence. "You put before me plans for making a definite, highly effective weapon," he said. "If such a weapon is going to be made, we must do it first. We can't afford not to. But I'm here to tell you, nothing significant will happen on a job like this unless we get into it with everything we've got."

He looked Lawrence in the eye. "Ernest," he said, "you say you are convinced of the importance of these fission bombs. Are you ready to devote the next several years of your life to getting them made?"

Conant was telling him point-blank to put up or shut up. Compton saw Lawrence freeze, his mouth half open in surprise. Ernest's hesitation lasted only a moment. Then he replied, "Jim, if you tell me this is my job, I'll do it."

It was the moment of truth, for both Lawrence and the bomb project. Bush and Conant had been skeptical about an atomic weapon in part because of the reluctance of eminent physicists to commit themselves to the work not only in word but deed. Conant declared later that he fully intended his question to Lawrence as a test—even a dare. "Lawrence was particularly vociferous about the need for mobilizing all scientific talent for the uranium program," he wrote. "I could not resist the temptation to cut behind his rhetoric."

Ernest met the challenge. The Chicago meeting was the watershed moment in the atomic bomb project. Before, Conant had leaned heavily against pursuing the bomb. Now his weight was on the other side. Shortly after the meeting, he reported the details to Bush (calling it an "involuntary conference to which [I] had been exposed," as though he had been caught unawares by its subject matter). On October 9, two weeks later to the day, Bush was back at the White House, bearing a copy of the MAUD report along with a page of talking points prepared by Conant. He delivered the MAUD conclusions with such conviction that Roosevelt agreed to launch without delay a comprehensive research program aimed at building the bomb. Only the actual construction of a uranium separation plant was be held in abeyance without a further order from the White House.

The president named a "top policy group" to make major decisions on the program. They, and only they, were to have full knowledge of all its details: Roosevelt himself, Vice President Henry Wallace, Secretary of War Henry Stimson, Army Chief of Staff George C. Marshall, and Bush and Conant. The quest for the bomb was now fully under way.

To Bush, the sequestering of knowledge and policy authority had one special virtue: it clipped Lawrence's wings. "Much of the difficulty in the past has been due to the fact that Ernest Lawrence in particular had strong ideas in regard to policy, and talked about them generally," Bush wrote Frank Jewett. He made Lawrence and Compton aware that henceforth their portfolio was to be exclusively scientific and technical. The debate over whether to build the bomb was over. Now they could move ahead—but under standards of military secrecy even more stringent than those of Briggs, about which Lawrence had groused constantly. He would soon come around.

Compton received the green light from Bush and Conant to review and coordinate American research on fission. Members of the scientific community were expected to "throw themselves into this exploration with everything they had," he recalled. "Lawrence's commitment of his own war years set the pattern." Compton scheduled a meeting with Lawrence and several other members of his technical committee for October 21 at the General Electric research facility in Schenectady. Lawrence responded to the summons by informing Compton, as a fait accompli, that he would be introducing a new participant to the technical roundtable: Robert Oppenheimer. He told Compton simply, "Oppenheimer has important new ideas."

Yet Ernest's decision to invite Oppie into the discussion had been a close call. Despite his continued admonishments to Oppenheimer to abandon his political "leftwandering," Oppenheimer found it hard to cease his engagement with liberal causes. Only a few weeks earlier, he had infuriated Ernest by hosting at his home an organizational meeting for a Berkeley branch of the American Association of Scientific Workers (AASW), a leftist union whose British branch had Communist associations. Worse, he had cajoled Martin Kamen and another Rad Lab scientist, Al Marshak, into attending. They arrived at Oppie's house to find fifteen employees of various Berkeley labs seated on the floor, listening to two labor organizers outline their goals. Finally, Oppie asked for the audience's views. When Kamen's turn came, he recalled, "I embarrassedly asked if permission had been obtained from E.O.L. for people such as Al and myself to be approached."

The question stopped Oppenheimer in his tracks, for, like every other habitué of the Rad Lab, he knew of Ernest's ingrained suspicion of anything that smacked of politics. The previous August, Lawrence had curtly rebuffed an invitation from Harold Urey to join with other American Nobel laureates in a "Federal Union of Democracies of the World"—a fairly innocuous antitotalitarian "campaign from the standpoint of ideas," in Urey's description, "one that I believe is just as important as any aid which we can give for national defense on the physical side." Lawrence replied that "the idea of a federation of democracies may have much practical merit, but . . . I would not think of using my position as a scientist in furthering such a political movement."

Oppenheimer, in his predilection for action, had foolishly overlooked Lawrence's mind-set. The next day, a shaken Oppie buttonholed Kamen at the lab and revealed that he had just reported the unionization meeting to Ernest, who "blew a gasket." Oppie angered Ernest even further by refusing to reveal which Rad Lab members had attended. "They'll have to come tell you themselves," he said.

Kamen went to Lawrence and confessed, explaining that he had been against the AASW idea from the start. His demurrer did not seem to register. "He urged me to 'get out,' " Kamen recalled, "whereupon I hotly asserted that I had never been 'in.' "

For the moment, Lawrence considered that Oppenheimer's talents as a theoretician outweighed his demerits as a political dabbler. He swallowed his misgivings, vouched for Oppie's scientific judgment to Compton, and escorted him to the Schenectady meeting. There Oppenheimer delivered an impressively detailed explication of atomic bomb physics, along with his own rough estimate that 100 kilograms of U-235—that is, 220 pounds—would be enough for a practical device. (This proved to be an overestimate: the core of the Hiroshima bomb, which comprised uranium enriched to 80 percent U-235, weighed about 140 pounds.)

Oppenheimer's exposure to the high-level deliberations in Schenectady impressed upon him, more directly than Lawrence's scolding, the wisdom of shedding his political entanglements. Here was an opportunity to make practical use of the theories in which he had marinated for years—and in the service of a fight against Fascism that he took seriously indeed. Anxious to signal that his involvement in the effort would be trouble-free, he wrote Ernest on November 12 to assure him that "there will be no further difficulties at any time with the A.A.S.W. . . . I doubt very much whether anyone will want to start at this time an organization which could in any way embarrass, divide, or interfere with the work we have in hand. I have not yet spoken to everyone involved, but all those to whom I have spoken agree with us; so you can forget it."

Compton's report of the Schenectady meeting to Bush asserted firmly the practicality of building a "fission bomb of superlatively destructive power" and of separating isotopes on an industrial scale for no more than $100 million. For the first time, Bush had received an unequivocal endorsement of the idea that an atomic bomb could be developed in time to influence the course of a war he knew was coming. He delivered the report to Roosevelt on November 27, with a covering letter stating that he was forming an engineering team and accelerating all other necessary research. It was understood that he would proceed unless and until an order came from the White House to stand down. But the only direct response from the White House to Bush would arrive nearly two months later, on January 19, 1942, when Compton's report came back to him with FDR's handwritten note: "V.B. OK—returned—I think you had best keep this in your own safe."

By then, American science had been mobilized. In early December, Bush had brought together a small group to act as the bomb project's civilian overseers: Conant, Briggs, Compton, Urey, and Lawrence. Compton was given the chairmanship of a new committee, known as S-1. (Bush considered, and then discarded, the idea of giving the chair to Lawrence, whom he ultimately considered too voluble. "The matter would . . . have to be handled under the strictest sort of secrecy," he confided to Frank Jewett. "That is the reason that I hesitate at the name of Ernest Lawrence.") S-1's task, as Compton described it later, was to determine within six months "whether atomic bombs could be made." If the answer was yes, the nation would provide virtually unlimited resources to make it happen.

At the committee's first meeting, they divided up the most important responsibilities. Lawrence would proceed with the magnetic separation technique he was already testing in the converted thirty-seven-inch cyclotron; Urey would develop a separation process based on gaseous diffusion, which exploited the different weights of vaporized uranium isotopes;

Compton was to assemble a team to begin work on the actual bomb design. They agreed to meet again two weeks later.

It was Saturday, December 6, 1941.

Compton returned to his Washington hotel and penciled out a budget of $300,000 to cover the next six months. Lawrence drove to the airport for a flight home to Berkeley. Before taking off, he was informed that the first microscopic quantity of U-235 had been separated from natural uranium in the thirty-seven-inch cyclotron.

The next day, as Compton was heading north by train from Washington to New York, where he was to meet with Fermi, a passenger boarding in Wilmington, Delaware, relayed the first sketchy radio reports of an attack on Pearl Harbor. Conant was already home in Cambridge, preparing with his wife to greet students arriving for a weekly four o'clock tea. Lawrence had disembarked from his flight and was home in Berkeley.

Glenn Seaborg, the discoverer of element 94, was relaxing in his room at the Faculty Club, listening to a football game on the radio. Suddenly the announcer broke in with a news bulletin. Seaborg and his team had been working on a project so secret that they had been forbidden to publish anything about their work, and he instantly understood the impact that the electrifying news from Hawaii would have on his life and those of his colleagues: "Before, we'd been jogging toward our goal. Now we would be at a dead run."

Chapter Twelve

The Racetrack

On December 18, less than two weeks after Pearl Harbor, the members of the S-1 Committee met again in Washington. They assembled in a very different spirit from that in which they had adjourned on December 6, their futures now clouded by the project to which they had all committed themselves on that final day of peace.

Conant presided, with Lawrence, Briggs, Compton, Urey, and Eger Murphree, the research director of Standard Oil of New Jersey, in attendance. Their task was to choose the most effective method for producing the fissionable fuel for an atomic bomb from among five options, none of which was appreciably better than the others. The most speculative option involved element 94, that purportedly hyperfissionable substance discovered in Lawrence's laboratory. Large-scale production of 94 would require a uranium chain reaction, which Fermi had not yet achieved. There were also four possible approaches for separating fissionable U-235 from natural ore. All relied on the weight disparity between isotopes 235 and 238, but all were experimental, and their potential for industrial-scale production conjectural. They were (1) gaseous diffusion, which involved passing uranium hexafluoride, a highly corrosive gas known unaffectionately to chemists as "hex," through a porous barrier; (2) thermal diffusion, which exploited temperature differentials to coax the isotopes into separating; (3) electromagnetic separation, which was based on the divergent paths followed by ions of different weights when beamed through a magnetic field; and (4) the use of a high-speed centrifuge.

Lawrence stunned his colleagues with the disclosure that the Rad Lab had isolated a microscopic quantity of U-235 on the very day before the attack on Pearl Harbor. Driven by frustration over the slow pace of the Uranium Committee's work, he had taken it upon himself to pull his best staff members off the 60-inch and 184-inch cyclotrons and assign them to a crash conversion of the 37-inch into a mass spectrograph, an electromagnetic device used to separate ions. The conversion, which was quietly financed from Rad Lab funds and a supplemental $5,000 grant from the Research Corporation, entailed replacing the machine's vacuum tank with a hastily designed new chamber fitted with an ion source of solid uranium chloride; new electrodes to ionize the vapor and accelerate the charged particles through the depressurized tank; and collectors on which the ionized isotopes would be deposited, hopefully in discrete clumps, after having traced divergent paths because of their different weights. The conversion had been completed on November 24, and the first ion beams hit the collectors exactly one week later, the two beams traveling semicircular trajectories about two feet in diameter and completing their journeys a fraction of an inch apart. By December 6, the spectrographers had accumulated their first measurable, if invisible, quantity of uranium 235—far from pure but at more than triple the concentration found in nature.

Lawrence emerged from the meeting with the first contract issued by the S-1 Committee: a grant of $400,000 to pursue his electromagnetic separation method as well as "certain experimentation on certain elements of particular interest involving cyclotron work." The oblique verbiage referred to element 94. When S-1 reconvened the next day, Compton was awarded responsibility for supervising the theoretical research on U-235 and plutonium. That gave him not only oversight of Fermi's reactor research at Columbia, for which the committee made a six-month appropriation of $340,000, but also jurisdiction over the spectrograph and plutonium projects at Lawrence's Rad Lab.

This was an inherently unstable arrangement, given the self-regard of the two scientists. Sure enough, conflict between them broke open at the end of January, when Lawrence appeared in Chicago with a proposal to centralize all the plutonium and isotope work, including the atomic pile, at Berkeley. This was not purely a power grab. Compton and Lawrence shared an uneasiness about leaving Fermi's reactor project at Columbia. Not only did they fear that New York City could be vulnerable to an attack launched from Europe, but also the university's resources were already stretched thin by Urey's gaseous diffusion research. Compton's preference, however, was to transfer the atomic pile to the University of Chicago, the administration of which was eager to host it and where, as a professor of physics, he could oversee the project from his office on campus. Lawrence countered that Chicago's experience with large-scale nuclear research was meager compared with that of the Rad Lab, which had support facilities to spare. And, of course, he felt he had more than enough administrative experience in large projects to oversee electromagnetic separation, plutonium, and the chain-reaction pile without overstretching himself.

Compton was seized with a vision of Lawrence supplanting his own authority by the sheer exercise of will, abetted by geography. Compton had been entrusted with the overall responsibility of managing the production of atomic fuel for the bomb, which could not be delegated—already he had drafted a project schedule, starting with a deadline to determine whether a chain reaction was feasible (July 1, 1942) and ending with the assembly of a working bomb (January 1945). He was determined to push back firmly at Lawrence, but at that moment, he was placed at a disadvantage by a miserable case of the flu. The decisive confrontation took place by his sickbed upstairs in his home. There it was witnessed by a nonplussed Luis Alvarez, whom Compton had brought to Chicago to help him organize the bomb project. Alvarez was torn by his loyalties to both men; he had received his doctorate at Chicago under Compton and then had become a trusted staff member at Lawrence's Radiation Laboratory. Now he was stuck in the middle of an increasingly acrimonious debate between the browbeating Lawrence and the bedridden Compton. The more Lawrence pushed, the harder Compton pushed back. "In all the years I had been Arthur's student," Alvarez would recall, "I had never seen him fight so hard for anything." Finally, Compton settled the argument by fiat: the pile would move to Chicago.

"You'll never get the chain reaction going here," Lawrence shot back. "The whole tempo of the University of Chicago is too slow."

"We'll have it going here by the end of the year," Compton replied.

"I'll bet you a thousand dollars you won't."

"I'll take you up on that," Compton said. At that moment, he and Lawrence realized that Alvarez, their student and protégé, was seeing them at their worst. Suddenly abashed, Lawrence said, "I'll cut the stakes to a five-cent cigar."

"Agreed."

Compton would win the bet, though he never received the cigar.

Lawrence was right about the Rad Lab's capabilities, however. The lab had been operating virtually on a war footing for two years. Up to September 1939, Ernest had expected the cyclotron to sit out any war, advising J. Stuart Foster, who was hoping to make a military case for building a machine at the University of Toronto, that it was "difficult . . . to suggest concrete practical applications of the cyclotron in warfare." But after the Nazi invasion of Poland and the family scare over John's passage on the *Athenia*, he set aside his doubts and conjured up ways to put the machine's capabilities to use in the war effort. Ever attuned to the ebb and flow of research funding, he recognized that the government stood to become as generous a financial backer as the medical foundations he had mined to build the sixty-inch. He became an assiduous collector of government contracts large and small, starting with the production of radioisotopes at a standard fee of $25 per hour of cyclotron time. By the summer of 1940, the flow of government money into the Rad Lab had grown into a torrent.

The most marked manifestation of the trend was an enhanced standard of living for Rad Lab scientists placed on the government payroll as consultants. Martin Kamen, who was supervising isotope production on the sixty-inch, was compensated for his high-pressure responsibilities with a government stipend of $5,000 a year, a "mind-boggling" sum for a scientist who previously had eked out a penurious existence grant to grant. He promptly took advantage of the windfall by purchasing a beautiful eighteenth-century Tassini viola, lest the money be snatched away as abruptly and mysteriously as it had arrived.

The lab's traditionally frugal approach toward overhead also disappeared. One day Kamen was summoned to the office of financial fussbudget Don Cooksey, who instructed him to prepare a detailed requisition for equipment to perform chemical assays—and not to worry about the cost. Kamen took him at his word. Paging through a stack of chemical supply catalogs, he chose lavishly, not excepting a "Podbielniak fractional distillation apparatus with gold-plated seals and ground joints," which he ordered mainly because he was curious to know what it looked like. Its price: $1,000.

Kamen submitted the list and stood by uneasily as Cooksey, who had been known to hold lengthy conferences over requisitions worth a single dollar, scrutinized every entry. "Martin, I don't think you understand the situation," he said finally. "Don't you think we should triple this order?"

As the pace of war planning intensified, it soon became impossible to tell where the Rad Lab ended and the bomb program began. Among the scientists who made the transition with ease was Glenn Seaborg, whose research would prove to be among the most important in the war.

Seaborg never lost his respect for the substance that would make his career. "Plutonium is so unusual as to approach the unbelievable," he would write a quarter century later. "Under some conditions it can be nearly as hard and brittle as glass; under others, as soft and plastic as lead. It will burn and crumble quickly to powder when heated in air, or slowly disintegrate when kept at room temperature . . . It is unique among all of the chemical elements. And it is fiendishly toxic, even in small amounts."

Seaborg's early life told the same quintessentially American story of immigration and assimilation as Ernest Lawrence's, although his upbringing was rather more insular and culturally constrained than that in the Lawrences' educated household. He was born in 1912 to Theodore Seaborg, a first-generation Swedish American, and his Swedish-born wife, Selma, in Ishpeming, a community in Michigan's upper peninsula where the dirt streets were tinted red by the iron ore mined from tunnels underlying the town in a subterranean latticework.

When Seaborg was ten, his parents fled the meager opportunities of Ishpeming, relocating to a small community just south of Los Angeles. There Glenn was raised in permanently straitened means, for his father never again found steady work. But Glenn was able to take advantage of the superb state university system of California, where tuition was free to residents. By 1933, he had received his bachelor's degree in chemistry from the University of California at Los Angeles, which only a few years earlier had relocated from downtown Los Angeles to a bucolic new campus west of the city, and moved on to Berkeley to pursue his graduate studies in chemistry under Gilbert Lewis. He soon gravitated to the twenty-seven-inch cyclotron, which occupied the ramshackle old Rad Lab next door to Lewis's domain, Gilman Hall.

Chemistry skills like Seaborg's were coming into demand at the Rad Lab. To the physicists, the separation and purification of radioisotopes seemed a black art; but it was well within the capabilities of a graduate chemist. In no time, Seaborg became one of the Rad Lab's reigning radiochemistry experts while also serving as Lewis's personal research assistant. Fortune had placed him in a position to plant one leg in each of the university's two most renowned departments, standing at the very spot where they came together to promote the most exciting research on earth.

The contrast between the two towering figures of Berkeley science was striking. Gilbert Lewis remained rooted in small science, happily puffing away on his acrid black cigars at a workbench festooned with hand-blown glassware. Lawrence, perfecting his style of Big Science, already was involved less in performing experiments than in managing his interdisciplinary staff. The Rad Lab was starting to operate on an industrial scale, with graduate students working in shifts to keep the cyclotron running round the clock. Ernest's genius, Seaborg perceived, was to draw into his orbit like-minded scientists in every field, not just physics, and imbue them with his own drive to build and perfect his magnificent invention. Without Lawrence, there would be no Rad Lab—none of the teamwork that combined the disparate knowledge and skills of physicists, chemists, biologists, physicians, and engineers into a new paradigm of science.

Like his colleagues, Seaborg never forgot where he was the moment he first heard about fission: in his case, at Lawrence's Monday Journal Club. He wandered the streets of Berkeley all that night, marveling at the discovery, cursing himself for not having recognized the phenomenon himself despite his hours of hands-on experience pulling isotopes from bombarded uranium, imagining the renown that would have come his way had the discovery been his, and hankering to join the full-scale assault on fission that Lawrence had ordered.

His opportunity came through Ed McMillan. As one of the lab's senior scientists, McMillan was at the hub of Ernest's campaign to augment the discovery of fission by bombarding uranium and logging the reactions. The result that especially piqued his interest was a 2.3-day "activity." (The term referred to the half-life of the isotope in question.) This activity, curiously, remained nestled close to the uranium target, unlike other fission products, which typically were driven some distance away by the energy of nuclear fragmentation. That suggested that the activity resulted not from fission but from some other reaction—most likely the absorption of a neutron by the uranium nucleus. If that neutron decayed into a proton, then the 2.3-day activity must be element 93, which had never been seen. It would also be the first element heavier than uranium ever found—the first transuranic.

The rules of physics dictated that an element with a half-life of 2.3 days had to be highly unstable. In turn that posed the possibility that it would decay rapidly into the next element up the line, element 94, which was likely to be especially vulnerable to fission. McMillan's initial observations suggested that 94 would be exceptionally long lived, which was both a drawback and, in terms of the war effort, a virtue. The longer an isotope's half-life, the more muted its radioactivity, and, therefore, the more difficult it was to detect. But a long-lived fissionable isotope, especially one that could be chemically separated from its uranium grandparent, might be especially suitable for a bomb.

From his bachelor's apartment in the Faculty Club, McMillan burbled incessantly about his work. Seaborg, who lived down the hall, was enthralled. At virtually every encounter—"whether at the laboratory, at meals, in the hallway, or even going in and out of the shower"—their conversation "had something to do with element 93 and the search for element 94," Seaborg recalled. McMillan was closing in on the elusive new transuranic by watching for alpha emissions in 93's decay products. One day he told Seaborg triumphantly that he had found an alpha emitter and had already ruled out that it could be an isotope of element 91, 92, or 93. It looked like element 94.

Shortly after that, McMillan disappeared.

Only six weeks later would his whereabouts become known, and then only by the bare bones: he was at MIT, working on an unidentified "war project." At the end of November, Seaborg reached him by letter with a proposal to collaborate on the search for 94 in McMillan's absence. Responding with traditional Rad Lab collegiality, McMillan wrote back that he did not expect to be back in Berkeley for quite some time (showing how little he credited Ernest's promise that his East Coast sojourn would be brief) and saying he would be pleased if Seaborg took over the project in the meantime. Seaborg joined with Joseph Kennedy, a newly appointed chemistry instructor with the gangly physique of a scarecrow and a gentle drawl that proclaimed his North Texas origins, and brought the plan to Ernest.

Lawrence viewed the quest for 94 through a double prism. The discovery of a fissile isotope that could be separated from its parent by chemical means might quell the skepticism still bogging down the uranium program. Then there was, as always, his desire to demonstrate the cyclotron's unique properties. If element 94 was to be found, it could be found only through the intense and repeated bombardment of uranium—and the only machine capable of reaching the required energies was the sixty-inch cyclotron. With Lawrence's endorsement, Seaborg's project was awarded top priority for time on the Crocker Cracker.

By January 20, 1942, Seaborg was confident enough to write McMillan with assurance that the bombardments had produced an unknown isotope of element 94. "Things look good for element 94," he wrote, adding the cautionary note: "No one else knows about the most important of these results except Wahl and Kennedy . . . The committee [that is, Briggs] will want us to keep the results *very secret*." He staked the team's claim to the discovery with a letter to the *Physical Review* dated January 28, 1941. "We felt like shouting our discovery from the rooftop, but the war had changed everything," he reflected later. In accordance with Lyman Briggs's secrecy rules, their letter would be withheld from publication until April 1946. By then, Seaborg would observe, the existence of element 94 had been revealed to the world "in the most dramatic form possible": with a detonation in the skies above Nagasaki, Japan.

Seaborg's next step was to test the new element's fission cross section—in other words, to determine whether it could sustain a chain reaction and make a bomb. This new phase of research required the irradiation of uranium on a heroic scale, with the goal of producing enough 94 to perform the necessary experiments. Seaborg, now working with the Italian refugee physicist Emilio Segrè, calculated that a week's bombardment of 1.2 kilograms of uranium (a touch more than 2.5 pounds) by the sixty-inch cyclotron would yield one microgram—a millionth of a gram—of element 93. A further calculation gave Seaborg pause: the irradiated uranium would be ferociously radioactive. The lab had been so blasé about the radioisotopes produced at the Crocker Lab that it routinely sent some of them through the US Mail; but this sample would have to be treated with sedulous respect, handled at greater than arm's length by personnel wearing lead gloves and goggles.

One day in early March, Seaborg and Segrè carried a hot sample of bombarded uranium out of the Crocker Lab in a lead bucket suspended from a long pole. Swaddled in leaded clothing, they crossed to Gilman Hall and crab-walked up two flights of stairs to a vacated laboratory. There they separated out 93 by repeating a tedious sequence of heating, evaporating, dissolving, precipitating, and centrifuging over three long days. They poured their final precipitate into a platinum dish smaller than a dime, boiled off the liquid, and slathered the residue with a layer of Duco Cement. The dish was glued to a piece of cardboard, labeled "Sample A," and left alone to decay into element 94. That took three more weeks. At last, by Seaborg's reckoning, they had one quarter of a millionth of a gram of element 94.

Then came what he called the moment of truth: they brought the sample to the thirty-seven-inch cyclotron, placed it in a neutron beam, and waited for their detector to announce the kicks that indicated fission. The sounds came instantly and unmistakably. Later tests would tell them that isotope 239 of element 94—that is, a nucleus pregnant with 94 protons and 145 neutrons—was nearly twice as fissile as U-235 (later experiments would refine this result to 1.24, still appreciably more fissile than the uranium isotope) and that its half-life was somewhere in the neighborhood of 30,000 years (the true value is 24,100 years). That was surely long enough to provide the stability required of a bomb core. By transmuting common uranium into a fissionable product that could be extracted chemically, Seaborg calculated, they could increase the supply of raw material available for a bomb by a hundredfold. It was May 1941. Around the end of that year, after McMillan had named his element 93 neptunium to commemorate the eighth planet from the sun, Seaborg followed suit and named 94 for the ninth planet, which had been discovered only in 1930. It became known as plutonium.

• • •

A few weeks after Pearl Harbor, Arthur Compton summoned Seaborg to Chicago for a heart-to-heart talk.

The subject boiled down to this: Could Seaborg devise a chemical process for separating element 94 from the stew of radioactive fission products in which it was mixed after the bombardment of uranium? Compton specified that the work would have to be done at breakneck speed and that the process would have to be scalable to an industrial level. Without hesitating, Seaborg answered yes. He would have many opportunities later to curse himself for this "rather hasty expression of confidence delivered by a young man not likely to admit to probable failure, who was too ignorant to realize the ultimate magnitude of the project or even his inexperience about the intricacies of large-scale production." He might have been even more unnerved had he known how skeptically his seniors regarded his headstrong pluck. Bush and Conant both advised Compton that in pressing ahead on plutonium, he was making a long-shot bet, for no plutonium separation process had even been demonstrated in the lab.

Compton stood his ground bullheadedly. "Seaborg tells me that within six months from the time plutonium is formed, he can have it available for use in the bomb."

Conant snorted dismissively. "Glenn Seaborg is a very competent young chemist," he said, "but he isn't that good."

But he was. Once he settled on a production method, Seaborg beat the deadline by four months.

Following the $400,000 contract awarded by S-1, the Rad Lab outgrew its quarters rapidly. As early as the beginning of 1941 the staff had expanded to nearly one hundred. The feel of a tight-knit group pulling together under Ernest Lawrence's paternal eye was slipping away. "It wasn't a cozy gang," Kamen lamented in a letter to McMillan after the Rad Lab Christmas party that year, attended by scores of strangers. To house his burgeoning staff, Ernest commandeered every vacant room in LeConte

Hall and, with Sproul's permission, completely took over the university's newest instructional building, then officially designated simply as the "New Classroom Building." (It would later be christened Durant Hall in honor of Berkeley's first president.) Kamen's strangers mixed with previously departed veterans, for Ernest was summoning his cyclotroneers back from their nationwide diaspora—back from Cornell and Princeton and Washington University in St. Louis, and back from Westinghouse and General Electric—to help him perfect the electromagnetic separation of uranium-235.

When a warm body was needed, Ernest would recruit almost off the street, like a Hollywood talent spotter dragooning a soda fountain girl into instant stardom. To manage the construction of the 184-inch cyclotron, he appointed Ed Strong, a Berkeley philosophy professor, whose home Ernest had helped finance with his Nobel Prize money as a way of investing informally in California real estate. After Strong described at dinner one evening his experiences supervising the carpenters, plumbers, and electricians building his house, Lawrence drove him up the hill to the Rad Lab site. The new cyclotron's 4,500-ton magnet, the largest in the world, was still standing in the open air and other parts were packed in huge crates. "I need somebody who on the one hand can work with the mechanics, who talks their language, and, on the other hand, won't be run over by the physicists and the engineers," he explained. Strong would spend the war years overseeing one of the largest scientific construction sites in the world. After the war's end, however, he turned down Ernest's offer to continue as the lab manager in order to resume the classroom teaching of Hegel and Marx.

Lawrence was once again making fresh footprints into a new world, an experience he had not enjoyed since the construction of the twenty-seven-inch cyclotron. Those who were working with him for the first time were instantly infused with his confidence, not to mention his sense of urgency. Even Vannevar Bush, who had spent his career surrounded by accomplished scientists, felt overcome by the "stimulating" and "refreshing" atmosphere in the lab when he paid a visit in February 1942.

The chief drawback to the lab's expanding relationship with the government was the growing burden of security. Briggs's rules prohibiting publication of fission-related articles soon expanded to a blanket ban on any discussion of the topic outside the confines of the laboratory. The scientists' initial reaction to these increasingly draconian strictures was to treat them whimsically. Seaborg's group started referring to element 94 as "copper." That worked for a few months, until they needed actual copper fittings for their equipment, at which point they referred to the genuine metal as "honest-to-God copper." The scientists were forbidden to ever refer to uranium or plutonium by name. ("All I knew was that *uranium* was a dirty word, and I wasn't supposed to use it," Molly Lawrence recalled.) Eventually they took to designating these elements by a code assembled from the final digits of their atomic number (92 for uranium and 94 for plutonium) and their isotopic weight, so that uranium-235 became known as "25" and plutonium-239 as "49." This code persisted through the war.

Having been upbraided by Cockcroft for the Rad Lab's publication of atomic research in 1940, Ernest soon accommodated himself to the demands of military secrecy. These included the regime of "compartmentalization" decreed by General Leslie Groves, the newly appointed head of the bomb project, to limit the knowledge any scientist could have beyond what was needed to perform his or her specific task. That does not mean that government security officials invariably found Lawrence cooperative, especially when their suspicions fell upon Rad Lab members he considered indispensable. "We had more trouble with Ernest Lawrence over personnel than any four other people put together," Lieutenant Colonel John Lansdale, Groves's security chief, was still complaining many years after the war. His specific reference was to a contretemps in August 1943 over Rossi Lomanitz, a twenty-one-year-old Rad Lab physicist who was valued as a protégé by both Lawrence and Oppenheimer. Lawrence planned to place Lomanitz in charge of the electromagnetic separation work at Berkeley, but the security men had other ideas. Viewing Lomanitz as a dangerous leftist, they plotted to remove him from sensitive nuclear research via the simple expedient of drafting him into the army. Lawrence demanded a face-to-face meeting with Lansdale over Lomanitz's fate. It was not a warm encounter. Lansdale recalled, "Ernest Lawrence yelled and screamed louder than anyone else about us taking Lomanitz away from him." But in the end, the army got its way—and got its man.

Lawrence's approach to security pressures in the bomb program soon evolved under pressure and the omnipresence of security officials in his life. "The crushing responsibilities of his later years," Martin Kamen would observe, "hardened his attitudes about people and increased his suspicions of their motivations."

Kamen was the most prominent victim of Lawrence's increasing rigidity on security matters, for it cost him his job at the Rad Lab and almost his entire career. Despite the importance of Kamen's work—his discovery of carbon-14 was a notable scientific achievement, and his indefatigable management of radioisotope production a contributor to the lab's reputation as a reliable supplier of research materials—he had never been part of Lawrence's inner circle. He was a chemist, not a physicist, with an intellectual temperament much closer to Oppenheimer's than to Lawrence's. A talented musician, he socialized with what he described as an "exciting group of leftist intellectuals and bon vivants" in San Francisco. Also, he was Jewish, and although Lawrence was not an anti-Semite, he could be oversensitive to ethnic distinctions, possibly because of his upbringing in an ethnically homogenous rural community. In describing Kamen to his friend Alex Allen, who was hiring at the Bartol Research Institute in 1937, Lawrence wrote: "He is Jewish and in some quarters, of course, that would be held against him, but in his case it should not be, as he has none of the characteristics that some non-Aryans have. He is really a very nice fellow." (Whether Allen offered Kamen a job is unclear, but in any event, Kamen stayed at Berkeley.)

By early 1943, Kamen's social contacts brought him under the eye of army security. He was aware that his home phone was tapped and his house watched by comically conspicuous security men, who would sit for hours in the front seat of a parked car with the motor running. Despite the surveillance, Kamen remained imprudently nonchalant about his social activities. One night he was spotted dining with Gregory Kheifetz, the local Soviet vice-consul, whom he had met at a party at the home of the violinist Isaac Stern. The dinner was Kheifetz's way of thanking Kamen for his having arranged radiation treatment for a Soviet diplomat's leukemia in John Lawrence's lab, but it was viewed with much greater gravity by the security men. Not long after the dinner, Cooksey called Kamen into his office. His face ashen, he wordlessly handed over a one-page typed document ordering Kamen to leave the lab immediately. He would never return.

With virtually every academic lab in the country working under government contracts, Kamen could not find a research job anywhere. He sat out the war as a technical inspector at the Kaiser shipyards in nearby Richmond, California. For years, the chemist harbored a bitter resentment of Lawrence, who had plainly dumped the task of firing him in Cooksey's lap and absented himself from the Rad Lab while the deed was done. "EOL thinks I told the Russians something," Kamen complained to Oppenheimer. "How he could have fallen for this cock-and-bull story is beyond my comprehension except that he may have wanted to." The cloud over Kamen's career did not lift until the onset of peace, when he was invited by Arthur Compton to supervise the construction of a cyclotron at Washington University in St. Louis, where Compton had been named chancellor. As it was later revealed, Compton had called Lawrence to inquire about Kamen, and received a glowing recommendation and the assurance that he "felt no doubt with regard to Kamen's essential loyalty to the United States."

The security restrictions also burdened Lawrence personally, and not merely because they added a new administrative concern to his expanding responsibilities. They also narrowed his options for respite from his care-laden work. Arthur Compton, who faced similar pressures and was accustomed to freely discussing his research with his wife, arranged to have her cleared by security; Seaborg, a newlywed whose bride—Ernest Lawrence's secretary, Helen Griggs—had been vetted before the war so that she could type Rad Lab scientists' technical reports, did not even need to take that step.

Ernest, however, had never been one to talk about his work at home. He was not inclined to start now, when it was more sensitive than ever. His resulting isolation only increased his fatigue and nervous tension. He was on the road almost constantly, swallowing his deep aversion to airplane travel merely to get from one meeting to another; for the first time in their married life, Ernest and Molly spent Thanksgiving and Christmas apart. The closest he came to confiding in her was to hand her a list of the stops on his itinerary, but if she asked why he was going to any particular place, he would snap, "It's none of your business." "So I got out of the habit of asking questions," she recalled.

As Lawrence expected, the skills of the mass spectrograph operators improved dramatically with experience, expanding the device's capabilities; by mid-January, a nine-hour run yielded 18 micrograms of uranium enriched to 25 percent U-235. The sample weighed about a thousandth of a grain of sand, but it was thirty-four times more concentrated with the fissionable isotope than was natural uranium. A month later, the lab had accumulated 225 micrograms enriched to 30 percent U-235. This was divided into three samples, two of which were dispatched to Compton at the bomb project's Chicago laboratory, known as the Metallurgical Laboratory, or "Met Lab," and to Mark Oliphant, who was in charge of the British government's uranium program at Cambridge University.

Lawrence continued to press ahead. Even as his staff was mastering the first-generation separator, he started sketching out a new one. This device was shaped like the letter C to accommodate the semicircular ion beam; it was smaller than the original tank, so its internal vacuum would be easier to maintain, and was fitted with an ion source ten times as powerful as the original. The upgrade also featured a new type of isotope collector, which was shaped like a box with separate compartments for each isotope and was water-cooled to keep it from melting under the torrent of energized ions. In acknowledgment of the University of California's tolerant hospitality toward a laboratory that was now devoted almost entirely to government work, Lawrence dubbed the new unit the calutron.

Within days of its installation between the magnet poles of the thirty-seven-inch, the calutron exceeded Lawrence's expectations. He telephoned Bush with the news, his enthusiasm coming through so loud and clear over the long-distance line that Bush dashed off a note to FDR declaring that "there is a possibility of production of fully practicable quantities of material by the summer of 1943 [fifteen months away]." That would beat Compton's timeline by six months.

Nor was Lawrence finished. So far the calutron's production was minuscule and its reliability imperfect, but he was already planning to replace its 10-milliampere ion source with one offering yet another tenfold increase in power. But that was itself a mere stepping stone to greater things. Having shown that the calutron would work in the thirty-seven-inch magnet, he was now prepared to sacrifice his crown jewel to the war effort: the enormous 4,500-ton magnet ordered for the he-man cyclotron and still standing in solitude on the hillside above campus.

The original planning for the 184-inch cyclotron had called for the magnet's construction to be completed by November, but that deadline had slipped with the arrival of war. Now Ernest launched a crash program to convert it into a gigantic mass spectrograph containing several calutrons, each with multiple ion sources and collectors. His goal was to demonstrate that electromagnetic separation could produce enriched uranium on an industrial scale.

Ernest estimated that the conversion program would cost about $60,000, mostly for labor on round-the-clock shifts. Vannevar Bush balked at the sum, complaining that it was too much for the government to spend on equipment that would remain the property of the University of California. Lawrence turned instead to his newest and wealthiest private patron, the Rockefeller Foundation. He solicited an emergency meeting in Washington with Warren Weaver, refusing to divulge the subject in advance but offering to arrange government clearance for Weaver if he desired a full discussion. Weaver demurred. "Don't tell me anything secret," he said. He could guess the reason for the meeting, anyway. "What else could you want from me but money?"

Lawrence's appeal to Weaver resembled his recruiting pitch for the MIT Rad Lab: he needed the foundation's help with a project of the utmost importance that he could not divulge; his reputation and his record would have to do. Weaver brought the blind request to Raymond Fosdick, disguised as a grant application "for expediting the construction of the giant cyclotron, and for the purchase of certain associated equipment." Trustees who inquired why Berkeley needed more money so soon after receiving one of the biggest grants in the foundation's history were told "the need was urgent but could not be stated in specific terms," Fosdick would recall. "And so the $60,000 was voted as a matter of faith."

Only after the bomb was dropped did Lawrence reveal to Fosdick what the latter recalled as "the vital part played by the Rockefeller Foundation in the development of the atomic bomb." The trustees' discovery that they had unwittingly funded a weapon of mass destruction was not regarded in the boardroom as grounds for pride. "In the whole history of the Foundation, no grant had ever been made for a destructive purpose," Fosdick observed unhappily in the foundation's official history, "let alone such a lethal weapon as this."

The expedited work was finished by the end of May, and a vast circular building near Strawberry Canyon originally designed to house the world's biggest cyclotron became the headquarters of an entirely different project. Compared with its predecessor facilities on campus, the new building was unimaginably spacious, but the ceaseless activity inside could make it seem cramped. Machine shops and arrays of heavy-duty electrical equipment were situated on the ground floor along the curved inner walls, facing the two-story arch of the magnet. A second-floor deck held offices and conference rooms. Between the magnet's fifteen-foot-wide pole faces yawned a six-foot gap, wide enough to accommodate at least two C-shaped vacuum tanks arranged back-to-back. The arms of the Cs faced out so the ion sources, electrodes, heaters, and collectors could be moved, tilted, pushed in, or drawn out by operators seated within the concave space, which was fitted with vacuum-proof windows through which they could monitor the beams.

Work on the calutrons proceeded twenty-four hours a day even while the magnet was energized, its powerful field a constant presence in the building. "We all knew not to wear a watch or carry keys," recalled Bill Parkins, a young recruit from Cornell, but the magnet's tug on the nails in his shoes produced the sensation of slogging through mud. The workers' tools were forged from a nonmagnetic alloy of beryllium and copper, though occasionally an iron nail fell within the magnetic field and shot to a pole face like a stray bullet, striking its target with a metallic *ping*. Large ferrous appliances, such as the wheeled metal canisters for liquid nitrogen known as Dewar flasks, had to be chained in place. Perched atop the magnet housing was a sofa on which technicians working long shifts could catch catnaps, the nighttime chill of the Berkeley hillside chased off by the heat generated by the humming behemoth below them.

Ernest and his crews tinkered with the setup through the spring and into the summer while Conant and Bush pressed for faster progress. On May 23 Conant summoned to Washington all the program chiefs—Lawrence, Briggs, Compton, Murphree, and Urey—in the hope of settling on one or two separation methods and scrapping the others. A few days earlier, he had outlined for Bush the ultimatum he would deliver to his cadre of highly competitive experts: they had to decide whether "the possession of the new weapon in sufficient quantities would be a determining factor in the war." If it was, speed would be of the essence, and funding on a heroic scale would flow. If not—if even a couple of dozen atomic bombs would be "not in reality determining but only supplemental" to a successful war effort—then there would be no rush, no need for hasty spending, and possibly no need for a wartime bomb program at all.

To a certain extent, Conant was bluffing, for the rising concern in Washington and the scientific community about a possible German bomb made an allied effort seem imperative, regardless of the prospects of success. The concern was provoked in part by news that the Nazis had commandeered a Norwegian heavy-water plant, which could only mean that the enemy was building an atomic pile moderated by heavy water—one of the options considered by Fermi before he settled on graphite to control neutrons. That suggested in turn that the Nazis had discovered plutonium, which they intended, like the Allies, to breed in a reactor. Earlier that spring, Eugene Wigner had spun a ghastly scenario for Compton: if the Nazis knew about element 94, it might take them only two months to produce 6 kilograms of it in a heavy-water pile. That would be sufficient to build six bombs by the end of 1942, beating Compton's timeline for an Allied bomb by two years. As was learned later, the Nazi bomb program never got off the ground; but obviously this could not be known at the time.

Despite the urgency, Conant was doomed to be disappointed in his hope for a firm decision on a separation process. None of the four options for uranium enrichment—electromagnetic or thermal separation, gaseous diffusion, or the centrifuge—had clearly outdistanced the others, so none could be ruled out. Fermi had yet to achieve a chain reaction, so the availability of plutonium was still in question. Yet proceeding at full speed with all the options meant an unnerving commitment of hundreds of millions of dollars.

Still, one method did show distinct promise: Lawrence's. With his customary bravado, he reported that the first calutron in the big magnet would start separating uranium isotopes within three weeks. By September, he promised, he would be producing 4 grams a day of U-235. He believed that a process capable of producing 100 grams a day was technically within reach. Assuming that the calutron could be made to enrich its output to 80 percent U-235 (though that was far beyond its capabilities at the moment), there could be enough material for one 30-kilogram bomb core in a year.

Conant knew Lawrence well enough to have confidence he was not drawing castles in the air. He toyed briefly with casting the government's die in favor of the calutron. But this was tantamount to calling Lawrence's bluff: at its current level of development, Ernest acknowledged, the calutron could produce enriched uranium fast but only in very small quantities. Whether it could shoulder all the demands of a full-scale bomb program was still in doubt. Conant, unhappy with his scientists' inability to settle on a single separation process, penciled out a proposal to spend $870 million for four individual pilot plants.

Over that summer, the Rad Lab continued to improve the calutron design. The most important advances involved manipulating the magnetic field to sharpen the ion beams; the problem harkened back to the imperfection of the magnetic field in the earliest generations of cyclotrons, and the solution proved to be the same: arrangements of metal shims to "shape" the field, much as an optical lens focuses light entering a camera to produce a sharp image. On August 13, at an S-1 meeting on his home turf in Berkeley, Ernest delivered another bravura presentation on the Rad Lab's progress. The remaining problems in calutron design were merely engineering, he declared, a cat he had skinned dozens of times. Electromagnetic separation now had widened its lead in development over the other methods. Gaseous diffusion and the centrifuge appeared to be feasible in theory for the large-scale production of U-235, but both were hobbled by unsolved technical problems, and neither had yet produced any U-235 to speak of. The main obstacle for the diffusion process being developed by Urey was its reliance on uranium hexafluoride gas, the detestable "hex," which lived up to its nickname by severely corroding every permeable barrier Urey tried. The centrifuge, which was being pursued by Lawrence's old friend Jesse Beams at the University of Virginia, persistently underperformed its expected enrichment yield; it would be the first method abandoned outright. Progress on the atomic pile gave reason for optimism, but Fermi—unhappily relocated to Chicago on Compton's orders—was still building only prototypes.

The key unsolved question was whether electromagnetic separation could produce more than a single uranium bomb core. Lawrence remained uncertain. Once again he advised Conant to keep his options open rather than "picking a horse for the long pull, for it seems to me quite likely that other methods will ultimately prove to be better." Meanwhile, he worked out a design for a large-scale calutron plant employing an array of vacuum tanks arranged in large ovals between the poles of huge electromagnets. There would be ninety-six tanks to each oval, each holding two calutrons back to back, designed to beam uranium ions in 180-degree arcs four feet in radius from source to collector. At the Rad Lab, the arrangement was known, after its ovoid configuration, as the racetrack. Despite Lawrence's own initial doubts, it would produce the fuel for the bomb that destroyed Hiroshima.

Chapter Thirteen

Oak Ridge

In June 1942 Vannevar Bush approved the purchase of a rural fifty-two-thousand-acre parcel about twenty-five miles from Knoxville, Tennessee. It was a long, flat valley shaded by wooded ridges on either side, sufficiently remote to serve as a secret facility for the manufacture of bomb-grade uranium but also close to the Tennessee Valley Authority electrical plants that would have to feed its prodigious appetite for power.

Preparing the site would require heavy construction, grading, and converting cow paths into four-lane highways. The job of finalizing the purchase and launching the work had been assigned to the US Army Corps of Engineers, but the Corps dithered, deciding that until the S-1 Committee settled on an enrichment technology, there was no need to acquire the site, much less get construction under way.

At the end of August 1942, Bush warned Army Chief of Staff George C. Marshall and Secretary of War Henry Stimson that the delay threatened to stifle a bomb program that had won the unanimous endorsement of "a group of men that I consider to be among the greatest scientists in the world." His stern words got the project jump-started. General Brehon Somervell, the head of the Corps of Engineers, placed the program—then still known obliquely as the DSM Project, for "Development of Substitute Materials"—under the jurisdiction of the Corps's Manhattan Engineer District, headquartered at 270 Broadway, and in the hands of his most trusted and efficient subordinate, Colonel Leslie R. Groves. Thus was born the Manhattan Project.

Groves, the son of an army chaplain, was a West Point graduate who tipped the scale at three hundred pounds and was not shy about throwing his weight around. The previous year, he had assumed command of construction of the Pentagon, then beleaguered by labor problems, material shortages, cost overruns, and miserable work conditions. He got the largest office building in the world completed within eight months. Groves joked later that when this complex and contentious job was done, "I was hoping to get to a war theater so I could find a little peace."

Bush learned that he had a new teammate from the army only when Groves, who had been promoted to brigadier general upon taking over the bomb project, showed up at his office on September 17. Furious at being blindsided by the appointment, Bush dodged Groves's brusque queries about the state of the project and dashed off a prickly letter to Stimson's assistant, Harvey Bundy. "Having seen General Groves briefly, I doubted whether he had sufficient tact for such a job," he wrote. "I fear we are in the soup."

He soon changed his mind. The two most pressing necessities for the bomb program were the acquisition of the Tennessee site and the securing of a top-priority classification for the work, both of which had been in limbo for four months. Groves met both goals within forty-eight hours of taking over.

Three weeks later, Groves was in Berkeley. He arrived at the Rad Lab skeptical about electromagnetic separation, having been convinced of the superiority of gaseous diffusion by his technical advisors, who came from a petroleum industry familiar with that technology. But they had not reckoned with the horrors of hex—or with the galvanizing personality of Ernest Lawrence.

Lawrence escorted Groves around the Rad Lab, showed him the calutron in operation, and explained its virtues so convincingly that the general was carried away. He measured Lawrence against the other program heads he already had met, and judged him the most capable. Lawrence's experience as an impresario of Big Science gave him an instinctive grasp of how to scale up a process from prototype to production. That was a critical talent for a project facing such tight deadlines that immense factories would have to be designed and built for processes that could not be tested in advance. After a few days with Lawrence, Groves recognized that his process was the only one with a demonstrated capability to produce appreciable quantities of U-235. On November 5 he asked Lawrence to draw up specifications for a sprawling electromagnetic separation plant based on the existing calutron design, which was to be frozen, or made final, so that commercial manufacturers could begin turning out the tanks for installation. The plant, code-named Y-12, would be located on 825 acres at the southeastern end of the new federal reservation in the Tennessee Valley. Five miles away, a residential community would be built to house the staff. It would be known as Oak Ridge.

One other important event occurred during Groves's visit to Berkeley: his first encounter with J. Robert Oppenheimer.

Since being tapped by Arthur Compton in June to supervise the theoretical work on bomb design, Oppie had been working out of his Berkeley office with a small group of trusted colleagues. One of them, Robert Serber, was present when Groves, fresh from his briefing by Lawrence, strode into the room trailed by his adjutant, Colonel Kenneth D. Nichols. "Groves walked in, unbuttoned his tunic, handed it to Nichols, and said, 'Take this and find a dry cleaner and get it cleaned,' " Serber recalled. "Treating a colonel like an errand boy. That was Groves's way."

Oppenheimer, who could tell they were witnessing a theatrical display of military authority, was unfazed. He spent the next few hours in a collegial discussion with Groves about the challenges facing the bomb designers. As Groves recalled later, he had been advised by his consultants that the device could be "designed and fabricated in a very short time by a relatively small number of competent men"—even by twenty scientists in three months or less. He was looking for validation of his instinct that this was a "dangerously optimistic" view. Oppenheimer provided it, warning that a multitude of theoretical and technical questions still remained unresolved; therefore, design work should be initiated immediately if the weapon were to be ready in time to influence the war.

Groves asked Oppenheimer to join him in Chicago a week later for further discussion. There they boarded the New York–bound 20th Century Limited passenger train and, crammed into a rattling roomette with Nichols, talked about how to organize a bomb design lab: "setting up an organization, building the needed facilities . . . and dealing with such expected issues as recruiting scientists and confining them in a laboratory in a remote area," Nichols recalled.

Groves originally had considered appointing Lawrence to head the bomb lab but decided to leave him in place overseeing the crucial electromagnetic separation work. Of the other possible candidates, Compton could not be spared from Chicago, and Harold Urey, a chemist, was not qualified to manage a physics lab. By the time Oppenheimer disembarked from the 20th Century in Buffalo to return west, Groves recognized that he was the best choice of all.

He did have several concerns, however. Oppenheimer had no administrative experience. Nor did he have a Nobel Prize. Groves, who thought the latter shortcoming deprived Oppie of the scientific prestige the project leader should possess, plainly was unaware of the scientist's towering reputation in the physics community. Another concern was Oppenheimer's leftist background. Citing Oppie's habit of dabbling in politics, the Manhattan Project's security apparatus, which was not yet entirely under Groves's control, refused to clear him for war work. To overcome that obstacle, Groves sought help from Lawrence, who delivered an enthusiastic personal recommendation and provided Groves with a letter intended to vanquish all doubts: "I have known Professor J. Robert Oppenheimer for fourteen years as a faculty colleague and close personal friend," Ernest wrote. "I am glad to recommend him in highest terms as a man of great intellectual caliber and of fine character and personality. There can be no question as to his integrity."

Lawrence further assured Groves that, if Oppenheimer failed in his task, he would step in himself. With that escape clause in his pocket, Groves ordered the Manhattan Project's security staff to award the necessary clearance "irrespective of the information which you have concerning Mr. Oppenheimer," and appointed him to head the desert laboratory that would be known as Los Alamos.

Glenn Seaborg arrived in Chicago to take up his assignment from Compton on Sunday, April 19, 1942. The blustery day was his thirtieth birthday. He understood the challenge ahead, for he and his colleagues had yet to produce a speck of element 94 detectable even by microscope. The work, as he put it wryly, involved "invisible materials being weighed with an invisible balance." The scientists could determine if plutonium was present only by reading its trace radioactive signature; this was science performed not by observation but by deduction.

Compton named Seaborg's group Section C-1 and gave it space on the fourth floor of the University of Chicago's Herbert A. Jones Laboratory, where the aging benches, sinks, and fume hoods reminded Seaborg of the laboratories he had haunted in his undergraduate years. Now he was the man in charge, however, and his first task was to fill out his team. Taking a page from Ernest Lawrence's recruitment handbook, he reached out to the best chemists he had known as classmates and fellow faculty members, and dangled before them the prospect of career-making jobs on a project that, alas, he was not at liberty to describe. "Unfortunately I cannot divulge to you the nature of the work but . . . you are in a fair position to guess," he wrote a college friend whom he knew to be unhappy in an oil company job. "It is the most interesting problem upon which I have ever worked."

It was a difficult sell. Seaborg had Lawrence's verve and enthusiasm but not his reputation, and he was not always successful at persuading his quarries that his mysterious summons warranted their dropping everything and moving to Chicago for an indefinite period. Still, within five months, he had assembled a staff of twenty-five; within a year of his arrival, it would be fifty, and at its peak, one hundred.

Their primary goal was to bombard uranium to obtain sufficient plutonium for study. There were two methods to choose from: a cyclotron-generated neutron beam or a chain-reacting atomic pile. But the first produced minuscule quantities; at a production rate of 1 microgram of plutonium per week of uranium bombardment, Seaborg calculated it would take the cyclotron twenty thousand years to make a single kilogram of element 94. A chain-reacting pile would be more efficient but still existed only in theory, for Fermi's most recent prototype had produced less than one neutron per fission—a definite fizzle. Even assuming that a chain reaction could be started and maintained, the task would remain of extracting the plutonium—concentrated at 250 parts per million, or a half pound per ton of uranium—from material so radioactive that the processing would have to be done from behind walls of dense concrete.

Luckily, Seaborg did not need a kilogram of plutonium just then, only a few micrograms. And for that, the cyclotron at Washington University in St. Louis, which Compton had built before moving to the University of Chicago, would do. Seaborg's group commandeered the cyclotron to bombard uranium twenty-four hours a day. Over the next eighteen months, this indefatigable machine would produce two thousandths of a gram of plutonium, a quantity about the size of a grain of salt. Seaborg monitored the work with only a single break in his routine: a quick visit to Berkeley to collect Helen Griggs, his bride-to-be, followed by a hurried wedding in a dusty Nevada town called Pioche and a wedding night in a hotel room on the second floor of the rail depot. From there the married couple continued on to Chicago by train. Soon after their arrival, the first shipment of irradiated uranium arrived from St. Louis by truck: three hundred pounds of it packed in plywood crates shielded by lead bricks. Some of the boxes had broken open on the way, spilling hot uranium over the truck bed; Seaborg advised his assistants to wear rubber gloves to sweep up the detritus.

Once they hauled the cargo upstairs to their lab, they let it cool for a week before beginning the tedious process of reduction, oxidation, precipitation, and extraction.

A breakthrough came on the morning of August 20. Seaborg's microchemists used hydrofluoric acid to reduce a solution made from the bombardment products and watched a minuscule quantity of pinkish material precipitate out: this was pure plutonium-239. As Seaborg reported in his journal, it was the first time that plutonium—indeed, any synthetic element—had been seen by the naked eye. "I'm sure my feelings were akin to those of a new father," he wrote. It had been twenty months since he and Joe Kennedy had identified element 94's alpha signature in their microscopic sample. With Fermi's chain-reacting pile still undergoing its difficult gestation, for all they knew this was the only plutonium that would be seen for months, or even years.

All that day and into the evening, Met Lab associates streamed into room 405 of the Jones Lab to peer through the microscope at the pink speck. A few weeks later, General Groves was ushered upstairs to view the plutonium sample for himself. He put his eye up to the microscope, where a researcher had set the sample on a glass plate with exquisite care.

"I don't see anything," he growled. "I'll be interested when you can show me a few pounds of the stuff!"

Groves's order to freeze the calutron design was honored only in the most general terms, and then only for the very first units to be shipped for installation at Oak Ridge. This was inevitable, for the separation process was so novel that no firm specifications could have been devised without continued experimentation by the Rad Lab, even after the start of construction of Y-12. The mass and purity of the U-235 collected from the calutrons varied widely depending on the pressure within the units, the strength and shape of the magnetic field, the voltage of the accelerating electrodes, and myriad other factors, all interrelating in ways not fully understood.

Lawrence originally calculated that two thousand calutron source-and-collector arrays—two thousand vacuum tanks, or about twenty racetracks—would yield 100 grams a day of U-235. By January, the calutron had been modified so that each one could be operated with pairs of sources and collectors. That would double their output or, to put it another way, reduce the requirement for 100 grams of daily production to only one thousand tanks. Having absorbed Lawrence's confidence that further efficiencies eventually would be achieved, Groves reduced the specification for Y-12 to five hundred tanks, or five racetracks. The plant was built with sufficient room to install more tanks if production failed to continue ramping up. Groves set a punishing pace for construction, demanding that the first racetrack be running by July 1, 1943, and all five by the end of the year. These deadlines left no time for idle theorizing. The Rad Lab returned to the days of cut-and-try; any configuration that produced a stronger or sharper beam was incorporated into the standard, even if the experimenters could not figure out why it worked.

Every piece of equipment had to be machined to the most demanding specifications and designed for the heaviest duty. Gaskets had to survive extreme temperature changes and maintain their airtight seal over long periods of constant operation. Vacuum pumps, a constant source of frustration for the Rad Lab staff even when the machinery operated at its best, had to be scaled up to gargantuan dimensions so they could rapidly reestablish vacuums in tanks that would be regularly opened for cleaning but had to be promptly returned to service. The collectors were repeatedly redesigned in a never-ending effort to keep the deposits of U-235 and U-238 apart.

For the Rad Lab, the most familiar design elements were the magnets. After all, they had already erected the largest magnet in existence for the 184-inch cyclotron. But that did not mean designing the Y-12 magnets would be without its challenges, for all the racetrack magnets combined would be one hundred times larger.

This led to the first major procurement crisis of Groves's tenure. The original magnet design called for using copper for the electromagnet coils and the massive transmission bus that carried electricity to the racetracks. But the entire national supply of copper had been requisitioned for other war needs, and not even Groves's bullishness could overcome that cold fact. As an alternative, the designers settled on silver. That metal was also in short supply—except for the US Treasury Department's monetary reserve, which was located in a vault at West Point. Groves demanded access to the hoard and got his way. Some 14,700 tons of silver worth more than $300 million were procured from the depository on the understanding that every ounce would be returned after the war. The appropriation of the silver bullion underpinning the US dollar was kept absolutely secret, its milling into electrical cables and its transport watched over by armed military guards until the finished product arrived at Oak Ridge. Groves kept his promise: the precious metal was all returned to the Treasury, though the last of it was not transferred until 1968.

In General Groves, Lawrence encountered—possibly for the first time in his adult life—someone who was even more driven than himself. Ernest now found himself in the unaccustomed position of being regularly outflanked and outrun on decision making; often Groves would make a show of consulting with him, sometimes even visiting Berkeley as part of the charade, only to disclose at last that he had settled unilaterally the question supposedly under joint consideration. Thus a debate over which contractor to place in charge of operating Y-12 ended with the appointment of Tennessee Eastman Corporation, a Kodak subsidiary with which Groves finalized a deal while Lawrence was still pondering a list of candidates. Equipment manufacturers were appointed in the same mysterious manner while supposedly awaiting Lawrence's approval. Instead, he would learn that the decision had been made and contracts signed, and that Groves was already on his way to Berkeley with representatives of the contractors—Westinghouse, General Electric, or the Allis-Chalmers Manufacturing

Company—demanding a full Rad Lab briefing about the process and the equipment they were expected to manufacture and install in Tennessee a month or two hence.

Yet Groves was fully cognizant of Lawrence's crucial role in the program. Without Ernest's energy, his serene self-confidence, and his charismatic command over the scientists and engineers inventing the separation process, there would be no Y-12. Groves placed Lawrence within the same bubble of personal security he ordered for Oppenheimer and a few other indispensable Manhattan Project personnel, notifying him by letter of "certain special precautions" he was to take for his personal safety. Among these was an absolute ban on "flying in airplanes of any description," except with Groves's explicit consent. Lawrence was not to get behind the wheel of an automobile for any distance longer than a few miles, or travel "without suitable protection on any lonely road" or after dark. Instead, he was provided with an armed chauffeur at government expense.

Nowhere was the awe-inspiring scale of the Manhattan Project—in spending, in manpower, in physical size—as evident as at Y-12. The electromagnetic separation plant was the largest and most complicated installation in the valley. By the end of the war, Groves calculated, its construction, engineering, and electricity would cost more than a half billion dollars, making it the single most expensive component of the Manhattan Project. During a two-week period of construction, 128 carloads of electrical equipment arrived by rail. Deliveries of concrete blocks filled 63 railcars; lumber—38 million board feet—another 1,585 cars.

When Lawrence visited Y-12 in May 1943, less than three months after ground had been broken, the sight of the transformed valley left him spellbound. He had been primed to expect big things after inspecting the bustling Westinghouse, General Electric, and Tennessee Eastman factories turning out equipment for Y-12. But the Clinton Engineer Works, the official name of the Tennessee site, looked to be in a class of its own. Four immense two-story buildings were rising in various stages of construction. Cow paths had been transformed into paved thoroughfares, railroad tracks crisscrossed the valley floor, and electric lines snaked over the ridges and connected up to a vast switchyard hulking with transformers. "When you see the magnitude of that operation there," he reported back to the Rad Lab staff, "it sobers you up and makes you realize that whether we want to or no, that we've got to make things go and come through . . . Just from the size of the thing, you can see that a thousand people would just be lost in this place." Great achievements always inspired Lawrence to think bigger, and Oak Ridge was no exception. "We've got to make a definite attempt to just hire everybody in sight and somehow use them, because it's going to be an awful job to get those racetracks into operation on schedule. We must do it."

Groves was anxious to advance Y-12 from the experimental phase to industrial operation quickly, despite warnings from the Rad Lab that what its scientists did not know about electromagnetic separation still outweighed what they could say for certain. Other players expressed similar concerns. When an executive at Tennessee Eastman fretted that his firm might not be capable of the scientific research needed to build a U-235 plant, the executive recalled, Groves snapped that he already had "so many PhDs that he couldn't keep track of them." What he needed from Eastman was its industrial expertise.

Lawrence relocated one hundred of his Berkeley staff to Oak Ridge to tune the calutrons personally. His recruitment method, perfected for the MIT Rad Lab, held firm: "Would you like to go to Tennessee?" he'd ask a candidate. Anyone who responded with even a conditional assent was told, "Good. You leave the day after tomorrow."

The Berkeley scientists arrived to find Y-12 mired in mud, chaos, and rock-bottom morale. With its bulldozer-graded streets and its rows of prefab houses, Oak Ridge looked like an "unfinished movie set," recalled Seaborg. Everything was encrusted in dense Tennessee clay. Those who braved the Oak Ridge cafeteria invariably suffered an introductory bout of intestinal distress dubbed "Clinton fever." The physical conditions were overlaid with military regimentation far beyond anything imposed at the

Rad Lab, even after its absorption into the bomb project. Portions of Y-12 were partitioned off so that the scientists could not cross from one end of its vast floor to the other without stopping repeatedly to show their passes.

Things seemed to run smoothly only when Ernest was on hand. This was discouraging, since the demands on his attention elsewhere were multiplying. Still, he had only to make an appearance at Y-12 for all the issues of logistics and construction staging and equipment design that had piled up in his absence to evaporate. There, as at Berkeley, he convened a policy meeting every morning at eight, marched into the room to its most comfortable chair, and launched the proceedings with a breezy "What's new?" He absorbed the details of every problem instinctively, formulated a solution or a path to the solution as he listened, and moved rapidly on to the next item. Lawrence's visceral feel for the mechanics and capabilities of the calutrons matched the mysterious affinity he had always shown for the cyclotrons. "He felt it in his bones," marveled Bill Parkins, the Cornell recruit.

Another malady suffered by the Berkeley transplants was severe culture shock. Primed to disdain the young women recruited from the Tennessee hollows to work as calutron operators—many of whom were most comfortable padding around the floor barefooted and conversing with one another in an impenetrable back-country drawl—the Berkeley PhDs could not get their minds around the thought that the success of Y-12 depended on these platoons of "essentially illiterate hillbillies," as Kamen labeled them. "Mass spectrometers were still lab instruments of extreme fragility and erratic performance, which could be operated only by highly trained technicians. To assert that they could be built on the scale needed and operate more or less on a push-button basis was to invite invitations to don a straitjacket."

Nor were the women themselves immune from culture shock. They were trained to perform a purely mechanistic job "watching meters and adjusting dials" from swivel chairs set before tall steel consoles marked only with numerals. They reported to work in vast factories that had appeared almost overnight—at least they seemed to be factories, but there was "no receiving dock and no loading dock," Parkins reported. "Nothing went in or out, but everybody was very busy." Women who inquired about their jobs were told "a cock-and-bull story about broadcasting radio signals that jammed the communications of our enemies" or, as one instructor told his training class, "I can only tell you that if our enemies beat us to it, God have mercy on us."

Yet the Tennessee women outdid the Berkeley scientists in one crucial quality: patience. Once Y-12 was operational, their equanimity made them superb operators. Indeed, Colonel Nichols, exasperated by the scientists' condescension toward his workforce, soon informed Lawrence that the women could outproduce the scientists and that he was prepared to prove it in a production race. Lawrence accepted, and lost. The reason, Nichols reckoned, was that the women were trained to follow without debate or explanation the instructions they were given, so they went ahead without distractions; the scientists, however, could not avoid investigating the cause of even the most minor fluctuation of their meters—and since they actually knew what every figure meant, they were constantly making trivial and unnecessary adjustments. As Nichols understood, once the plant entered the production phase, what was needed was unquestioning operation, not ceaseless tweaking.

Lawrence regarded Y-12 as his personal fiefdom (notwithstanding the eleven thousand Tennessee Eastman employees who made it run). On his frequent visits, he could be spotted chatting amiably with workers or engineers, sometimes in the company of less patient companions, such as Groves. One time he gave a lengthy tour to a tall, stoop-shouldered man with a lugubrious expression on his long face, whom Lawrence introduced with unusual deference as "Dr. Nicholas Baker." This was the code name of Niels Bohr, whose gloom reflected his misgivings about the horrific weapon being brought forth by his own discoveries. Lawrence, of course, displayed no such doubts, which may only have deepened "Dr. Baker"'s dismay.

Many issues raised at Lawrence's briefings resulted from the decision to begin construction of Y-12 before the central technology was perfected. An alteration in calutron design that seemed trivial in Berkeley could set off an avalanche of costly and time-consuming changes downstream at Oak Ridge. New designs kept flowing out of Berkeley: by the end of the program, 71 different types of ion sources and 115 designs for the receivers and collectors had been prototyped and tested, with the results sent on to Groves. Some design changes arose from Lawrence's own indefatigable quest for the next big technical advance. He implemented the doubling of the source-and-collector arrays in the calutrons without much fuss, but even before they were installed, he was pondering four arrays, and then eight, and then sixteen—each upgrade requiring new studies of beam interference and power consumption.

Through early 1943, the encouraging design experiments at Berkeley, the rapid progress of construction at Oak Ridge, and Ernest's infectious enthusiasm combined to make electromagnetic separation appear to be the best bet for producing core material for the uranium bomb. Urey's gaseous diffusion plant at Oak Ridge, designated K-25, was falling far behind schedule: in the spring of 1943, no one yet had any idea what material for a diffusion membrane would best resist the corrosive properties of the malevolent hex, and no one could say how soon an answer might be found. Urey quarreled repeatedly with John Dunning, a Columbia colleague who was a specialist in the process. More seriously, he clashed with Groves. "The problem of where to concentrate Dr. Urey's energy is still unsolved," Colonel Nichols scribbled in his notes after a fraught meeting over K-25.

By contrast, plans for Y-12 kept expanding even as the plant rose from the valley floor. The newest idea percolating through the hallways at the Rad Lab called for a two-stage separation process using a new variety of calutron to further enrich the product of the original racetracks. The originals were to be designated the Alpha racetracks; the new, smaller racetracks, which would have thirty-six tanks arranged in a rectangle, with two ion sources each, were labeled Betas.

Groves signed off on the new system in March, launching another round of frenetic experimentation in Berkeley. The Betas would process less material but at higher intensities, leaving less flexibility for loss. Some material always spattered over the interior of the Alphas, much of it recoverable through washing and scrubbing, but the partially enriched uranium serving as the Betas' feedstock was too valuable for even a speck of it to be left behind on the ion sources, electrodes, or vacuum tank walls.

By late spring, construction was well along on five Alpha racetracks, two Betas, and a host of outbuildings for chemical processing. Lawrence predicted that each Alpha would be capable of producing 300 grams of enriched uranium a month. The first racetrack was to be ready to start producing by September.

Then this one great step forward was pushed two steps back. The Los Alamos lab, recently established under Oppenheimer's leadership, tripled to 40 kilograms its estimate of the quantity of U-235 needed for a bomb. Hearing the news in Washington, Conant despaired of obtaining so much bomb material out of Oak Ridge by any method. But Lawrence saw opportunity in the discouraging new figure. Arguing that Y-12 was the most reliable option for meeting the new specifications, he pressed for a major expansion of the plant, encompassing additional Betas and a new, more productive Alpha design. After a test Alpha at Oak Ridge ran successfully for the first time on August 17, Groves approved the Alpha expansion to nine racetracks from five.

In early November, the first racetrack was being readied for the production phase. But disaster struck when it was energized on November 13, 1943. Nichols would remember the day, his thirty-sixth birthday, without fondness. For the racetrack failed almost instantly. The powerful magnets jolted the fourteen-ton vacuum tanks out of place, which threatened in turn to pull apart the superstructure of pipes, pumps, and cables. Even the tanks that stayed put sprung vacuum leaks, microscopic holes that had to be hunted down one by one and sealed by hand.

The magnets kept shorting out for reasons the engineers were able to diagnose only after opening one up, when they discovered its cooling oil had been thoroughly contaminated by rust, metal shavings, and sediment. The only option was to ship all the magnets back to the manufacturer, Allis-Chalmers, for cleaning and radical reconditioning. At least a month's delay was in store.

The fiasco placed heavy pressure on Lawrence, whose outwardly sunny disposition concealed how close he had been brought to physical collapse by the weight of his responsibilities and the strain of constant traveling. Things only got worse when, in the midst of the frenetic effort to track down the causes of failure, an incensed General Groves arrived on the scene. Groves blamed Lawrence personally for the delay, especially after learning that the Rad Lab had experienced a similar flaw with its cyclotron magnets and therefore should have anticipated the problems. He aimed a stream of "caustic comments" (in Nichols's surely understated words) at Lawrence, and finally Lawrence broke. He made his way to Chicago for an S-1 meeting at the Met Lab, so tormented by back spasms that he had to be carried into the session in a chair. Immediately afterward, he checked himself into the University of Chicago Hospital and summoned Luis Alvarez from the Met Lab.

Alvarez was aghast at Lawrence's appearance. "I had never seen the guy so completely beaten in my life," he recalled later. "He was just exhausted, he was depressed, he was just in very poor spirits." Lawrence feared that the failure at Y-12 would destroy the army's confidence in the entire bomb project, with billions of dollars—indeed, the entire war effort—hanging in the balance. Adding to his misery was his feeling of impotence; for perhaps the first time in his career, he was confronted with an engineering crisis he could not fix himself. The repairs were dependent on the manufacturer's engineers pulling the magnets apart, flushing the oil lines, and putting them back together properly. "There was nothing he could do but sit there and sweat it out," Alvarez recalled. Ernest was fighting a chronic sinus infection, his back was aching, and an orthopedic brace that had been made for him fit so poorly it aggravated rather than relieved the pain. Alvarez hung around to "hold his hands for several days," at which point Ernest staggered home to Berkeley to await the completion of repairs.

The crisis proved to be temporary. At length, Allis-Chalmers achieved the necessary fixes. The successful launch of the Alpha II array in mid-January 1944 helped dispel Ernest's gloom. The new calutrons were plagued by their own flaws—electrical failures, cracked insulators, leaky vacuum tanks, a cascade of trivialities that spelled long halts for dismantling and reassembly—but millimeter by millimeter, the engineers and operators mastered the temperamental machines. By the end of February, the Alpha II racetracks had produced 200 grams of material enriched to 12 percent U-235. The volume was disappointing and the enrichment far below what was needed for a bomb core, but it was also far more than any other method had produced. Some of it was shipped to Los Alamos for experimentation, the rest stored as feedstock for the Betas.

Lawrence, now back on his feet, felt encouraged enough to outline an appeal for yet another expansion of Y-12. He pressed his case in a letter to Conant, reminding him that Y-12 stood alone as validation of the feasibility of the bomb program. "The primary fact now is that the element of gamble in the overall picture no longer exists," he wrote. "The electromagnetic plant is in successful operation and the experimental developments at Y [that is, Los Alamos] leave no doubt that the production can be used as an overwhelmingly powerful explosive. It is only a question of time."

In mid-1944, Groves addressed the problem of K-25's disappointing output by scaling back the gaseous diffusion program. Diffusion worked as a "cascade"—in steps, with each cycle of diffusion incrementally increasing the concentration of U-235 in the feedstock produced by the previous cycles. Groves's order reduced K-25 to a supplemental source of raw material for the Beta calutrons of Y-12. In connection with the new arrangement,

Lawrence received approval for his expanded Alpha II plant, which would supplement the feed from the diffusion process in a final push to produce uranium enriched to weapon-grade.

There was one last element to the process. Working on a navy contract quietly and virtually single-handedly at the Philadelphia Navy Yard, Lawrence's former graduate student Philip Abelson had perfected a way of enriching uranium by thermal diffusion. In April 1944, amid persistent doubts that Oak Ridge could supply a bomb core, Groves learned of the work and dispatched a reconnaissance team to Philadelphia. They found Abelson laboring in a forest of one hundred towering cylindrical columns, in which uranium hexafluoride with a high concentration of U-235 was accumulating. Three weeks later, Groves ordered the erection of a 2,100-column thermal diffusion plant at Oak Ridge.

By January 1945, the three-headed process was working. The Alpha tanks produced more than 250 grams per day of uranium enriched to 10 percent U-235. The Betas mixed that with the output of K-25 and the thermal columns to turn out 204 grams of uranium per day enriched to 80 percent U-235. This was bomb-grade fuel. At that rate, one bomb core would be produced by July 1, 1945. This core, manufactured largely by a process essentially cobbled together out of the bits and pieces of Ernest Lawrence's precious cyclotrons, would go into the device known as Little Boy, which detonated above Hiroshima on August 6, 1945.

While the refinement and construction of the uranium plants were taking place, Glenn Seaborg ran his own race to perfect a plutonium extraction process. An important breakthrough occurred in a squash court under the west stands of the University of Chicago's Stagg Field, where Enrico Fermi had erected an atomic pile. On December 2, 1942, he achieved a chain reaction. Arthur Compton was on hand to watch Fermi's men draw the control rods from a towering pile of graphite blocks within which the uranium was arranged, and to hear the telltale clicks from counters measuring radioactive output from the world's first controlled atomic reaction. Elated—perhaps in part because he had won his January bet with Lawrence that the chain reaction would be going by the end of that year—Compton called Conant in Cambridge. "Jim," he said, "you'll be interested to know that the Italian navigator has just landed in the new world."

"Is that so?" Conant replied. "Were the natives friendly?"

"Everyone landed safe and happy."

Seaborg got the news at his office across campus. He recorded the moment in his journal, leavening his relief with the one overriding concern that had lent urgency to the work of all the atomic scientists of those years. "Of course, we have no way of knowing if this is the first time a sustained chain reaction has been achieved," he wrote. "The Germans may have beaten us to it."

On June 1, six months after Fermi's achievement, Seaborg met with executives of DuPont, which had been cajoled by Groves into contracting to build a pilot plutonium separation plant at Oak Ridge. Their goal was to finalize the plant's design. Its chain-reacting pile went live at the beginning of November and within six weeks was producing plutonium by the milligram, and soon by the gram. This was shipped back to Chicago for further study by Seaborg's team, which shortly would abandon its techniques of microchemistry and work with quantities they could see—but which required new safety precautions to protect against breathing or ingesting their "fiendishly toxic" godchild. Meanwhile, DuPont began work on the next stage of the manufacturing cycle: a full-scale production plant located on a bend of the Columbia River in central Washington State, near the little town of Hanford.

Seaborg was as awestruck by the immensity of the Hanford plant as Lawrence had been on witnessing the transformation of Oak Ridge. It had been four years since he and Emilio Segrè had crossed the Berkeley campus with a bucket of ether infused with microscopic quantities of plutonium and suspended from a long wooden pole. All his work since then had culminated in a factory that could irradiate two hundred tons of uranium at a time to produce a half pound of plutonium every two hundred days. From the slugs of irradiated uranium, fissile plutonium would be extracted at a second plant ten miles away, eventually to be sent on its way to Los Alamos. From that material, the Los Alamos bomb makers would create the device known as Fat Man, the plutonium bomb dropped over Nagasaki three days after the Hiroshima bombing.

Chapter Fourteen

The Road to Trinity

The course of the war had outrun work on the bomb, rendering the Allied scientists' concerns about a Nazi bomb program increasingly irrelevant. The entry of America into the European war, with its stupendous resources, made an Allied victory seem almost inevitable by mid-1943, notwithstanding earlier tactical setbacks in North Africa and the Balkans. The Normandy invasion launched on June 6, 1944, heralded a final Allied push toward Berlin, interrupted chiefly by the six-week winter counterattack by German forces known to Allied military historians as the Ardennes Counteroffensive and in the popular mind as the Battle of the Bulge.

For the physicists of the atomic bomb program, Germany's surrender on May 7, 1945, complicated the moral and ethical issues connected with the device they had invented. Those issues had seemed comparatively straightforward while the Allied war effort remained primarily focused on the Nazi regime. The terrifying thought that Adolf Hitler might beat the Allies to exploitation of the atom's destructive capacity had prompted many eminent scientists, including numerous refugees from Nazi Germany, to participate willingly in the Manhattan Project. In 1942 and 1943, the uncertain course of the war pushed doubts about building and using an atomic bomb to the background. Given the existential threat posed by a German bomb, few harbored any qualms about the Allies using theirs first.

Japan presented an entirely different case, at least to the scientists.

Japan's technological capabilities appeared to be far inferior to those of Germany, and the Japanese regime's threat to the world of a much more limited order. Although only a handful of the Manhattan Project scientists were fully aware of the state of progress on the bomb—chiefly those working on the device itself at Los Alamos and those with high-level clearance, such as Ernest Lawrence—the experts at the program's far-flung laboratories perceived that a successful conclusion to their work was drawing near. This sharpened the feelings in the physics community and among the program's civilian leadership that discussions about the deployment of the bomb and postwar management of its technology should be urgently stepped up.

Leo Szilard felt deeply the difficulty of balancing the imperative to build the bomb with humanitarian concerns about its use. The refugee Hungarian physicist had placed the very notion of atomic weaponry on the federal government's radar in 1939, when he prompted Albert Einstein to alert Franklin Roosevelt to the military potential of nuclear fission. Szilard's views on the program he midwifed had traveled a tortuous path. For four dispiriting years, he had hectored government officials to place the program on the fast track. Yet by May 1945, he was urging those same officials that the best hope for averting a postwar nuclear arms race lay in "not using the bomb against Japan, keeping it secret, and letting the Russians think that our work on it had not succeeded." After the German surrender, he brought one more argument to the table: using the bomb on Japan would fatally compromise America's moral and humanitarian standing, undermining any efforts to create a workable international control regime. Yet the hour was getting late. The planned test of the plutonium bomb over the sands of Alamogordo, New Mexico, an event known as Trinity, was nigh; deployment of the bomb in war could be only a few weeks further off.

Discussions about the control of nuclear arms and use of the bomb on Japan had been going on since 1943. The center of debate over postwar planning, as well as of opposition to dropping the bomb, was the Metallurgical Laboratory—the Met Lab—headed by Arthur Compton at the University of Chicago, where Glenn Seaborg had successfully separated plutonium. There were several explanations for the Met Lab's prominence in the debate. One was that its direct work on the bomb had been completed by early 1944, for as soon as Seaborg and his colleagues perfected the separation process in Chicago, large-scale plutonium production using atomic piles was moved to Hanford. The Met Lab scientists therefore had the time and leisure to ponder the greater implications of nuclear weaponry and its control. Then there was the presence at the lab of James Franck, a German physicist of unassailable reputation who had devoted a great deal of careful thought to the social and political implications of atomic energy. These conditions did not exist at the other important research centers within the Manhattan Project universe: Los Alamos remained frantically busy through the Trinity test and up to the bombings, and Berkeley marched to the tune set by one man, Ernest Lawrence, whose hostility to the distractions of "politics" was crystal clear—though he himself would soon be drawn into exactly the sort of discussions he abhorred hearing in his own laboratory.

Sharpening the tenor of the discussions in Chicago were signals that the government was considering shutting down the Met Lab. In July 1944, General Groves told Arthur Compton to plan a personnel cutback of as much as 75 percent by September 1. The prospect dismayed the staff. Having been immersed in the development of the atom's destructive potential, they were hoping to shift as a team into work on the generation of electricity and other such peacetime aspects of the embryonic science. As Franck would observe later, they were uneasy that the government's interest in atomic energy might begin and end with its effectiveness as a weapon. The fear was that if the bomb failed, government support for all other work in the field would cease, and research into the atom's benefits to society would go fallow. On the other hand, if the bomb succeeded, the demand for fundamental research on atomic energy would only increase, though it would be likely focused on more and better weapons. But whether the future of nuclear technology lay in war or peace, dispersing an effective team of experienced researchers and leaving them without financial support would be a foolhardy course for the government.

Compton managed to bring these concerns before Groves and his civilian counterparts, Vannevar Bush and James Conant. He proposed several long-range projects to carry the Met Lab through 1945, among them supporting Hanford and Los Alamos with research and development assistance, and launching research into new technologies such as advanced nuclear reactors and applications of radiation for industry, medicine, and the military. Compton also moved quietly to establish the University of Chicago as a postwar center of nuclear research by trying to lure Fermi from Columbia University with a professor's chair at the university and the directorship of the Manhattan Project's Argonne National Laboratory west of Chicago. His attempted poaching of Fermi succeeded only in provoking an outraged protest by Columbia to Bush, who forbade the move.

The episode was a reminder for Bush and Conant that if they did not begin outlining a postwar research program within the Manhattan Project, its members would start taking matters into their own hands. They also understood that, beyond keeping the program's scientists mollified, there were other reasons to think about postwar policy. One was that the secret of atomic weapons was bound to get out one way or another, whether through deployment of the bomb itself, or the natural march of basic research, or through leaks from the huge contingent of scientists who had worked on the project. Estimates of how long it might take the Soviets to develop their own bomb ranged from three or four years (the view of Oppenheimer, Bush, and Conant) to twenty (the opinion of Groves, who saw the matter through the prism of his own efforts to manage the enormously complicated program). The Soviet Union was a crucial ally in the war effort, but the leaders of the US and Great Britain mistrusted Stalin and feared that the Soviet leader's thirst for European domination would make for an uneasy postwar peace. Divulging the bomb to the Soviets and enlisting them in an international control regime prior to a public demonstration of the weapon's power, Bush and Conant reasoned, might be the best way to avoid an arms race that would threaten the world.

Ideas for international control had been making their way to Roosevelt and British prime minister Winston Churchill for months, albeit haphazardly. In August, Supreme Court Justice Felix Frankfurter, an old friend of FDR's, had brought Niels Bohr to the White House with a proposal for immediate disclosure of the bomb secret as a prelude to establishing an international control regime. The proposal landed on the agenda of the September 1944 Quebec summit meeting between Roosevelt and Churchill, but they decided instead to maintain the bomb exclusively as an Anglo-American technology even after the war, in effect establishing the two countries as permanent stewards of world peace and excluding the Russians.

Bush and Conant, who learned of this decision from Secretary of War Stimson, responded on September 30 with a memo pointing out that news of the atomic bomb was likely to reach the world by August 1945 either through a nonmilitary demonstration or deployment in war. The weapon in development, with the power of about ten thousand tons of TNT, had the capacity to dramatically alter the nature of war and peace; but that was modest compared with the potential of a thermonuclear "Super" bomb, which was already the subject of theorizing by Manhattan Project scientists and might be a thousand times more powerful. The United States and Britain could not hope to keep a monopoly on nuclear technology for more than a few years, they warned. Bush and Conant also were aware that as long as the atomic bomb was a secret, the talks already taking place on postwar diplomacy—specifically the 1944 Dumbarton Oaks conference, which would set the foundations for the United Nations—would be proceeding in a vacuum. As James Franck would write in early 1945, scientists "know in our hearts that all these plans are obsolete, because the future war has an entirely different and a thousand times more sinister aspect than the war which is fought right now."

The efforts by Bush and Conant yielded little in terms of policy movement. The White House was gearing up for the conference of the Big Three—Roosevelt, Churchill, and Stalin—in Yalta, scheduled for the beginning of February 1945 amid indications of Stalin's growing intransigence over a strong Soviet Union role in postwar Europe, especially Eastern Europe. Military planning, meanwhile, was proceeding at full speed. Groves had begun training of the airborne unit that would deliver the bomb to its target in Japan; on December 30, 1944, he received approval from FDR to inform key combat officers of the army, air force, and navy about essential details of the mission. In this frenzied atmosphere, the inchoate ideas of Bush and Conant about international control made barely a ripple.

After Yalta, however, opportunity seemed to beckon. The White House was euphoric over what it considered to be a successful summit. More to the point, the bomb program was moving toward a concrete conclusion, with plans laid for a decisive test in the desert. As Stimson recalled the moment, "it was considered exceedingly probable that we should by midsummer have successfully detonated the first atomic bomb . . . What had begun as a well-founded hope was now developing into a reality." Bush and Conant urged upon Stimson the idea that serious preparatory steps needed to be taken before the bomb was revealed to the world: the drafting of public statements and legislation addressing the question of international control, and of domestic management of a postwar technical program. The world would move from the prenuclear age to the nuclear era with the very first atomic fireball; if there were a vacuum of national or international control at that moment, the result could be "something akin to mass hysteria."

Stimson brought these warnings to a meeting with FDR on March 15. His nominal goal was to end backbiting about the bomb program from Roosevelt's director of war mobilization, James F. Byrnes, a former senator and Supreme Court justice. The courtly South Carolinian New Dealer, who knew almost nothing about the program, had been plying FDR with alarming rumors of extravagant spending by the Manhattan Project and suggestions that, with the cost having grown to $2 billion, "Vannevar Bush and Jim Conant have sold the president a lemon on the subject and ought to be checked up on," as Stimson paraphrased him. In a memo to FDR, Byrnes had urged that an outside panel of eminent scientists give the program a once-over—"rather a jittery and nervous memorandum and rather silly," Stimson recalled. At the White House, he dismissed Byrnes's concerns by pointing out that the scientific team behind the bomb included Ernest Lawrence and three other Nobel laureates, along with "practically every physicist of standing." Then he addressed the lingering issues of future control and the necessity of having some sort of statement at hand to release to the world along with the first news of the explosion. Roosevelt agreed that those issues should be settled in advance, but no action ensued. The meeting was Stimson's last with FDR. Less than four weeks later, on April 12, the president was dead of a massive stroke at the age of sixty-three.

Stimson's next visit to the White House occurred on April 25, when his task was to explain the cataclysmic implications of nuclear energy to a new president "whose only previous knowledge of our activities was that of a senator who had loyally accepted our assurance that the matter must be kept a secret from him." President Harry S. Truman, Stimson observed with relief, accepted his new responsibility "with the same fine spirit that Senator Truman had shown before in accepting our refusal to inform him." Stimson did not minimize the international and political importance of the weapon. As his advance notes for the meeting stated, without robust principles in place for sharing knowledge of the bomb and placing it under control, "the world in its present state of moral advancement compared with its technical development would eventually be at the mercy of such a weapon. In other words, modern civilization might be completely destroyed . . . On the other hand, if the problem of the proper use of this weapon can be solved . . . the peace of the world and our civilization can be saved." Stimson's immediate goal was to obtain Truman's permission to establish a committee devoted to these postwar issues. He received it on the spot.

By May 1, the Interim Committee was established—so named on the assumption that Congress would appoint a permanent committee as soon as the war was over—with Stimson as chairman and Byrnes, Bush, Conant, and Karl Compton as its members. Not long afterward, Stimson named a scientific panel to advise the committee: Ernest Lawrence, Robert Oppenheimer, Arthur Compton, and Enrico Fermi. They were all leaders of Big Science, convened to contemplate its most important creation.

The scientific panel's role was never precisely spelled out. Conant, who had come up with the idea, thought it could serve as a conduit to the Interim Committee for the views of bomb scientists who had become restless about the implications of their work. It is doubtful that the other members agreed with that role; Compton was the only one who oversaw a lab where the scientists' ambivalence was expressed openly.

In formal terms, the scientific panel would have the opportunity to give its advice on any subject brought before the interim committee. This appeared at the outset to be an expansive portfolio. In the event, however, both committees soon got drawn into what Conant later described as "the most important matter on which an opinion was to be recorded. This was the question of the use of the bomb against the Japanese."

The Interim Committee met with its scientific panel for the first time on May 31 at the Pentagon, with Groves and Army Chief of Staff George C. Marshall in attendance. It had been three weeks since the fall of Germany, and attention was focused on Japan. As Compton recalled the discussion, Japan "was an overpowered nation, but she was fighting desperately, unwilling to acknowledge defeat . . . The great danger was that the fanatical military group would retain such control of Japan that surrender would be impossible." The historical debate over the likely intensity of Japanese resistance and therefore the possible cost of an invasion of the home islands in Allied lives has continued to this day, but there can be little doubt that for decision makers in 1945, the magnitude of possible losses—one million lives was a not-uncommon estimate—looked horrific. The May 31 discussion unfolded against that backdrop.

Compton placed the concerns of rank-and-file scientists about the social and political implications of the bomb on the committee record, via a memo from the Met Lab's Franck. The document warned that the use of a weapon that could kill thousands of people in a single strike would mean "moral isolation" for the United States; if America was hoping to outlaw the use of atomic weapons by international treaty, Franck wrote, their prior use would put the country "in a weak position to recommend their prohibition."

Instead, Franck proposed "a demonstration before the eyes of all the world on some barren island." This was by no means the first time the idea of a nonmilitary demonstration had been heard; in fact, it had been in the wind for years. Bush and Conant had proposed something of the sort to Stimson in September 1944: "complete disclosure" of the technology, omitting manufacturing and military details, "as soon as the first bomb has been demonstrated." They specified, "This demonstration might be over enemy territory, or in our own country, with subsequent notice to Japan that the materials would be used against the Japanese mainland unless surrender was forthcoming."

But the May 31 meeting was the first time that the idea of a demonstration was placed before a decision-making body of the US government. Even so, it was presented delicately, not as part of the official agenda, and only after the committee addressed several other formal agenda items. These included America's supposed head start on the rest of the world in atomic weaponry. (In context, "the rest of the world" meant the Soviet Union.) Compton estimated the American lead on the atomic bomb at no more than six years. As for developing and producing the next-generation weapon, a thermonuclear bomb, Oppenheimer reckoned that would take the United States three years—not a comforting timeline for anyone concerned about the proliferation of weapons capable of menacing civilization. Byrnes, who up to then had viewed the atomic bomb largely as an abstract device useful for bargaining with the Russians, would recall that Oppenheimer's astonishing figures left him "thoroughly frightened."

When the discussion turned to the future of the American technical program, Lawrence took the opportunity to secure a foothold for postwar Big Science by lobbying for a "vigorous program of plant expansion and stockpiling" with government support. This was an idea endorsed by both Arthur and Karl Compton, who had large academic institutions of their own to nurture. Lawrence appeared untroubled by the prospect of continued military oversight of the program, which was where Oppenheimer drew the line. Oppie thought such conditions could only stifle basic research; there was a fundamental difference, he said, between basic science and the work that had taken place at Los Alamos and the other Manhattan Project labs, which had only "plucked the fruit of earlier discoveries." To keep American science in full flower, he asserted, what was needed was "a more leisurely and normal research environment." On that point, he won the assent of Bush.

Then it was time for lunch. All during the morning session, Compton would recall, "it seemed to be a foregone conclusion that the bomb would be used. It was regarding only the details of strategy and tactics that differing views were expressed." Yet during the morning, Lawrence had alluded approvingly to the possibility of a nonmilitary demonstration. At lunchtime, Byrnes asked him to expand on it. "It was discussed at some length," Lawrence remembered, "perhaps ten minutes."

As Lawrence and Compton both recalled, the general reaction was negative. "An atomic bomb was an intricate device, still in the developmental stage," Compton recounted the conversation. "We could not afford the chance that one of them might be a dud . . . Though the possibility of a demonstration that would not destroy human lives was attractive, no one could suggest a way in which it could be made so convincing that it would be likely to stop the war." Oppenheimer was uncomfortable with the entire discussion, for he considered the question of the bomb's deployment to be outside the scientists' expertise. "We didn't think that being scientists especially qualified us as to how to answer this question of how the bombs should be used or not," he recalled years later, taking it upon himself to articulate the sense of the entire scientific panel. "We didn't know beans about the military situation in Japan. We didn't know whether the invasion was really inevitable . . . We did say that we did not think that exploding one of these things as a firecracker over a desert was likely to be impressive." But as he observed in recalling the discussion, that was before they actually had exploded their firecracker, which yielded a display that turned out to be very impressive indeed.

Oppenheimer's point made a vivid impression on Lawrence. "Oppenheimer felt, and this feeling was shared by Groves and others, that the only way to put on a demonstration would be to attack a real target of built-up structures," he related in late August to his friend Karl Darrow. As the general explained, this narrowed the targets to cities hosting large munitions plants surrounded by worker housing.

The other telling argument against a demonstration, Lawrence recalled for Darrow, was that "the number of people that would be killed by the bomb would not be greater in order of magnitude than the number already killed in fire raids." To the extent this represented a quantitative judgment, it was certainly true; yet in qualitative terms, it may be viewed as a desperate rationalization of a foregone conclusion. Oppenheimer estimated for the committee that the bomb might take twenty thousand lives; later estimates place the death toll from the blast itself, not counting subsequent deaths from radiation exposure, at sixty thousand to eighty thousand in Hiroshima and forty thousand to fifty thousand at Nagasaki. These figures are indeed within or even somewhat below the range of estimated deaths from the firebombings of Tokyo, the raids by low-flying aircraft carrying incendiary bombs, which had commenced that February and climaxed on the night of March 9–10. Yet the comparison ignores the fact that almost everyone involved in the Manhattan Project and the bombing decision recognized that a nuclear bomb was transformative, certain to introduce a new type of warfare and new relationships among nations; that was the point of establishing the Interim Committee in the first place. Stimson acknowledged this himself in fragmentary handwritten notes he prepared for the May 31 meeting:

"We don't think it *mere* new weapon . . . *infinitely greater*, in *respect* to its *Effect*—on the ordinary affairs of man's life. May *destroy* or *perfect* International Civilization . . . *Frankenstein or* means for World Peace." If it were just a matter of counting casualties, the decision to drop the bomb would have been much simpler. Stimson wrote later that at the May 31 meeting, he and Marshall both expressed the view that atomic energy "could not be considered simply in terms of military weapons but must also be considered in terms of a new relationship of man to the universe." But he remained silent about the incompatibility of that view with the cold calculation of acceptable casualties.

The May 31 meeting broke up with the Interim Committee and its scientific panel at cross-purposes. After lunch, Compton recalled, the scientific panel was asked to prepare a report "as to whether we could devise any kind of demonstration that would seem likely to bring the war to an end without using the bomb against a live target."

The scientists therefore were justified in thinking that a final decision on dropping the bomb would await their further input. But matters were moving ahead without them. One day after the Pentagon meeting, the Interim Committee adopted three recommendations for President Truman: (1) the bomb should be used against Japan "as soon as possible"; (2) it should be used on a "dual target," such as a military installation or munitions plant surrounded by "houses and other buildings most susceptible to damage"; and (3) there should be no prior warning. The vote was unanimous, although one member, navy undersecretary Ralph Bard, presently dissented from the third recommendation, resigned his post, and proposed to Stimson that the Japanese receive two or three days' warning. His views made it all the way to the White House where, plainly, they were disregarded.

Struggling to meet their final assignment from the Interim Committee, the four members of the scientific panel convened at Los Alamos on June 15. It was a painful, emotionally fraught gathering. "We thought of the fighting men who were set for an invasion which would be so very costly in both American and Japanese lives," Compton reflected. "We were determined to find, if we could, some effective way of demonstrating the power of an atomic bomb without loss of life that would impress Japan's warlords. If only this could be done!"

"The last one of our group to give up hope," Compton wrote later, was Ernest Lawrence. Ernest's colleagues observed his "obvious distress" all weekend long; Compton suggested that he was excessively swayed by the friendships he had forged with Japanese physicists who had worked on the cyclotron. This was tactless but not necessarily inaccurate, for two Japanese scientists, Ryokichi Sagane and Tameichi Yasaki, had worked for more than a year at the Rad Lab before building the very first cyclotron outside the United States at Tokyo's Institute for Physical and Chemical Research, copying Lawrence's machine so slavishly that they even incorporated mistakes that the Berkeley cyclotroneers had already discovered and rectified.

The panel, to its own dismay, could conceive of no alternative to bombing Japan. Oppenheimer, writing for the group, dispatched to Washington a dispirited one-page memo on June 16 memorializing their conclusion. The memo suggested that Britain, Russia, France, and China all be informed in advance of the progress of the work and the likelihood of the bombs' deployment, and that the allies be solicited for suggestions "as to how we can cooperate in making this development contribute to improved international relations." On the question of deployment versus demonstration, "we can propose no technical demonstration likely to bring an end to the war; we see no acceptable alternative to direct military use." To what extent Ernest Lawrence concurred with that baleful conclusion is not documented, for no notes by him have been found. Oppenheimer's memo acknowledged that "the opinions of our scientific colleagues on the initial use of these weapons are not unanimous," but he might well have been referring to the range of thought within the entire scientific community, not merely among the four committee members. In any event, Lawrence put his name to the statement on June 16, and thereafter never admitted publicly to any misgivings about the result.

By then, the die had been cast. The 509th Composite Group under the command of Paul Tibbets, a twenty-nine-year-old lieutenant colonel from Illinois, had been domiciled with its Boeing B-29 Superfortress heavy bombers on Tinian Island in the Northern Marianas, about 1,500 miles south of Japan, since the week before the scientific panel's last meeting. At Los Alamos, the most important task remaining was to test a bomb using plutonium as its core, for its implosion design—in which a hollow plutonium sphere would be crushed into a supercritical ball by an outer shell of explosives, painstakingly arranged and timed to deliver a symmetrical shock wave—was so novel that no one could be absolutely sure it would work. By contrast, the bomb carrying the fissile uranium produced by Lawrence's racetracks in Oak Ridge needed no testing, for its design was judged to be a trivial engineering feat; in any event, the output of enriched U-235 from Oak Ridge was so meager that there was only enough for one device, to be carried on Tibbets's aircraft, the *Enola Gay* (named after his mother), to Hiroshima. The reactors at Hanford, however, were breeding plutonium in sufficient volume to make a test feasible and provide for an unlimited number of plutonium bombs afterward.

The Trinity test was to take place under nerve-wracking scrutiny. Present at the Alamogordo test range would be almost every high-level official of the Manhattan Project, from General Groves down. At Los Alamos, the almost insupportable tension fostered a sort of mania. "It was hard to behave normally," recalled Elsie McMillan, Edwin McMillan's wife and Molly Lawrence's sister. "It was hard to think. It was hard not to let off steam. It was hard not to overindulge in all the natural activities of life."

Meteorologists at Los Alamos scanned their weather reports, searching for a period of clear skies under which to stage the test at Alamogordo, some 250 miles to the south. Finally, they settled on a span of a few days in mid-July. On July 5 Lawrence received a confirmatory cable from Oppenheimer in Los Alamos: "Anytime after the fifteenth would be good for our fishing trip. Because we are not certain of the weather, we may be delayed several days. As we do not have enough sleeping bags to go around, we ask you please do not bring any friends with you. Let us know where in Albuquerque you can be reached."

Groves arrived in Berkeley on July 13, with Bush and Conant in tow. That night, they were feted by Ernest at Trader Vic's in Oakland, where the fare ran to spareribs eaten with the fingers and washed down with the restaurant's signature mai tai cocktails. They all continued on the general's plane to Albuquerque, where Groves was infuriated to discover eminent physicists gathered in clumps at the main hotel. On the reasoning that the recognition of any of them by an outsider would breach security, he ordered them dispersed to other hotels around town. At eleven o'clock on the night of the fifteenth, a government sedan picked up Lawrence for the bumpy three-hour ride to Compania Hill, a watching post twenty miles from the bomb site. Reaching his destination at two in the morning, he encountered McMillan, Robert Serber, Edward Teller, the newly knighted Sir James Chadwick, and a young Caltech physicist named Richard Feynman, who exercised his radio skills to pester the post's balky shortwave to life. It instantly began broadcasting conversations between ground control and the B-29 observation planes. In the distance, Lawrence could see the glow of spotlights illuminating the spindly hundred-foot-tall test tower from which the plutonium "gadget" was suspended.

The test shot had been postponed from four o'clock to five thirty to allow a pelting thunderstorm to pass. Each observer chose to fight off the tension in his own way. Lawrence placed wagers on the outcome and yield, as was being done at the main viewing post less than ten miles from the tower; the bets ranged from the equivalent of forty thousand tons of TNT down to zero. Teller slathered his face with suntan lotion. Others fidgeted with their protective goggles—McMillan had brought along a welder's mask with the darkest visor he could find.

At the base camp the observers were instructed to lay prone in ditches excavated to protect them from the blast, with their feet pointed toward the tower. The precautions at Compania Hill were less stringent. As the countdown squawked from the radio, Lawrence was unable to keep still. As he reported to Groves:

> *I decided the best place to view the flame would be through the window of the car I was sitting in, which would take out ultraviolet, but at the last minute decided to get out of the car (evidence indeed I was excited!) and just as I put my foot on the ground I was enveloped with a warm brilliant yellow white light—from darkness to brilliant sunshine in an instant, and as I remember I momentarily was stunned by the surprise. It took me a second thought to tell myself, "this is indeed it!!" and then through my dark sun glasses there was a gigantic ball of fire rising rapidly from the earth—at first as brilliant as the sun, growing less brilliant as it grew boiling and swirling into the heavens. Ten or fifteen thousand feet above the ground, it was orange in color and I judge a mile in diameter. At higher levels it became purple, and this purple afterglow persisted for what seemed a long time . . . This purple glow was due to the enormous radioactivity of the gases. (The light is in large part due to nitrogen in the air, and in the laboratory we occasionally produce it in miniature with the cyclotron.) . . . It was a grand spectacle . . .*
>
> *A little over two minutes after the beginning of the flash, the shock wave hit us. It was a sharp loud crack and then for about a minute thereafter there were resounding echos from the surrounding mountains . . . like a giant firecracker set off a few yards away—or perhaps like the report of 37 mm artillery at a distance of about one hundred yards.*

Standing next to Lawrence, Serber had trained his unprotected eyes directly on the fireball at the moment of detonation. The flash blinded him for many long seconds, during which he could make out only subtle changes in color and feel the heat of the blast on his face. McMillan perceived the same purple glow as Lawrence through his welder's glass, and similarly attributed it to the ionization of atmospheric gas. All the observers shared the unsettling sensation of having witnessed a cataclysmic event. "The immediate reaction of the watchers was one of awe rather than excitement," related McMillan.

Oppenheimer's recollection of his own reaction has been oft repeated; he claimed that a verse from the Sanskrit Bhagavad Gita leaped into his mind: " 'Now I am become death, destroyer of worlds.' I suppose we all thought that, one way or another." There is reason to believe that recollection was an ex post facto elaboration; those on the scene recall a sensation more suggestive of relief and euphoria in the air.

In any case, Ernest Lawrence, a man normally not given to introspection in the Oppenheimer style, struggled to set down his emotions on paper. He incorporated the reactions he had witnessed in the others. "The grand, indeed almost cataclysmic proportion of the explosion produced a kind of solemnity in everyone's behavior immediately afterwards," he wrote. "There was restrained applause, but more a hushed murmuring bordering on reverence in manner as the event was commented upon[.] Dr. Charles Thomas (Monsanto) spoke to me of this being the greatest single event in the history of mankind, etc. etc.

"As far as all of us are concerned, although we knew the fundamentals were sound and that the explosion could be produced, we share a feeling that we have this day crossed a great milestone in human progress."

The outcome of the test was communicated promptly to Stimson, who had accompanied Truman to the conference of the Big Three in Potsdam, outside Berlin. The message from George Harrison, a prominent banker and member of the Interim Committee, read: "Operated on this morning. Diagnosis not yet complete but results seem satisfactory and already exceed expectations . . . Dr. Groves pleased."

The successful test brought into sharp relief the issue of what to disclose to America's allies. Although the scientific panel had proposed informing Britain, France, China, and Russia, in practical terms the real question boiled down to what to tell the Soviets. In consultation with Stimson, Truman decided to wait until the final day of the Potsdam conference and then treat Stalin to a bare-bones report. As Truman later recalled the episode, he wandered around the baize-covered conference table to the Russian side, leaving behind his own interpreter, and ambled up to Stalin: "I casually mentioned to Stalin that we had a new weapon of unusual destructive force. The Russian Premier showed no special interest. All he said was that he was glad to hear it and hoped we would make 'good use of it against the Japanese.' "

Truman may have considered himself crafty in giving Stalin just enough information to forestall Russian complaints of having been left in the dark, without telling him enough to be useful. Stalin's indifferent reaction could have been the product of Truman's studied nonchalance. But it may also have reflected the fact that he already knew about the Alamogordo test, informed by his network of spies in the West. To Oppenheimer, Truman's refusal to even try to forge a genuine partnership with the Soviet Union for the control of nuclear weapons was a tragic lost opportunity. "That was carrying casualness rather far," he remarked later.

After Trinity, the drive toward the bombing of Japan proceeded without further consultation with the scientific panel. But the ferment among the bomb scientists continued—centered, as always, at the Met Lab in Chicago. Szilard circulated a petition arguing on moral grounds against any deployment of the bomb at all. (He later revised the text to condone the bomb's use after giving "suitable warning and opportunity for surrender under known conditions.") The petition carried the signatures of more than sixty Met Lab scientists when Compton delivered it to Washington. But it was not the only petition accumulating signatures at the Met Lab, where opposition to use of the bomb was by no means unanimous; one other petition read in part: "Are not the men of the fighting forces . . . who are risking their lives for the nation entitled to the weapons which have been designed? . . . If we can save even a handful of American lives, then let us use this weapon—now!"

The precise timing of the Hiroshima mission was known only to a few physicists on the bomb team. Among them were Alvarez and Serber, who had been sharing a tent on Tinian for two months, preparing for their assignment tending gauges to be dropped by parachute with the bomb itself. At two forty-five on the morning of August 5, Alvarez climbed into the *Great Artiste*, the B-29 accompanying Tibbets's *Enola Gay* on the bomb run. Almost exactly six and a half hours later, they were over Hiroshima. The *Enola Gay* dropped its payload, the uranium bomb christened Little Boy. Alvarez watched his three gauges waft down behind the bomb, and as his plane veered around in a turn of two g-forces to outrace the shock wave, checked his receiver to make sure the gauges were acquiring data. He felt the blast forty-five seconds after the bomb was released.

"Suddenly a bright flash lit the compartment, the light from the explosion reflecting off the clouds in front of us . . . A few moments later, two sharp shocks slammed the plane." Then the *Great Artiste* circled back over Hiroshima. "I looked in vain for the city that had been our target," Alvarez recalled. "My friend and teacher Ernest Lawrence had expended great energy and hundreds of millions of dollars building the machines that separated the U-235 for the Little Boy bomb. I thought the bombardier had missed the city by miles . . . and I wondered how we would ever explain such a failure to him." But the bomb had reached its target. Hiroshima could not be seen because it had been destroyed.

Two days later, Alvarez was preparing for the second run that would drop the plutonium bomb, "Fat Man," on Nagasaki. Taking a break in the Tinian officers' club with Serber and Phil Morrison, a theoretical physicist from Berkeley, he remembered Ryokichi Sagane and his two-year sojourn at the Rad Lab. Perhaps this personal connection could be used to press for an end of the war—one tiny personal step the three American scientists might take to counterbalance their role in raining destruction down upon Sagane and his countrymen. Hurriedly they drafted a message to be placed in envelopes taped to the three gauges launched into the maelstrom:

> *We are sending this as a personal message to urge that you use your influence as a reputable nuclear physicist, to convince the Japanese General Staff of the terrible consequences which will be suffered by your people if you continue in this war . . . Within the space of three weeks, we have proof-fired one bomb in the American desert, exploded one in Hiroshima, and fired the third this morning.*
>
> *We implore you to confirm these facts to your leaders, and to do your utmost to stop the destruction and waste of life which can only result in the total annihilation of all your cities if continued. As scientists, we deplore the use to which a beautiful discovery has been put, but we can assure you that unless Japan surrenders at once, this rain of atomic bombs will increase manyfold in fury.*

They signed it "From three of your former scientific colleagues during your stay in the United States."

The letter, which was recovered from the rubble of Nagasaki, did not reach Sagane until after the Japanese surrender. Many months later, Wilson Compton, a brother of Arthur and Karl traveling in Japan, received a copy from Sagane. He passed it on to Alvarez, who put his signature to it and returned it to Sagane in 1949 as a mordant keepsake.

Amid the public fascination with the "secret weapon" following its spectacular disclosure and the euphoria of the Japanese capitulation announced by Truman five days after Nagasaki, the bombings occasioned a surge of soul-searching among the scientists of the Manhattan Project. Lawrence was sympathetic to the displays of doubt shown by some of his closest friends and colleagues. But he also was impatient with the second-guessing of a decision that, as he saw it, ended this war and, conceivably, all war. "I am sure the whole world will realize that war is no longer possible in human affairs," he wrote to Lewis Akeley, the South Dakota professor who so many years earlier had set him on his career path in physics.

Ernest fended off numerous efforts to draw him into the debate over the bomb. On August 9, in a moment of indiscretion following Nagasaki, he had confided to Karl Darrow that he had proposed a nonmilitary demonstration to the Interim Committee. Darrow, who hoped to head off public backlash for the scientists' complicity in the creation of a lethal new technology, leaped at the chance to disseminate an example of ambivalence at the summit of the scientific community. "I hope you will publicize the fact that you made this plea . . . mainly because of the possibility of public opinion harmful to science," he wrote Lawrence. "Some people go so far as to blame scientists for the consequences of their discoveries. I think it is not far-fetched nor absurd to conjecture that in time to come, people will be saying, 'Those wicked physicists of the "Manhattan Project" deliberately developed a bomb which they knew would be used for killing thousands of innocent people without any warning . . . Away with physicists!' It will not be accepted as an excuse that they may have disapproved in silence."

Lawrence would have none of it. After recapitulating the counterarguments raised at the Interim Committee against a demonstration, he advised Darrow, "I am inclined to feel that they made the right decision. Surely many more lives were saved by shortening the war than were sacrificed as a result of the bombs. Further, it goes without saying that . . . the world must realize that there can never be another war. As regards criticism of physicists and scientists, I think that is a cross we will have to bear, and I think in the long run the good sense of everyone the world over will realize that in this instance, as in all scientific pursuits, the world is better as a result."

He displayed the same impatience with Oppenheimer, who expressed his personal torment over the bomb while delivering the Arthur D. Little Memorial Lecture at MIT in 1947. "In some sort of crude sense which no vulgarity, no humor, no overstatement can quite extinguish," he declared, "the physicists have known sin; and this is a knowledge which they cannot lose."

Oppenheimer's point was more nuanced than it might have seemed on the surface; he was calling for introspection among scientists that had been lacking—and perhaps unnecessary—before the power of fission was unleashed. But Lawrence responded brusquely to what sounded like Oppie's assumption of guilt on behalf of scientists everywhere. "I am a physicist," he snapped, "and I have no knowledge to lose in which physics has caused me to know sin."

Yet bonds of friendship, colleagueship, and shared travail in the bomb project made it difficult for Lawrence to dismiss Oppenheimer's disquiet entirely. The weekend after Nagasaki, Ernest visited Los Alamos. He found Oppie wracked with self-doubt and struggling to draft another communiqué for the Interim Committee. This one addressed the "scope and program of future work in the field of atomic energy." Oppie was attempting to place before the committee his "profound" thoughts about the future of atomic weaponry, including what he called "the superbomb." Oppenheimer reported the scientific committee's conclusions that there could be no "effective military countermeasures for atomic weapons," that the United States could not be sure of maintaining "technical hegemony" in atomic weapons, and that "such hegemony, if achieved, could [not] protect us from the most terrible destruction." That could be done not through scientific and technical expertise but only by eradicating war. Lawrence and Oppenheimer agreed on that point. Where they disagreed was that Lawrence believed that goal had been reached by the bombing of Japan; Oppenheimer, by contrast, feared that the bombs of August had made it more remote than ever.

Lawrence largely approved of Oppenheimer's draft of the communiqué but requested one telling change. Oppenheimer had written, "We are most doubtful that even a profound strengthening during the coming years of our technical position in the field of atomic weapons could make an essential contribution to the problem of ending war." Lawrence proposed a substitute: "It is needless to say that as long as our nation requires strong armed forces we must stockpile and continue intensive development of atomic weapons, and it is probable that we can thereby retain supremacy for a number of years. However, we are certain that . . . other powers can produce these weapons in a few years . . . We consider it imperative, therefore, to take determined steps towards international arrangements that will make such developments improbable, if not impossible."

To Oppenheimer, this sounded like a marketing pitch for government funding of Lawrence's laboratory, and he persuaded Ernest to back down. The scientific committee's memo, as finally prepared for Stimson, acknowledged that the development of more effective nuclear weapons "would appear to be a most natural element in any national policy," but emphasized that national security could be "based only on making future wars impossible. It is our unanimous and urgent recommendation to you that . . . all steps be taken, all necessary international arrangements be made, to this one end."

Yet when Oppenheimer arrived in Washington with the letter, Stimson was absent—a man of seventy-seven trying to recover his strength at an Adirondacks resort after the superhuman exertions of the summer. Instead, Oppenheimer was shunted off to George Harrison, who relayed his thoughts to Jim Byrnes, now Truman's secretary of state. Byrnes instructed Harrison to give Oppenheimer a blunt reply: "For the time being, his proposal about an international agreement is not practical and . . . he and the rest of the gang should pursue their work full force"—work, that is, on the Super, the next-generation weapon predicted to be thousands of times more powerful than the bombs dropped on Japan.

Back in Los Alamos, Oppenheimer recorded his thoughts in a disconsolate letter to Lawrence. In Washington, he wrote, "it was a bad time, too early for clarity." He had tried to communicate that the national interest did not lie in simply continuing atomic bomb work, and even suggested banning atomic weapons by international convention "just like poison gases after the last war." (The allusion was to a prohibition on poison gas signed at the Geneva Convention in 1925.) But no one was listening. "I had the fairly clear impression from the talks that things had gone most badly at Potsdam, and that little or no progress had been made in interesting the Russians in collaboration or control," he told Ernest. "I don't know how seriously an effort was made." (In fact, no effort had been made.) Oppie reported two "gloomy" developments in Washington: one being Byrnes's order to continue the bomb research; the other being a "ukase," or edict, from Truman "forbidding any disclosures on the atomic bomb . . . without his personal approval." It seemed to Oppenheimer that the opportunity to keep atomic weapons under control via international agreement was slipping away, perhaps never to be recovered. More disturbingly, scientists' ability to control their own professional destinies seemed to be slipping away. He confessed to feeling "a profound grief, and a profound perplexity about the course we should be following."

The final paragraphs of Oppie's letter to Ernest concerned his return to Berkeley. The last few weeks' discussions about the future course of the bomb program, conducted in an atmosphere of shared emotional and physical exhaustion, had driven a wedge between the two friends, though by no means was it the only wedge. Oppenheimer had doubts about how well he might fit in on campus—"any fruitful future in Berkeley would have to depend . . . on a certain mutual respect for nonidentical points of view."

He was intimating that physics at Berkeley might have to continue without Robert Oppenheimer. This was only one change that Lawrence knew he would face as war gave way to peace. There were challenges ahead, but also opportunities—massive opportunities. And no one in science would be better positioned to exploit them than he was.

Chapter Fifteen

The Postwar Bonanza

American physicists emerged from the war torn between two contradictory sensations. In the aftermath of Hiroshima and Nagasaki, they were hailed as heroes; even portrayed bearing the "tunic of Superman" (in the words of *Life* magazine). Their pronouncements on scientific, social, and even political issues were eagerly sought and widely published.

But many felt burdened by the consequences of their work. "In the past, scientists could disclaim direct responsibility for the use to which mankind had put their disinterested discoveries," wrote James Franck, the Met Lab physicist who had emerged as the intellectual leader of the debate over the social consequences of nuclear research. "We cannot take the same attitude now because the success which we have achieved in the development of nuclear power is fraught with infinitely greater dangers than were all the inventions of the past."

For the moment, the American public and its political leadership were indifferent to these concerns. President Truman, who felt he had shown a valorous decisiveness in ordering the bombs to be dropped, had no tolerance for retrospective moralizing. That soured his first meeting with Oppenheimer, who was invited to the Oval Office on October 25, 1945, to discuss legislation for domestic control of the atom. Oppie immediately got off on the wrong foot with Truman by declaring, "Mr. President, I feel I have blood on my hands." In Truman's version of the meeting, he offered Oppie a handkerchief and replied, "Here, would you like to wipe them?"

Recounting the episode to Dean Acheson, then his undersecretary of state, he labeled Oppenheimer a "bellyaching" scientist and groused about how he had "come into my office . . . wringing his hands and telling me they had blood on them because of the discovery of atomic energy." He instructed Acheson, "I don't want to see that son of a bitch in this office ever again."

Unburdened by Oppie's metaphysical brooding, Ernest Lawrence eagerly accepted public accolades for his role in ending the war. These included the Presidential Medal for Merit, then the government's highest civilian honor, presented by General Groves and Bob Sproul at a Berkeley ceremony in early 1946. Consulting contracts were tendered by Eastman Kodak, General Electric, and the American Cyanamid Company, and invitations flowed in to serve on government committees, to lecture, to testify before Congress. Regent Jack Neylan, seeing that his protégé was inclined to accept them all and that the flood of commitments was sapping Ernest's health, stepped in as a buffer "to protect him from being marauded," as he put it later. Ernest, somewhat abashed at telling supplicants that he would have to check with Neylan before accepting their invitations, called him his "father confessor."

Lawrence had expected that the Rad Lab, like other scientific institutions, would be placed on a crash diet after the war as the torrent of military funding dried up. The demands of the Manhattan Project had swelled the complex on the hill above the Berkeley campus to thirty buildings filled with a professional workforce of 1,200, but as early as mid-1944, Lawrence had advised Sproul that it soon would shrivel into a modest outpost of the Physics Department, its budget cut by some 99 percent to $85,000 a year.

But he made that prediction before the bombs fell—indeed, it was during that grim period when the failure of the Alpha racetracks had laid him out in a Chicago hospital room. Now Lawrence's expectations for growth and funding turned positively libidinous. He mapped out a spectacular expansion of the laboratory to be funded jointly by the university,

General Groves's Manhattan Engineer District, and the Rockefeller Foundation, without devoting much thought to how the bills might be divided up among them. "Our stock was so high at that time that it didn't make any difference," Bill Brobeck would recall. "The money would come from somewhere." At the outset, most of it came from Groves, who committed to supporting Lawrence's laboratory at the level of more than $3 million a year.

What drove the new program forward was the return of Lawrence's gifted team of physicists from their wartime assignments. Their publication of new discoveries had been halted during the war, but they had not stopped thinking—or absorbing the knowledge gained from work on the physics of the bomb. McMillan and Alvarez returned from Los Alamos impatient to try out new approaches they had conceived to address that old bugaboo, the relativistic barrier to higher energies. Bethe had not been wrong about this obstacle, merely premature; but if the cyclotron were to move beyond 30 million electron volts, the issue had to be faced now.

McMillan dubbed his idea "phase stability." It was based on a realization that had come to him "lying in bed one night" at Los Alamos. Particles could be driven to ever higher energies in the cyclotron, he realized, by accelerating them not in streams but in pulses, with the accelerating voltage oscillating in frequency to keep them in phase as they approached the speed of light. The process would yield less current—that is, fewer particles—but give them higher energies. The insight, which would be implemented in an accelerator known as the synchrotron, was a breakthrough on a par with Lawrence's own recognition of the cyclotron principle. "If I had the true historical sense," McMillan recollected, "as soon as I woke up in the morning I would have written down in a notebook: 'I made a big invention last night.' " Instead, he wrote a letter to the *Physical Review*, which, in accordance with the rules of wartime secrecy, kept it locked away until it could be published (in September 1945). Only then did McMillan learn that he had not been the first physicist to conceive of phase stability; that honor went to Vladimir Veksler of the Lebedev Physical Institute in Moscow, who in 1944 published two papers on the subject that had not reached McMillan when he wrote his own. Veksler protested in a letter to the *Review* that McMillan's paper had made "no reference . . . to my investigations." He had a point, as McMillan acknowledged in a published apology. The synchrotron principle has ever since been attributed to McMillan *and* Veksler, who discovered it nearly simultaneously but certainly independently, working half a world apart.

Alvarez, meanwhile, pondered a similar physical limitation governing the acceleration of electrons. His thoughts turned to linear acceleration, the technology that Lawrence had been developing with David Sloan at Berkeley until the cyclotron demonstrated its superiority in 1931. In physics, however, that was ancient history. Alvarez concluded that at the much higher energies attainable in 1945, the linear accelerator might be more efficient than the cyclotron for certain purposes; his inspiration came from advanced oscillators that had been invented for radar at the MIT Rad Lab. Known as SCR-268s, these had been rendered militarily obsolete before war's end. Consequently, three thousand devices were collecting dust in army warehouses. At Alvarez's behest, Lawrence requisitioned them from Groves. Soon 750 surplus oscillators were on their way to Berkeley, where Alvarez proposed to use them to accelerate electrons to speeds sufficient to produce artificial muons, subatomic particles that thus far had been detected only in cosmic rays.

As Lawrence had perceived toward the end of the war, peacetime offered the opportunity to make Big Science much bigger. The ambitions of McMillan and Alvarez alone called for three huge new machines: a synchrotron capable of accelerating electrons to 300 million and eventually to 1 billion volts; a linear accelerator to drive protons to 140 million volts (Alvarez had reconfigured his machine to accelerate protons once he recognized that McMillan's phase-stability synchrotron would work better for electrons); and the 184-inch cyclotron, redesigned after the war to exploit phase stability and eventually rechristened the synchrocyclotron.

That was not all. Glenn Seaborg, anxious to return to Berkeley from

Chicago, proposed building a "hot lab" on the hill to continue his work on transuranic elements, which he believed were lying in wait to be discovered in abundance. For this, he needed equipment to manage intense radioactivities and, as a capstone, a nuclear reactor—the one nuclear technology that had eluded Lawrence's grasp. To pressure Berkeley to meet his requirements, Seaborg took a page from Lawrence's book: he let the university know that Arthur Compton had offered him a salary of $10,000 and the authority to hire a dozen scientists if he remained in Chicago. Lawrence pried the necessary salary and staff approvals from Sproul, and committed himself personally to finding money to build Seaborg's hot lab and reactor.

Last but not least was the cherished research in biomedicine, which Ernest did not propose to abandon. His plans included the continuation of John Lawrence's work with radioactive tracers and therapeutic isotopes as well as John's research in the biological effects of fission and the metabolism of plutonium and other transuranics by the human body, conducted in collaboration with Joseph Hamilton of the medical school. In 1945 the regents had finally addressed the medical school's hostility to this research by creating a Division of Medical Physics within the Department of Physics. At its inception, the division had a roster of four faculty members, including John, who was appointed an assistant professor of medical physics and mostly paid out of the Physics Department budget.

The Rad Lab budget for 1946 and beyond climbed to more than $2 million a year—not including construction of the new machines. Building the synchrotron and the hot lab and refurbishing the 184-inch would cost at least another $605,000. As for Alvarez's linear accelerator, Lawrence did not even submit an estimate—the device was too novel even to calculate a rough construction budget. No one knew better than Ernest that his expansion plans would outstrip the resources of the university, the Rockefeller Foundation, and the Research Corporation, his customary financial backers. He also knew that his fund-raising would face intense competition from traditional university rivals as well as such new bidders as the Manhattan Project labs at Chicago, Oak Ridge, and Los Alamos, which all hoped to establish themselves as permanent research centers in peacetime.

Keeping ahead of the competition would demand the full measure of Lawrence's fund-raising prowess. On September 19, a few weeks after the Japanese surrender, he was in Raymond Fosdick's New York office, painting word pictures of the marvelous 184-inch cyclotron. His goal was to obtain a new grant to finish the construction that had been halted by the war. In the years since the first grant was made, he explained, the cyclotron's original design had been outrun by new technological possibilities. The $400,000 remaining from the Rockefeller Foundation's initial $1.15 million was insufficient to finish the job.

In making his pitch, Ernest committed a grave faux pas. Laboring under the misimpression that the foundation would surely be proud of its unwitting role in the Manhattan Project—the $60,000 grant it had made blindly to convert the 184-inch for uranium separation—he assured Fosdick that "if it hadn't been for the RF, there would have been no atomic bomb," as Fosdick repeated to Weaver the next day. Lawrence was unaware how tactless this sounded to Fosdick, who had been brooding over the foundation's investment in what had turned out to be an instrument of unparalleled death and destruction. Fosdick considered that his foundation had been hoodwinked into a deal with the devil. The organization had funded the 184-inch in the name of "man's hunger for knowledge," as he had written in the foundation's 1940 annual report; five years later, in its annual report for 1945, he declared mournfully that this once-honorable quest had "brought our civilization to the edge of the abyss . . . The pursuit of truth has at last led us to the tools by which we can ourselves become the destroyers of our own institutions and all the bright hopes of the race." Fosdick had no good answer to the dilemma now facing science. "In the long run there is probably no method of sifting out the bad from the good in scientific research . . . The mighty imperative of our time, therefore, is not to curb science but to stop war . . . Science must help us in the answer, but the main decision lies within ourselves."

Yet even as Lawrence's words were reminding Fosdick of the foundation's guilt, they cracked open a door to its redemption. By outlining the cyclotron's vast potential for advancing the peacetime goals of basic science, Ernest convinced Fosdick that the machine just might serve as a beneficent scientific counterbalance to Hiroshima. Entranced again by Lawrence's "thrilling" optimism, as he described it to Weaver, Fosdick came away from the meeting anxious that the Rad Lab might not need the foundation's funds after all. Lawrence had mentioned the possibility that he might look to the army for the funding necessary to complete the cyclotron, even if a government grant would mean devoting the machine chiefly to military research.

The thought of ceding to the army a project in which the foundation had such an immense stake stirred Fosdick's proprietary blood. "This cyclotron is our baby, and it is going to be one of the jewels in our crown—that is, if we don't use the bomb to blow the world to pieces with," he told Weaver. "If Lawrence needs more money and is unable to get it from the government under the conditions which will give him complete freedom of action, I would be in favor of our stepping in." This was not the first time, nor would it be the last, that Lawrence would play multiple sponsors against one another.

Lawrence perceived better than the other Manhattan Project lab directors that General Groves was likely to remain a key patron of postwar physics research for years to come, since the transfer of nuclear research from the military to a civilian body was being hotly debated in Washington and was by no means assured. He had cultivated Groves assiduously, even sponsoring the general for an honorary Berkeley doctorate, conferred in 1945. Groves returned the favor by delivering the keynote at the presentation of Lawrence's Presidential Medal in 1946, declaring, "We bet a hundred million on him and won." His confidence in Lawrence's administrative and scientific abilities, which dated from their very first meeting, had only been strengthened by the success of the electromagnetic separation process. During 1944 and 1945, open criticism of Groves's regime surfaced at the Chicago Met Lab and other Manhattan Project outposts. But nothing of the sort was heard at the Rad Lab, thanks to Lawrence's "very drastic and successful efforts to prevent any expression of opinion on the part of members of the laboratory," as one staff member recalled. It was not by chance that Lawrence's personal recollection of the Trinity test was the only one Groves selected for direct transmission to the Pentagon on the day after the blast.

Ernest's determination to keep on Groves's good side showed in his muted response to an incident that outraged the physics community and inflamed public opinion. This was the army's wanton destruction in November of five Japanese cyclotrons. The machines, which were dismantled by occupation personnel and dumped in the Pacific Ocean, included the identical copy of Lawrence's sixty-inch cyclotron built in Tokyo. Adding to the insult, the Tokyo lab's director, Yukio Nishina, already had received permission from occupation officials to restart his cyclotron for biological and medical research. Instead, army engineers arrived unexpectedly on November 24 with sledgehammers and blowtorches and reduced his precious accelerator to scrap. Army officials rationalized their act as essential to the Allied effort to destroy the Japanese war machine, but this demonstration of crass insensitivity to the scientific value of the cyclotrons, which had played no part in the war, undermined the Pentagon's claim to be a reliable steward of nuclear science.

The incident placed Lawrence in a delicate position, for he was then awaiting Groves's approval of a budget for the Rad Lab's ambitious postwar program. As a prewar patron of Nishina, who had sent his assistants Ryokichi Sagane and Tameichi Yasaki to the Rad Lab for training, Lawrence might have been expected to register a powerful objection to the army's behavior. Instead, he remained offstage, leaving the campaign of condemnation to Karl Compton, who was serving as scientific advisor to the occupation forces and publicly labeled the cyclotron destruction "an act of utter stupidity." Compton was even more infuriated to learn that the army had invoked his name to justify its action. Nishina, appearing at occupation headquarters to demand an explanation, had been told that Compton had specifically endorsed the destruction order. The truth was exactly the opposite: Compton had drafted the order allowing the Tokyo machine to be restarted for biomedical research.

While Groves sustained volleys of political fire for the episode, Lawrence consoled him by agreeing that "the present time is hardly appropriate for any move to restore the cyclotrons, and the matter probably should be deferred." In a note to Vannevar Bush, he was somewhat more solicitous of Nishina's needs: "I hope very much something will be done to rectify this error . . . The right thing to do would be at least to re-equip Nishina's laboratory." But he did not say so publicly.

Lawrence's careful groundwork with Groves bore fruit, for the general gave freely of the army's surplus bounty, including the radar oscillators for Alvarez and a supply of capacitors for McMillan's synchrotron, worth $203,000. In December Groves granted the Rad Lab $170,000 from Manhattan Engineer District funds to complete the 184-inch cyclotron, authorized the launch of the synchrotron project under the army's wartime contract with the University of California, and allocated $2.2 million for construction on the hillside and operating expenses. Recalling the Rad Lab of that era, Alvarez exclaimed delightedly, "We ran it with a big barrel of greenbacks."

Lawrence thereby presided over a transformation of American science as profound as any change inspired purely by scientific discovery: the launch of peacetime government patronage. In 1936 the federal government had spent $33 million on research and development, accounting for 15 percent of the $218 million advanced by all sources, including industry, academia, and philanthropic foundations. From 1941 through 1945, the government's share rose to $500 million a year, or 83 percent of the annual total. War's end failed to flatten the trend line of spending: by 1947, the federal research budget had risen to $625 million, still representing more than half of a much-expanded total national investment in scientific research and development.

The government's enormous role in research, especially in nuclear physics, made many scientists uneasy. In 1946, wrote Philip Morrison, "for every dollar the University of California spends on physics at Berkeley, the Army spends seven." At that year's meeting of the American Physical Society in Berkeley, half the delivered papers carried the disclosure that they had been funded by the army's Manhattan Engineer District or its aggressive new service rival, the Office of Naval Research. Some thirty universities were performing nuclear physics research on navy contracts, which in some cases accounted for 90 percent of their research budgets.

Lawrence did not share his colleagues' misgivings. One reason was his rapport with Groves, which helped him fend off any constraints on the Rad Lab's spending of government funds. The Rad Lab occupied a special place among the Manhattan District labs for other reasons. It was the only lab to boast a prewar history and, of course, the only one with a preexisting management. Consequently, whenever Groves made plans to slim down the Manhattan Project's lab network, he left the Rad Lab alone. "The Berkeley laboratory, I felt, would continue as long as Ernest Lawrence lived, provided it received proper financial support from the government," he wrote later.

Lawrence's trolling for government funds took place against the backdrop of a fierce battle in Washington over legislation to bring atomic energy under civilian control. At issue was a bill introduced by Representative Andrew J. May of Kentucky and Senator Edwin C. Johnson of Colorado following a presidential message to Capitol Hill. Truman's message—drafted by State Department lawyer Herbert Marks with input from Oppenheimer—acknowledged that "in international relations as in domestic affairs, the release of atomic energy constitutes a new force too revolutionary to consider in the framework of old ideas." The Truman administration committed itself to seeking "a satisfactory arrangement for the control of this discovery in order that it may become a powerful and forceful influence towards the maintenance of world peace instead of an instrument of destruction."

At first, the scientific community was inclined to look with favor upon the May-Johnson Bill, especially since it was endorsed in a statement written by Oppenheimer and cosigned by Lawrence and Fermi—endorsed, that is, by three of the four members of the Interim Committee's scientific panel.

But support for the May-Johnson Bill evaporated once scientists got around to reading the text, which, unlike Truman's message, had been drafted by War Department lawyers. The bill ceded a large measure of control over atomic research to the military. The penalties for violating security regulations covering information judged to be sensitive had a distinctly militaristic cast, with fines of up to $300,000 and jail terms as long as thirty years for "willful" disclosures.

Over a period of weeks, criticism mounted steadily in volume and color; in one interview, Harold Urey labeled the measure the "first totalitarian bill ever written by Congress." He added morosely, "You can call it either a Communist bill or a Nazi bill, whichever you think the worse." The scientists' distaste soon fell upon the bill's eminent defenders. As Chicago physicist Herbert Anderson wrote to William A. Higinbotham, chairman of the Association of Los Alamos Scientists: "I must confess my confidence in our own leaders Oppenheimer, Lawrence, Compton, and Fermi . . . is shaken. I believe that these worthy men were duped—that they never had a chance to see this bill."

Anderson was being unfair to Compton, who wisely postponed signing Oppenheimer's statement until he could read the bill, and refused to support it once he had. But he was correct about the others: not even Oppenheimer had read the actual text before drafting the message to which Lawrence and Fermi had blindly signed their names. All three soon withdrew their support. The May-Johnson Bill would meet its legislative doom thanks to opposition from the scientists and public skepticism about military control, energized by public missteps such as the destruction of the Japanese cyclotrons.

Following his ill-fated endorsement of May-Johnson and his subsequent repudiation, Lawrence absented himself from the debate over domestic control. Anyway, his closest personal advisors were discouraging him from playing an active role in Washington. His brother, John, fretted constantly about Ernest's susceptibility to bronchial and sinus infections, back conditions, and other manifestations of exhaustion and stress; these had grown worse toward the end of the war and immediately afterward, when Ernest suffered a bout of viral pneumonia. Jack Neylan was concerned that the demands of Washington service would expose Berkeley's faculty star to the no-win politics of the nation's capital. When Secretary of War Robert Patterson invited Ernest to Washington to give his views on successor bills to May-Johnson, Neylan issued a strong veto even though Ernest seemed honored to be asked. "He was kind of boyish about it," Neylan recalled. "Came over to see me and said, 'What does it mean?' I said they're looking for a fall guy and you're nominated. He said, 'Do you think they'd do that?' I said you just don't know politicians." Alfred Loomis, who shared Neylan's uneasiness, deputized San Francisco lawyer H. Rowan Gaither, who had helped manage the MIT radar lab early in the war, to assist Neylan in screening Ernest's invitations to advisory commissions and corporate boards.

Lawrence's withdrawal from the policy debate also reflected his long-held conviction that science and politics made for an unwieldy mix. He had been willing to lay aside that conviction during the war, when national security required his participation in the highest councils of government planning. But peace had transformed the Rad Lab back into a civilian research institution pursuing its own interests; Lawrence's instincts told him that under the circumstances, it would be unwise for him to meddle in impenetrable policy battles in which he could find himself suddenly and unpredictably isolated in the wrong camp.

Nevertheless, appeals for his involvement in the public debate kept arriving at the Rad Lab. In February 1946 the Federation of American Scientists, a coalition of more than two dozen scientists' groups at universities and government labs, sought his endorsement of a new atomic energy bill introduced by Senator Brien McMahon of Connecticut. The McMahon Bill excluded the military from control of atomic energy, except as the technology was directly applicable to weapons, propulsion, and other military needs. Civilian supervision was vested in an Atomic Energy Commission comprising five presidential appointees. The irksome security provisions of May-Johnson were liberalized, with penalties reduced to maximum fines of $20,000 and prison terms of up to five years.

There is no written record that Lawrence replied to the overture, although around the same time, he registered a private opinion of the federation—and of scientific activism in general—in a letter to Weaver. "My own feeling is that this political activity of many of our atomic scientists is unfortunate in many ways," he wrote. "It is particularly a great pity that they are frittering away so much time and energy on political problems, when they could be devoting themselves to scientific pursuits."

Truman signed the McMahon Bill into law on August 1. His five appointees to the new Atomic Energy Commission were approved by the Senate before the end of the year: David Lilienthal, a former chairman of the Tennessee Valley Authority, as chairman; and as members Robert Bacher, a Los Alamos physicist; Republican financier Lewis Strauss; businessman Sumner T. Pike; and Des Moines, Iowa, newspaper editor William Waymack. The commission's most important bequest from the Pentagon's nuclear energy apparatus was the Manhattan District's network of atomic labs, which Groves had kept intact after the war.

This had been a difficult task, for they all suffered brain drains as their leading scientists returned to prewar academic posts or accepted jobs stemming from their newfound eminence. The lack of clarity about their postwar roles as the handover of policy authority to civilian control approached had further diminished morale among the staff members who remained. Lawrence did what he could to lend stability to the pieces of the network he could control: the Rad Lab, of course, remained his personal fiefdom and never suffered the doubts of the labs that had been created purely for the bomb program. He also pressed for the University of California to renew its government contract to manage Los Alamos. This brought him into rare conflict with his mentor Jack Neylan, who had rammed through a vote on the Board of Regents to terminate the deal. Ernest, who had initiated the contract himself, finally persuaded Neylan that its termination was tantamount to "running out on a duty" owed to the government—and that it might sever a critically important pipeline for the government patronage on which the continuing expansion of Big Science depended. Neylan agreed to extend the contract only after Lawrence pledged not to spend any time on direct management of the Los Alamos lab, though he would have "free run" to consult on its research program. Lawrence's instincts were right, for the Los Alamos contract helped to keep Berkeley at the center of government-funded nuclear research—so much so that other universities competing for government grants began to complain about the "University of California Atomic Trust," as it was labeled by Columbia's Isadore Isaac "I. I." Rabi, a Nobel laureate and veteran of the MIT Rad Lab and the Manhattan Project.

In the vacuum created by the long transition to civilian control of atomic energy, Lawrence and his lab prospered. McMillan's synchrotron and Alvarez's linear accelerator were under construction. Seaborg had accepted an appointment at Berkeley as a full professor, with the authority to hire four assistant and associate professors and twelve salaried graduate fellows. His hot laboratory was funded with the $400,000 left over from the Rockefeller Foundation's original cyclotron grant, for the 184-inch accelerator, now officially a synchrocyclotron, had been completed with the grant of $170,000 from the Manhattan District, supplemented with $132,000 from University of California funds.

The 184-inch, nestled within an enormous, circular red-domed building resembling an enclosed carousel on the hilltop above campus, was still the Rad Lab's showpiece. The machine first produced 200-million-volt deuterons before dawn on November 1, 1946, an event celebrated quietly by Lawrence and the lab crews who had worked long into the night to produce the most powerful particle beam in the world. The machine's formal launch, scheduled for Monday, November 18, was preceded by a three-day retreat hosted by Alfred Loomis at the Del Monte Lodge in Monterey for Bush, Conant, the Comptons, and Lawrence—a replay of the 1940 retreat at which he originally had brought them together. The guests then drove to Berkeley, where they were met by the new members of the Atomic Energy Commission.

Ernest hosted a dinner for them all at Trader Vic's, featuring his customary personal menu—"a kind of smoked sparerib, eaten 'by hand,' and a coffee plus brandy," David Lilienthal recorded in his personal journal. The AEC chairman described the physicist as "a rather fabulous figure in the nuclear field—indeed, in research generally . . . a very youthful man—big, red-faced, full of vitality and enthusiasm. Looks nothing at all like the picture of a great scientist—not at all. You get a sense of drive in talking with him, and that impression is in accordance with the facts." Throughout the visit, Ernest's bonhomie worked overtime, but with a serious purpose: his goal was to inculcate the commissioners with an understanding of atomic energy's potential for peaceable uses such as generating electricity, a field in which the Rad Lab expected to play a central role. The subtext was the necessity of government funding for more nuclear reactors—including one, it was hoped, at Berkeley. "He hammered this hard," Lilienthal recorded.

For all that he felt drawn into Lawrence's field of attraction, Lilienthal was no innocent. He was a battle-hardened bureaucratic fighter whose career in federal service had been launched with his appointment to the

New Deal's Tennessee Valley Authority. There he had to battle wealthy and influential utility magnates hoping to throttle the TVA at birth. Lilienthal had beaten them back, despite their able leadership by Wendell Willkie, a future Republican challenger to Franklin Roosevelt (and a business protégé of Alfred Loomis), and had established the TVA as one of the most effective institutions of the New Deal.

Lilienthal was to become amazed at the number and variety of Lawrence's patrons. One was Colonel Robert R. McCormick, the idiosyncratic right-wing owner of the *Chicago Tribune*, who conceived a friendship with Lawrence based on his role in developing the bombs that had destroyed Hiroshima and Nagasaki and transformed war "from irrationality to idiocy." Lawrence confided to Lilienthal that they were so close that he stayed at the McCormick country mansion when visiting Chicago. He worked assiduously to forge a relationship between the New Dealer Lilienthal and McCormick, whose *Tribune* had been Franklin Roosevelt's fiercest critic during the New Deal; he arranged for McCormick to visit the Argonne and Hanford labs and even tried to persuade Lilienthal to join the publisher on one of the trips—a potentially unpleasant encounter from which Lilienthal wisely begged off.

In the first week of January 1947, the Atomic Energy Commission appointed its most important body of technical counselors, the General Advisory Committee. Chaired by Oppenheimer, the GAC was composed of Conant, Fermi, Seaborg, Rabi, DuBridge, former Los Alamos physicist Cyril Smith, and industrialists Hartley W. Rowe and Hood Worthington. On the surface, this appeared to be another victory for Lawrence, for five of the members were his friends or former colleagues. When the GAC convened in early 1947 to apportion nuclear research funds among the old bomb labs and new university-based aspirants, Oppenheimer decreed that the committee should "pass no judgment on the work of the University of California, which has a special history." A contemporary witticism recalled by Brobeck told the story more succinctly. The AEC charter established that in the case of a disagreement with the director of one of the atomic labs, the commission could remove the director. "Somebody asked, 'What about the Radiation Laboratory at Berkeley?' The answer was, 'Oh, in that case, the director removes the commission.' "

A few months after the cyclotron launch, the opportunity arose for further Rad Lab lobbying. This time it was a joint meeting in Berkeley of the AEC, the General Advisory Committee, and the heads of the government labs. Again playing host, Ernest arranged four full days of wining and dining—what AEC historians recorded as "good, big dinners with plenty of red meat"—at Bohemian Grove, the rustic hideaway for the Bay Area elite in the Northern California woods. He had Jack Neylan on hand to assure the commission of the University of California's commitment to Lawrence and Los Alamos, and, undoubtedly, to forge a personal link between the powerful Lilienthal and the persuasive and influential Neylan.

From Berkeley's standpoint, the meeting came off as a roaring success. What helped was that it provided the commissioners with a chance to get out of Washington, where the atmosphere was thick with red-baiting security inquiries and a never-ending policy debate over whether atomic energy research should be directed more at munitions or peacetime applications. After the guests had departed, Don Cooksey informed Alfred Loomis that "members of the committee . . . and several of the directors enlarged to me on the wonderful progress they had made in coming to a mutual understanding of each other's problems. My own belief is that those four days may be marked as of inestimable value to the country." Two months later, the commission appropriated $15 million for new accelerator construction, much of it destined for Berkeley.

Yet the Rad Lab's standing with the AEC was not unassailable. For one thing, the relationship between Lawrence and Oppenheimer had become seriously frayed. Ernest had pressed Oppie to return to Berkeley after the war, but their wartime experiences had widened their personal differences into a gulf. Oppenheimer in particular had evolved; he no longer was the aloof, self-absorbed intellectual who had gone off to Los Alamos anxious to prove himself in the world beyond theoretical physics. Ernest would find it very difficult to deal with an Oppenheimer who had acquired a supreme self-confidence in the crucible of war. "He regarded me as potentially a very good physicist, a widely read man," Oppie reflected years later, "but in a certain sense not worldly, not experienced, and not very sensible. And this feeling is very hard to change once it's established in a friendship. And when the circumstances change, there is a little wrench."

Oppenheimer had spent the war embroiled in nearly constant battles with Berkeley administrators over the management of Los Alamos, and he was hurt that Lawrence failed to appreciate why that might have soured his feelings about the university. "It may seem odd and wrong to you that the lack of sympathy between us at Y [Los Alamos] and the California administration . . . could make me consider not coming back," he wrote Ernest a few weeks after the bombings. He chalked up Ernest's lack of empathy to the latter's having settled into the role of big man on the Berkeley campus: "It would not have seemed so . . . hard to understand if you remembered how much more of an underdogger I have always been than you." But Lawrence's insensitivity kept surfacing. When Oppie called to tell him he would rejoin the Berkeley faculty after all, Lawrence responded, "Good, I can clip your wings a little." Years later, Oppenheimer would still remember that as "an awful moment."

Oppenheimer's tenure at Berkeley would not last long. During a visit to the Rad Lab in November 1946, AEC commissioner Lewis Strauss, a trustee of the Institute for Advanced Study, in Princeton, New Jersey, pulled him aside to offer him the institute's directorship. The institute, an independent, privately funded research center, was known to the public largely as the home of Albert Einstein, but in general, its scientific reputation was mediocre—a condition the trustees thought Oppenheimer could remedy. Oppie irritated Strauss by dithering over the offer for months without replying. But in the end, he concluded that relocating to the East Coast would make it easier for him to participate in the important policy debates taking place in Washington while he created a world-class research mecca in Princeton. His appointment was announced in April.

Another challenge to the Rad Lab's domination of government patronage came from a new university lab under development at the eastern end of New York's Long Island. The Brookhaven National Laboratory was conceived as a powerful regional counterweight to Berkeley. It had been established in 1946 with the blessing of General Groves and under the guiding scientific spirit of I. I. Rabi of Columbia. Rabi stitched together the new lab's sponsoring consortium, known as Associated Universities, from nine large Eastern research institutions that would have been hard pressed to compete individually in the multimillion-dollar world of postwar high-energy physics.* Rabi would become as powerful a guardian of Brookhaven's interests as Lawrence was of Berkeley's, for as a member of the General Advisory Committee, he was perfectly placed to keep an eye out for efforts by Oppenheimer and Seaborg to tilt research grants toward the Radiation Laboratory.

If Associated Universities committed a misstep, it was to wait for the handover of power from the Manhattan District to the AEC before acquiring the lab's rural site, a decommissioned US Army base known as Camp Upton. The delay allowed Lawrence to exploit Groves's largess without competition: by the time Brookhaven was chartered, the foundations for Lawrence's three new accelerators already had been laid. The Brookhaven team, which included Stan Livingston, recovered quickly under Rabi's energetic guidance, however. By the time of the Bohemian Grove retreat, they were ready with a proposal for their own big accelerator.

The Rad Lab's entry in what became a two-way battle was a machine designed by Brobeck and labeled the Bevatron, referring to its goal of producing particles at energies over 1 billion electron volts—in fact, *10* billion electron volts. Brobeck had designed the Bevatron around McMillan's

*The nine universities were Columbia, Cornell, Harvard, Yale, Princeton, MIT, Johns Hopkins, the University of Pennsylvania, and the University of Rochester.

phase stability principle; his idea was to combine the synchrotron's variable magnetic field and the synchrocyclotron's variable frequency in one hybrid accelerator, exploiting both principles to drive energies to a new frontier.

Initially, the competition was amicable. Lawrence, placed in a companionable spirit by Loomis's hospitality at Del Monte, happily showed off Brobeck's designs to Rabi. He seemed pleased at the chance to expand the Cyclotron Republic once again. Anyway, explained Brobeck, "they needed something at Brookhaven." But trouble arose once it became clear that the AEC's accelerator budget of $15 million would not be sufficient to support two Bevatrons—in fact, it might not be enough to build even one. That locked the two labs into a zero-sum game that only one could win unless both compromised significantly. At McMillan's suggestion, Lawrence shrank his proposal to a 6-billion-volt machine budgeted at just under $10 million. Yet even that was so ambitious that the AEC was forced to consider whether Ernest was thinking too far ahead and whether it made any sense at all to build two similar machines at opposite ends of the country. On the GAC, Fermi emerged as a prime critic of Lawrence's scheme, suggesting that it was premature to build an accelerator in the billion-volt range even before the three new Rad Lab machines had been put through their paces. It would harm science, he argued, to "endorse what appears to be an unthoughtful program." Rabi, meanwhile, was urging his team to think on a Lawrencian scale. As Rabi's handpicked accelerator builder, Stan Livingston had sketched out a 750-million-volt synchrocyclotron. But measured against the Bevatron, that energy range seemed paltry. "Take something that's a bigger challenge," Rabi needled Livingston. "Make something big."

The two labs' plans landed at the AEC almost simultaneously in February 1948: Berkeley's for a machine rated at 2.8 billion volts and Brookhaven's at 2.5 billion. (To distinguish it from the Bevatron, it was christened the Cosmotron.) The rivalry weighed heavily at the Rad Lab: having learned that Brookhaven was about to submit its proposal, Lawrence gave Brobeck and his design staff a scant two weeks to finish theirs. Their task was eased somewhat by Lawrence's unassailable reputation as an accelerator builder. "What we submitted was largely a cost estimate," Brobeck recalled, "because you didn't have to convince these people that Lawrence could build an accelerator . . . It was a very brief proposal; it didn't have a lot of scientific justification." The cost estimate was $4.5 million.

When the GAC weighed the two proposals, questions of morale and laboratory politics, not scientific merit, came to the fore. Rabi, citing Berkeley's three existing accelerators, suggested that Brookhaven should be given a chance to catch up. That idea was parried by Oppenheimer, who argued that the "discouragement of the Berkeley group would result in the loss of something valuable to the national scientific health." Yet overall sentiment on the GAC was drifting in Brookhaven's direction.

Then Lawrence pulled a rabbit out of his hat. For the better part of a year, the Rad Lab had been trying to produce mesons, the most sought-after and elusive subatomic particles of the moment, in the restored 184-inch cyclotron. Their existence had been predicted in 1934 by Japanese physicist Hideki Yukawa, who speculated that they carried the force that binds the atomic nucleus together, counteracting the mutual electromagnetic repulsion of its positively charged protons. (The meson quest had been one of the goals Ernest presented to Warren Weaver in support of the original Rockefeller Foundation grant for the 184-inch.) Mesons had been found only in cosmic rays, which made them the natural quarry of high-energy physicists. The 184-inch synchrocyclotron was one of the few machines that could accelerate alpha particles to the energies required to produce them. Yet through 1947, Lawrence's attempts to capture an artificially produced meson on photographic film were unavailing.

Then, in February, a gifted young Brazilian experimentalist named Cesare M. G. Lattes arrived in Berkeley. Within days, he had remedied the experimental shortcomings that had prevented Lawrence from producing mesons, and before the end of the month had captured tracks of the first artificial mesons ever seen. Lawrence was summoned to the phone at Trader Vic's to receive the news. He bolted out of the restaurant, leaving his nonplussed dinner guests behind, to view the evidence for himself.

To Ernest, the discovery was more than a triumph of applied physics; it was a means of reminding the GAC of the Rad Lab's standing as the preeminent high-energy laboratory in the world. When the committee reconvened in April, there was no longer any question of denying it a piece of the Bevatron pie. The members decided to fund two machines of different sizes, each optimized to produce different particles; the only question left was which lab would build which machine. In the end, Berkeley got approval for a Bevatron to produce 6-billion-volt protons, satisfying Lawrence's predilection for building as big as he could. Brookhaven settled for a cosmotron rated at 2.5 billion volts, though it received the GAC's explicit promise that it could come back to the well for bigger machines in the future. Lawrence's reputation and his nimble management of this newest patron had carried the day.

Chapter Sixteen

Oaths and Loyalties

"E. O. Lawrence personally never quite demobilized after the end of World War II," the Rad Lab physicist Wolfgang "Pief" Panofsky would observe many years later.

The German-born scientist was right. While the Rad Lab quickly moved back into basic nuclear research, Ernest Lawrence maintained his personal interest in military projects. He continued working on the calutron, determined to increase its efficiency even after 1948, when Oak Ridge was fully converted to gaseous diffusion for uranium enrichment and the Atomic Energy Commission canceled funding for calutron development. He consulted with Admiral Hyman Rickover on the development of nuclear-powered submarines, advising that the service would have to invest "real cash" in the effort—not the cheeseparing $2.5 million budgeted for preliminary studies, but $100 million. ("To be credible, the project would have to be big," Ernest counseled in true Big Science style, "and if it were big, it would attract good people.") And he toyed with the how-tos of radiological warfare, a project that perverted the health-giving radioisotope research he had so long pursued with his brother John into a search for death-dealing tactical weaponry. "RW," as it was known, was "a subject which was very close to Professor Lawrence's heart," Alvarez recalled. But mainstream figures in science and the military disdained RW as ineffective, impractical, and immoral.

Ernest seemed to be trying to preserve the atmosphere that had helped

Big Science win the war. As the debate over the social and political implications of the bomb intensified during the first uneasy years of peace, he assumed an increasingly obstinate position in favor of more weapons research and less introspection about it. National security was paramount, he argued. Its protection meant pursuing the most advanced nuclear technology that could be exploited for the purpose. That would come to mean the Super: the thermonuclear, or hydrogen, bomb.

Lawrence's concern with the threat to national security from outside forces, specifically Communism, dated back to his earliest involvement with wartime research, when security issues deprived the Rad Lab of Rossi Lomanitz and Martin Kamen. But after the war, his outlook became colored more by his intimate association with wealthy conservatives such as Jack Neylan, whose opinions about the cunning of Communists and their determination to subvert Western civilization were one-dimensional to the point of self-caricature. Ernest was soon to become entwined in the increasingly supercharged atmosphere of suspicion and accusation that anticommunism brought to the Berkeley campus.

The hunt for secret communists at the Radiation Laboratory began in 1947, when a publicity-hungry state senator named Jack Tenney, chair of the California legislature's Un-American Activities Committee, staged an investigation into what he asserted was suspiciously lax security at Berkeley. Like many red hunts of the day, the Tenney hearing had a Gilbert and Sullivan comic-opera flavor. The star witness was the committee's chief investigator, who told of having ambled up to the Rad Lab's hillside perimeter with a flashlight. He crawled under the fence and wandered around without ever being challenged, thereby exposing an enormous hole in the security fabric of the nation.

These revelations failed to inspire an outcry by the press or the public, and the committee presently ended its investigation. The nationwide miasma of anticommunism, however, was not to be dispelled so easily. In 1948 the Atomic Energy Commission responded to pressure from the red-baiting House Un-American Activities Committee (HUAC) by forming regional personnel security boards around the country to vet employees of AEC contractors, such as the University of California, whose political affinities had come under question. The chairman of Berkeley's board was Neylan, who had been specifically recommended to the AEC by Ernest Lawrence. Admiral Chester Nimitz and Major General Kenyon Joyce, two renowned local war heroes, rounded out its membership, though they generally sat by silently during hearings while Neylan conducted the inquisitions.

The Neylan board's first case involved a former Los Alamos chemist named Robert Hurley, who had fallen under suspicion because his Latvian-born wife purportedly had leftist sympathies. Neylan subjected Hurley to a withering examination at his San Francisco law office, with only Hurley's mentor, Berkeley chemistry dean Wendell Latimer, present to offer counsel and support. Hurley sarcastically parried Neylan's questions about his associations with liberal organizations while Latimer quietly seethed at the injustice of a proceeding based on hearsay and FBI files that the accused was forbidden to see. Neylan concluded that Hurley was prevaricating—"he was the smart-alec type," he observed later—and ordered him fired, only to learn subsequently that Latimer had quietly rehired him. Neylan ordered him fired again, and this time the dismissal stuck.

Latimer brought his complaints about the board's dual role as prosecutor and jury to David Lilienthal. The AEC chairman referred Latimer's objections back to the board, prompting General Joyce to ask Neylan "if some kindly but pointed advice from Ernest Lawrence would not be beneficial in getting Latimer on a more conservative and less emotional track." Neylan responded that he already had approached Lawrence, who had assured him that Latimer was merely "overworked."

The board's appeal to Lawrence showed the confidence of its members in his ability to see things their way and balance the interests of the emerging security state and the scientists it employed. That confidence paid dividends when the board turned its attention to Robert Serber, who had been Oppenheimer's right-hand man in theoretical physics at Los Alamos. After the war, Serber had become a valued member of the Berkeley faculty, recruited by Lawrence to succeed Oppie as its chief theoretical physicist. Serber had come under attack because of the leftist associations of his friends and his wife, Charlotte. Both Robert and Charlotte Serber had held security clearances at Los Alamos, where Charlotte served as the lab's indispensable librarian. That was not good enough for the AEC, which denied her a security clearance to serve in the same capacity at Berkeley and placed her husband under a cloud.

Serber's character witness before the Neylan board was Lawrence himself, as good a guarantee of clearance as he could have wished. Ernest's handling of the hearing differed dramatically from Latimer's, producing a much different outcome. As Neylan recollected, Lawrence said, " 'I hope you'll give him the benefit of the doubt because he is a fine fellow.' Ernest was betting his life on Serber and was proved to be right." Ernest had learned how to play the security game. Unlike Latimer, he accepted the rules, focusing his efforts on a defense of the accused rather than an attack on the system.

The carefully orchestrated proceeding faltered only once, when Neylan asked Serber a dangerous hypothetical question alluding to the case of Haakon Chevalier. A member of the Berkeley French faculty and a close friend of Oppenheimer's, Chevalier had approached Oppie about passing information from the bomb program to the Soviets. Oppenheimer rejected the overture on the spot, though the encounter would create a political problem for him down the line.

Neylan's question to Serber was: "If somebody sympathetic to Russia were to ask you to get some secret information from Ernest Lawrence and retail it back to a Russian agent, would you put that matter up to Lawrence?"

Serber replied, "Yes, I think so."

"*What*?" Lawrence exclaimed.

But Serber had misunderstood the question. With Lawrence's careful guidance, he explained that he meant he would report the overture to Lawrence as a security breach, not as an invitation to espionage. Neylan accepted the explanation and cleared Serber, declaring him "frank and forthright."

To Serber, the entire episode was deeply unsettling. He pictured himself like a defendant in a movie court-martial, interrogated by three humorless inquisitors about suspicious acquaintances he barely remembered. He never received a formal notification that he had passed the ordeal, though Oppenheimer later mentioned that in a copy he had seen of the board's report, Serber had been "glowingly praised." That was cold comfort, Serber reflected: "I had found the experience humiliating and frightening, and resented having been put through it."

A year later, Lawrence again had to intercede to protect a Rad Lab colleague. This time Neylan's target was Melvin Calvin, a future Nobel laureate who had worked at the Metallurgical Lab in Chicago and then moved to Berkeley to research peacetime applications of medical radioisotopes with John Lawrence. Ernest stepped in at once, advising Neylan that his own investigation of the charges against Calvin had "strengthened and reaffirmed" his confidence in the chemist's loyalty. "Quite apart from his great scientific talents," he told Neylan, "I have always regarded him as a man of fine personal qualities of character, loyalty, and integrity." Lawrence's statement took the wind out of the security inquiry, and the board passed Calvin after a brief interrogation.

A more complicated case, though one that unfolded outside the AEC's security process, involved Frank Oppenheimer, a talented physicist who had received his PhD from Caltech but lacked his older brother's theoretical profundity. Unlike Robert, Frank had joined the Communist Party with his wife, Jackie, in 1937, when they harbored the impression that the party shared their own goals of social justice. They would resign in disillusionment in 1940. But Frank would not shed his political beliefs; nor was he as willing as Robert to suppress them in the name of career advancement. On accepting a job in the Rad Lab in 1941, however, he promised Lawrence not to embarrass the lab with labor activism or other causes. He had been carefully coached by his brother, who had experienced Ernest's hostility to politics in the Lab firsthand. "I warned him that Ernest would fire him if he was not a good boy," Robert recalled later.

Frank played an invaluable role at the lab, including work on the calutrons. But his leftist politics never lay far beneath the surface, posing a continual irritant for Lawrence. "Why do you fool around with these things?" Ernest chided Frank during a train trip to Oak Ridge in 1944. "Good scientists aren't like people who just want to eat, sleep, and make love. You're not like people who can't get anywhere. You don't need that." Lawrence's concern, Frank recalled, was that politics might cause dissension and distraction on the laboratory floor—creating what he called "inhomogeneity" in the lab.

In Ernest's view, Frank failed to keep his promise to avoid causing political trouble for the lab: shortly after the war's end, Frank explained to a newspaper reporter that he had moved a speech to a small hall because Negroes had been barred from a larger one. "Now look what you've done," Lawrence upbraided him. "You've brought race relations into the lab!" Still, when Frank came under consideration for a tenure-track job at the University of Minnesota after the war, Ernest recommended him glowingly as "one of the most useful members of our staff" and praised his "originality and soundness of scientific thinking."

Frank got the job. He left Berkeley in 1946 with assurances from Ernest that he would always find the doors of the Rad Lab open. But two and a half years later, everything had changed. In October 1948, as Frank was preparing to accompany physicist John Williams to Berkeley to discuss the construction of an accelerator at Minnesota, Lawrence abruptly informed Williams that Frank would not be welcome. The unexplained refusal shocked Williams, who scurried to find a traveling companion who "meets with your approval," as he wrote Lawrence. For Frank Oppenheimer, who had been looking forward to the trip as a long-deferred homecoming, it was even more stunning. That very day, he dispatched to Ernest an indignant cri de coeur:

> *Dear Lawrence:*
>
> *What is going on? Thirty months ago you put your arms around me and wished me well, told me to come back and work whenever I wanted to. Now you say I am no longer welcome.*
>
> *Who has changed, you or I? Have I betrayed my country or your lab? Of course not . . . Does anybody think that I ever let any classified information leak out, intentionally or unintentionally? . . . You do not agree with my politics but you never have, and there are no new rumors about my distant past floating around . . .*
>
> *I am really amazed and sore because of your action.*

For all its undoubted sincerity, Oppenheimer's letter displayed a brazen disingenuousness, perhaps its author's chief character flaw. In 1949, under investigation by the House Un-American Activities Committee, Frank would finally confess publicly to his Communist Party membership. At that point, he was fired by the University of Minnesota. But he knew well at the time of his letter to Lawrence that new rumors about his past had been "floating around"; a Washington *Times-Herald* reporter had badgered him about them in 1947, prompting him to deny his party membership for the record. The reporter's source was undoubtedly an FBI file on Robert Oppenheimer, which disclosed Frank's Communist affiliation. The dossier was being shown around Washington by Senator Bourke Hickenlooper of Iowa—curiously enough, as part of his case against David Lilienthal's confirmation to the Atomic Energy Commission. Numerous top-level scientific leaders, including Lilienthal, Conant, and Bush, were familiar with the file. There is some likelihood that Lawrence had heard about it too, in which case he would have known that Frank Oppenheimer's repeated denials of party membership were false. That could have been enough to prompt him to bar Oppenheimer from his lab.

In a final indignity, Lawrence asked Edwin and Elsie McMillan to rescind a dinner invitation to Frank. That hurt, for Frank's friendship with McMillan dated back to the mid-1930s, when they shared many companionable holidays camping and riding with Robert at his ranch in the New Mexico desert. Lawrence's interference in a private friendship only added to the growing rift between him and Robert and Kitty. "When we ran into Ernest at one of those big parties somebody else was giving [in Berkeley], I said something about it," Robert recalled. "I don't think Ernest minded that, but as often was the case, my wife said something sharper, and I think maybe he minded that."

The AEC personnel board hearings foreshadowed a conflict over security and politics that would have a historic impact on the Berkeley campus and the Rad Lab. The conflict involved the University of California's loyalty oath.

In early 1949 Jack Tenney, who had staged the earlier hearing into Rad Lab security, resurfaced with a package of thirteen bills targeting suspected communists at the university and elsewhere in state government. Hoping to head off wholesale interference in university affairs by conservative legislators, Robert Sproul asked the regents on March 25 to amend a 1942-vintage oath of office required of all university appointees, adding a clause stating that the appointee was not a member or supporter of "any party or organization that believes in, advocates, or teaches the overthrow of the United States Government." The regents went further, compelling university employees to explicitly disavow the Communist Party.

This was not Sproul's finest moment. He led the regents to believe that his addition had been approved by the faculty senate, which, in fact, was unaware of the proposed change. In one stroke, his misstep fractured his relations with the regents and alienated the faculty. The resulting mistrust would keep the loyalty oath controversy boiling for more than two years, during which Berkeley's reputation as an academic sanctuary was immeasurably damaged.

The question of whether to sign the oath split the faculty. A majority opposed the oath but chose to sign nevertheless, especially after the regents decreed that those who refused would face dismissal. But for many faculty members, the significance of the oath went much deeper. Even among those who signed, the experience of being forced to affirm one's political loyalty was so repugnant that the real issue became whether to stay at Berkeley at all. In that conundrum would lie the deeper injury the controversy caused the university in general and the Rad Lab in particular.

Lawrence's mentor Jack Neylan was also damaged by the controversy. Neylan originally opposed Sproul's version of the oath, on the curious grounds that it would simply encourage its targets—the radicals assumed to be swarming over the university—to commit perjury. "I was convinced that a Communist would swear to anything," he said later. Soon, however, he seized the reins of the affair as leader of the regents bloc pressing for dismissal of nonsigners. As his position became more controversial, Neylan became ever more obstinate. While Sproul may be judged as having started the uproar, it was Neylan who carried it to its institutionally destructive conclusion. On campus, he would be seen as the outstanding villain of the affair, and not without reason: at his behest, the regents would fire thirty-one nonsigners in 1950. Two years later, the California Supreme Court ordered them all reinstated. One of them, David Saxon, an MIT-trained physicist, would become president of the University of California in 1975.

Lawrence's friendship with Neylan desensitized him to the moral quandary faced by members of the Rad Lab, especially its European-born scientists. Even the most ardent anticommunists among them regarded the oath as an uncomfortable reminder of the impositions on academic freedom they had suffered in their homelands. To Ernest, this attitude was incomprehensible; he dismissed their objections as little more than "byzantine quibbles," recalled Emilio Segrè. Moreover, he took a firm stand against the holdouts, as in the case of Gian-Carlo Wick, an outstanding Rad Lab theoretician. Learning that Wick had refused to sign, Lawrence summoned him to his office for a bitter interview. Unless Wick changed his mind, Ernest declared, he could "get the hell out of the Rad Lab, as far as I'm concerned." When Wick stood his ground, Ernest demanded his security pass.

Luis Alvarez took it upon himself to step into the fray. Alvarez had been willing to sign the oath, but he understood that Lawrence had wildly exceeded his authority; at the time of the Wick confrontation, the regents had still not decided whether to dismiss nonsigners. As Alvarez recounted the episode, he hastened to Lawrence's office to advise him that as long as Wick remained in good standing as a faculty member, Ernest had no authority to fire him. "Ernest grunted and groaned and made a lot of noises for a while," Alvarez recalled, "and eventually calmed down and agreed that I was right." Alvarez visited Wick and asked him to forget about the whole encounter, explaining, "Ernest sometimes acts emotionally." Wick got his pass back, but he could not easily overlook the affront to his intellectual independence. A few months later, he quit Berkeley for a post at Pittsburgh's Carnegie Institute of Technology.

That was the beginning of an outflow of talent from the Rad Lab. Among the others to go was the genial Berlin native Pief Panofsky, a brilliant thirty-year-old particle physicist. Gifted with a quick theoretical mind and an exceptional talent for practical experimentation, Panofsky had been personally recruited from the Manhattan Project by Alvarez, who considered him "my secret weapon at Los Alamos and at Berkeley." Panofsky, who had risen precociously in his two years at Berkeley to the rank of associate professor, detested the very idea of the oath, but signed—"reluctantly," as he put it later. He considered himself a hardened veteran of security theater; scientists like himself who held security clearances during the war, he observed, were inured to "the kind of irrationality and lack of privacy inherent in personnel security measures." But once it became clear that the regents would enforce the oath with the threat of dismissal, he decided that continuing to serve at Berkeley was impossible. Armed with job offers from several elite universities, he informed Lawrence of his decision to resign.

"Don't do anything until you hear the regents' side," Ernest told him, and scheduled a meeting for the physicist with Neylan. Ernest and Pief motored together over the Bay Bridge and down the peninsula to Neylan's estate in the woodsy enclave of Atherton. There Neylan asked Panofsky superciliously, "Young man, what is bothering you?" Panofsky replied that he was disturbed by the regents' intolerance. "Now, listen, my boy," Neylan rejoined, launching into a two-hour monologue about the history of the oath, the duplicity of Robert Sproul, and the faculty's disrespect for the board of regents. On the return drive, Panofsky told Lawrence that he had not changed his mind. A few weeks later, he accepted a job at Stanford, where he served until the end of his life in 2007.

The exodus continued, fueled by growing dismay in the Rad Lab at Lawrence's refusal to oppose the oath. The entire episode underscored the fruitlessness of his hostility to politics in the lab, for it was now plain that there never was a way to keep the political world excluded; Lawrence's maxim that political discussion was irrelevant to the science performed within the lab's walls sufficed only to widen the political rift on the staff. Berkeley's physicists, whether in the department or the Rad Lab, ended up divided into two camps. On one side were Lawrence, Alvarez, and a few other supporters of the oath; on the other side were several equally distinguished scientists. Those who spoke out vigorously against the oath eventually felt themselves becoming marginalized: Jack Steinberger, a nonsigning Rad Lab staff scientist, bristled at being "lectured by Alvarez on the evils of the fifth column of communist 'sympathizers,' in whose traitorous ranks he probably included me." Alvarez's caution to Lawrence about his treatment of Wick had yielded to his intense loyalty to his boss; he barred Steinberger from access to the 184-inch cyclotron, which Steinberger needed to complete an important experiment. Soon after that, Steinberger was turned down for a faculty post for which he had been nominated by Segrè and McMillan; finally, he found a note on his desk stating that since he had refused to sign the oath, he was no longer welcome at the lab and must leave the premises before nightfall.

By the time the controversy ended, six physicists had departed the lab, including all four of its theorists. "For theory, it was a body blow," Segrè reflected. What remained behind was a pervasive dismay that the leaders of the lab, who were among Berkeley's most eminent figures—among them Lawrence, Alvarez, Seaborg, McMillan, and Segrè—had kept aloof throughout the controversy despite repeated efforts by their colleagues to enlist them to fight the oath.

Lawrence was not the only professor whose feelings about the oath as an issue of academic freedom were inextricably bound up with questions of loyalty to his faculty colleagues and his friends on the board of regents. But he was certainly among the most prominent, and his failure to speak out weighed the more heavily for that. Wrote David P. Gardner, the affair's first historian: "One can only speculate if the Regents would have dismissed, for not signing, men such as these on whose reputations the University's scientific fame in substantial part lay."

Some of them sought to rationalize their behavior after the fact, as though fearful that history would judge them for their silence. Segrè, several of whose closest professional friends lost or relinquished their jobs over the matter, deprecated the oath as "meaningless," one of a sequence of "transient lunacies" he was willing to accept with a mental reservation that made it a nullity. "I calculated that I had sworn my allegiance to king, Mussolini, party, constitutions, and institutions at least fifteen times." What could one more oath matter?

Glenn Seaborg, whose work with plutonium was still conducted under strict government security, was more candid about his capitulation. "While I thought that the oath was an extremely unwise policy, the extent of my protest against it was to sign it on the last possible day," he explained in his memoirs. "I saw nothing to be gained by refusing; getting fired wouldn't make it go away. There was also nothing to be gained by alienating Ernest Lawrence, who had been turning increasingly to the right after the war. I believed in saving my political capital for more productive fights, such as working quietly to get rid of some of the unneeded secrecy rules."

Lawrence left no public statement about his views of the affair. These can be gleaned only from the recollections of his associates, who uniformly described him as insistently supporting Neylan's position. "Ernest felt emotionally coupled to this situation through his friendship with Neylan," observed Alvarez.

Edward Teller detected the same emotional bond, to his intense distaste. When the oath controversy erupted, Teller, a brilliant theoretician, had just agreed to leave the University of Chicago for a faculty post at UCLA. He thought Lawrence's endorsement of a policy that deprived his lab of its entire theoretical physics staff disgraceful. Discovering that Neylan was trumpeting his UCLA appointment as proof that the loyalty oath did not hamper the recruitment of distinguished scientists to the university, he furiously withdrew his acceptance. Hoping to soften the blow by delivering his decision in person, Teller received a sympathetic hearing from Sproul and a few other administrators. "There was one exception," he wrote a friend: "Ernest Orlando Lawrence. Since the days of the Nazis, I have seen no such thing. I had talked sufficiently gently and generally so that Lawrence did not attack me personally. But he did use threats and he was quite unwilling to listen to any point of view except to the one of Nylan [*sic*]. I felt somewhat sick when I left his office." As harsh as his feelings were, however, Teller's split with Ernest would not last long. It would be the Super that brought them back together.

The loyalty oath affair initiated a subtle transformation in the Rad Lab's reputation as a haven for pure science. Instead, it began to seem a place where one's views on the fraught politics of national security and weapons development loomed over one's career prospects. Politics had come through the door after all, more or less at Lawrence's invitation. "Outstanding people left because the atmosphere at the Radiation Laboratory . . . did not make people who dissented feel they were welcome," observed David Saxon, the future university president. He believed the Rad Lab was destined to face inevitable decline as physics progressed and competitive laboratories emerged—many of them built on the model Lawrence had pioneered. "But I think its decline was accelerated and amplified because of the loyalty oath."

Chapter Seventeen

The Shadow of the Super

On September 23, 1949, a news flash turned up the political heat in the debate over nuclear policy. Ernest Lawrence spotted the startling headlines from his car when he pulled up at a corner newsstand in Merced, on his way to a business meeting at Yosemite National Park. President Truman had announced that an "atomic explosion" had been detected in the Soviet Union. His words were carefully chosen, but the implication was clear: America's monopoly over the atomic bomb was over. It had taken Joseph Stalin's physicists just four years to reach nuclear parity with the United States—about the time frame predicted by Robert Oppenheimer.

Back on the Berkeley campus, the news about Joe-1, as the blast was dubbed in the US, electrified Luis Alvarez. To Alvarez, who was chafing in his peacetime harness after spending four postwar years doing basic research, the Soviet bomb represented not merely a crisis but also an opportunity. As far as he knew, the American program for the Super, the thermonuclear bomb, had stagnated. If Soviets were prepared to move on from the atomic to the Super on their own, they might well beat the Americans to the prize. The big accelerators being installed at the Rad Lab with millions of government funds might as well have been lying fallow for all the good they were doing for national security; the scale of the Super project, however, was perfectly suited to the resources and ambitions of Big Science. The next day, he bearded Lawrence in his office with the words "We have to do something about this." Ernest required no further prompting. With Alvarez standing by, he put through a call to Edward Teller, who then was still at Los Alamos weighing UCLA's offer to join its physics faculty.

At Teller's invitation, Lawrence and Alvarez flew into Albuquerque, landing before dawn and arriving in Los Alamos at ten in the morning. Their conversations with Teller validated Alvarez's impressions about the lack of progress on the Super. The program, Teller told them, "had essentially not been of any magnitude worthy of the name." To hear Teller talk, it was only because of his personal efforts that there was any program at all.

Teller, irrepressibly voluble on his favorite subject, joined them for the drive back to Albuquerque, prattling all the way about the essential requirements of a program to build the Super. In simple terms, the bomb worked by unleashing the energy created by the fusion of light isotopes of hydrogen, but it required an enormous shot of energy to get the reaction started—perhaps a small conventional atomic explosion. The technological issues, Teller acknowledged, were far from being solved; indeed, it was unclear they even could be solved. But he made one compelling point that showed Lawrence and Alvarez how they might restore the program to Washington's front burner. Teller believed a promising option to fuel a Super bomb would be to use tritium, a superheavy isotope of hydrogen. (The tritium nucleus has one proton and two neutrons, one more neutron than deuterium, hydrogen's more familiar heavy isotope.) Tritium could be created by bombarding deuterium with copious neutrons in a heavy-water reactor. But government-sponsored reactor development had slowed nearly to a halt since the end of the war. There lay the key, Lawrence and Alvarez recognized: push the Atomic Energy Commission to sponsor production-scale heavy-water reactors and move to a tritium-fueled Super bomb from there.

Their mission now clarified, the two Berkeley scientists emplaned for Washington at three-thirty that morning. After landing, they began making the rounds of congressional and Atomic Energy Commission offices, concluding their first day in town at dinner with Alfred and Manette

Loomis, newlyweds who had extricated themselves from their previous marriages via divorces (Manette's in Nevada). Loomis eagerly approved of the Super project. Lawrence and Alvarez, egged on by their mutual mentor, continued their campaign the next morning.

On the whole, the reception was positive. At the AEC, they met first with Commissioner Lewis Strauss and research director Kenneth Pitzer, a Berkeley physicist who had taken a temporary leave for government service. Both were enthusiastic, as was Robert LeBaron, the AEC's Pentagon liaison officer, whose breakfast meeting with the visitors was interrupted by a telegram from Berkeley bearing the news that Ernest had just become a father for the sixth time (a daughter, Susan).

Then came the most important meeting of their trip: lunch with Senator Brien McMahon and other members of the Congressional Joint Atomic Energy Committee. Lawrence and Alvarez filled the lawmakers' heads with a calculatedly sinister vision of Russia's quest to rule the world by thermonuclear terror. They "even went so far as to say that they fear Russia may be ahead of us in the competition," scribbled the committee's hawkish executive director, William Borden, who was on hand to take notes. "They declared that for the first time in their experience, they are actually fearful of America's losing a war."

The two physicists received less hospitable treatment when they circled back to the AEC to see David Lilienthal, its chairman. Lilienthal had been brooding all day over a pending decision by President Truman about expanding the nation's nuclear arsenal, a $319 million proposition pushed assiduously by the Pentagon. "A whopping big [program]," Lilienthal grumbled to his diary. "More and better bombs . . . We keep saying, 'We have no other course'; what we should say is 'We are not bright enough to see any other course.' " His mood only darkened with the arrival of Lawrence and Alvarez, whose pitch left him so disaffected that he swiveled his chair around to turn his back on them. "The day has been filled . . . with talk about supers, single weapons capable of desolating a vast area," Lilienthal recorded sourly, his earlier admiration for Lawrence shattered by the latter's enthusiasm for this loathsome new weapon. "Ernest Lawrence and Luis Alvarez in here drooling over same. Is this all we have to offer?"

Lawrence and Alvarez were equally disgruntled by their encounter with the AEC chairman. Alvarez declared himself to be "shocked about his behavior . . . He turned his chair around and looked out the window and indicated that he did not want to even discuss the matter. He did not like the idea of thermonuclear weapons, and we could hardly get into conversation with him on the subject." This was their first hint of how the debate over the Super would divide the AEC, with Lilienthal and Lewis Strauss, the agency's leading enthusiast for the Super, glaring at each other across a thermonuclear policy gulf.

The Washington visit launched Lawrence into the role of the hydrogen bomb's most prominent and credible promoter. His efforts on behalf of the Super would bring Berkeley vast new patronage from the government, including the funding of an entire new weapons lab that would double the size of the nation's nuclear program and give the US technological momentum for decades to come. But his efforts would link his name forever with the issue of nuclear proliferation and cast a shadow over his scientific legacy.

Lawrence's campaign placed his friends and colleagues in a quandary. Most felt that the development of a US heavy-water reactor was a perfectly sound idea, given its potential to advance nuclear knowledge. But many viewed more equivocally the Super bomb program in which the reactor program was embedded. The spectacle of Lawrence's lending his highly developed talent for persuasion to a project of such dubious morality was more painful still.

Contemplating the grand tour of Lawrence and Alvarez and the resurgent electioneering of Teller for the Super, Oppenheimer aired his doubts to Conant in a letter affectionately addressed to "Uncle Jim." (Conant, then fifty-six, was nine years Oppie's senior.) "A very great change has taken place in the climate of opinion" in Washington, he advised Conant. "Two experienced promoters have been at work, i.e., Ernest Lawrence and Edward Teller. The project has long been dear to Teller's heart; and Ernest has convinced himself that we must learn from Operation Joe that the Russians will soon do the super, and that we had better beat them to it." As for the plan to build neutron-producing heavy-water reactors, Oppenheimer favored it. "For a variety of reasons, I think we must say amen," since the reactors would serve purposes other than the Super. But he doubted that science could overcome the technical challenges of building the Super, and he was disconcerted by the underlying politics of the debate: "I am not sure the miserable thing will work, nor that it can be gotten to a target except by ox cart . . . What does worry me is that this thing appears to have caught the imagination, both of the congressional and of military people, as *the answer* to the problem posed by the Russian advance . . . That we become committed to it as the way to save the country and the peace appears to me full of dangers."

After Washington, Lawrence and Alvarez planned to fly to Ottawa for a look at the Canadian government's heavy-water reactor in nearby Chalk River, a design they thought might be adapted for tritium production. Learning that seats to Ottawa were unavailable, they instead invited themselves to visit I. I. Rabi, a member of the AEC's General Advisory Committee, at Columbia University. Rabi, who believed they were making a social call, was surprised to hear them launch into a promotional spiel for a huge program to develop a hydrogen bomb.

As Alvarez recollected the meeting, Rabi praised their proposal as just what the doctor ordered. "Rabi was worried about the Russian explosion, too, and liked our plans," he wrote later. " 'It's certainly good to see the first team back in,' he told us. 'You fellows have been playing with your cyclotrons for four years. It's time you got back to work.' "

This would have been a significant endorsement from an influential scientist if it were true, but Rabi's recollection of the meeting was very different. He agreed that the US program should be ramped up to recover the American lead in nuclear weaponry, but he thought his visitors' fancies had far outrun the practicality of the Super. "They were extremely optimistic," he recounted later. "They are both very optimistic gentlemen . . . Dr. Teller gave them a very optimistic estimate about the thing and about the kind of special materials which would be required. So they were all keyed up to go bang into it."

Rabi did his best to bring them down to earth. "I generally find myself when I talk with these two gentlemen in a very uncomfortable position . . . Those fellows are so enthusiastic that I have to be a conservative. So it always puts me in an odd position to say, 'Now, now. There, there,' and that sort of thing."

Ernest's new cause revived his old indefatigable self. He sent Alvarez home to Berkeley to assemble a reactor design team and spread the word that the Radiation Laboratory had a new mission. Meanwhile, he returned to Washington to meet with Kenneth Nichols, Groves's former military adjutant and now the head of weapons development at the Pentagon, to urge that the Joint Chiefs of Staff designate the hydrogen bomb as a military requirement, which would help to obtain funding on Capitol Hill. Nichols briefed the Joint Chiefs the next day, asserting that the entire scientific community endorsed the project. This prompted the chairman, World War II hero General Omar Bradley, to declare that "if it can be done . . . it would be intolerable to have us sit on our butts."

Teller kept up a similar pace, albeit with less success. In October he visited in Chicago with Fermi, who had just flown in from Italy. Plainly skeptical but unwilling to engage in a colloquy with a wheedling Teller, Fermi pleaded that he was too tired from the flight to give him a hearing, much less a commitment. Teller moved on to Ithaca, New York, to enlist Cornell's Hans Bethe in the Super project, and afterward reported to Alvarez that he "felt he could count on Bethe."

This was Teller's self-delusion talking, not his judgment. Bethe had no intention of signing on until he had weighed the undertaking carefully. But he was predisposed against it. Teller's talk filled him with "the greatest misgivings," he would recall. "I never could understand how anyone could feel any enthusiasm for going ahead." It did not escape his notice that Teller, Lawrence, and Alvarez had papered over any moral qualms they might have had by considering the Super largely as a scientific challenge—"namely, how to overcome the technical obstacles." This was not sufficiently enticing for Bethe. He brought his misgivings to Oppenheimer, who confided that he shared them and showed Bethe his "Uncle Jim" letter, explaining that Conant also opposed the Super. Bethe brooded with another friend, physicist Victor Weisskopf, over the inevitable outcome of a thermonuclear war. "We both had to agree that after such a war, even if we were to win it, the world would not be . . . like the world we want to preserve." He turned Teller down.

But to Lawrence, seized with unrestrained enthusiasm, the project's future seemed assured. At home in Berkeley, he spent a weekend scouting locations for a heavy-water reactor, convinced that he was on the verge of securing the project as an adjunct to the Rad Lab. His preferred site was Suisun Bay, an estuary off San Pablo Bay north of San Francisco, which he judged to be distant enough from densely inhabited areas to be safe, yet convenient enough to Berkeley for frequent oversight visits. In his optimistic mood, he guaranteed Alvarez the appointment as director of the Suisun Bay project. Rather more cautious, Alvarez confided to his diary: "I am going on almost full-time as director of a nonexistent laboratory on an unauthorized program."

Indeed, it was not long before the Super's apparent lack of progress in Washington had Lawrence and Alvarez feeling uneasy. When they had left the capital in September 1949, things seemed to be moving at jet speed, but it had been weeks since they had heard an encouraging word. "There seemed to be a lack of enthusiasm suddenly pervading the scene," Alvarez recalled, "and we were worried about . . . whether it was a change in climate in Washington." They consulted Jack Neylan, who phoned his close friend Lewis Strauss to take the temperature of the capital. Neylan reported back that things were percolating along, albeit under the surface. Congress had signaled its enthusiasm for an expanded AEC program by allocating the agency a healthy budget for research, he explained. "Keep your shirts on, boys," Neylan said. "It's going to be all right."

But the scientists could not shed the suspicion that someone was throwing sand in the gears, and they thought they knew who: Oppenheimer. Lawrence quietly dispatched Robert Serber to Princeton to sound out Oppie. Serber had been Oppenheimer's scientific deputy at Los Alamos and author of *The Los Alamos Primer*, a thin volume handed to every scientist arriving at the bomb lab to bring him up to speed on the theoretical physics underlying the bomb design. He arrived in Princeton without a firm idea of the purpose of his trip, for he had been so deeply immured in the Rad Lab echo chamber that he assumed East Coast physicists were just as enthusiastic about the Super as Lawrence, Alvarez, and Teller. Oppenheimer promptly set him straight. At Berkeley, skepticism about the Super, much less Oppenheimer's conviction that it should not be pursued at all, "would have been unthinkable." But as Serber now discovered, "The East was evidently a completely different world from California."

The next day, Serber and Oppie boarded a train to Washington for a GAC meeting, at which Oppenheimer would preside and Serber would give a presentation on the proposed heavy-water reactor. In effect, the agenda concerned the Super's fate. When they arrived, they found Alvarez haunting the lobby, prepared to relay the GAC's decision to Lawrence as soon as it was made. Barred from the closed meeting, he had to cool his heels for hours, watching "my friends and any number of famous military men go upstairs to a closed GAC session." He felt almost as though his own destiny was being decided.

The GAC spent the morning probing the views of the military. As Lilienthal recorded the scene, the thought of the Super made the Pentagon officers' "eyes light up." But they also were aware that deploying a weapon of such enormous power would not be a military action; the Super's value would be strictly "psychological," General Bradley declared. Oppenheimer ran the meeting in his typically neutral manner, not taking sides during the discussion, but the committee seemed to see things his way. As he described the session later, "quite strongly negative things on moral grounds were being said."

At lunchtime, Oppie led Serber and Alvarez to a small, dark café a short stroll from the agency's offices. For the first time, he informed Alvarez directly of his opposition to the project. "The main reason he gave," Alvarez would recount, "was that if we built a hydrogen bomb, the Russians would build a hydrogen bomb, whereas if we didn't build a hydrogen bomb, then the Russians wouldn't build a hydrogen bomb." Alvarez tried desultorily to argue that Americans would find it hard to see the logic of Oppie's position. Later, testifying at Oppenheimer's security trial, he delivered a blunter assessment: "Pretty foggy thinking," he called it.

But Oppenheimer's position accurately reflected the GAC consensus. The report issued by the committee following the October meeting was as uncompromising a brief against the Super as any that would emerge from a government panel. Oppenheimer's preamble to the document delivered the gist: "No member of the Committee was willing to endorse this proposal," he wrote. The hydrogen bomb's unlimited explosive power was its most horrific characteristic: once the technical problem of initiating the fusion reaction was solved, ever-larger explosions could be arranged simply by adding more deuterium, which was easily available and cheap. "It is clear that the use of this weapon would bring about the destruction of innumerable human lives; it is not a weapon which can be used exclusively for the destruction of material installations of military or semimilitary purposes. Its use therefore carries much further than the atomic bomb itself the policy of exterminating civilian populations."

Oppenheimer's theme was picked up in two concurring opinions issued with the report. A majority bloc composed of Conant, Cyril Smith, Lee DuBridge, and the industrial scientists Hartley Rowe and Oliver E. Buckley warned that deploying the weapon "would involve a decision to slaughter a vast number of civilians . . . A super bomb might become a weapon of genocide . . . In determining not to proceed to develop the super bomb, we see a unique opportunity of providing by example some limitations on the totality of war and thus of limiting the fear and arousing the hopes of mankind." In a separate statement, Fermi and Rabi repeated the majority's premonitory tone and language: "Necessarily such a weapon . . . cannot be confined to a military objective but becomes a weapon which in practical effect is almost one of genocide." They further proposed that President Truman "tell the American public, and the world, that we think it wrong on fundamental ethical principles to initiate a program of development of such a weapon. At the same time it would be appropriate to invite the nations of the world to join us in a solemn pledge not to proceed in the development or construction of weapons of this category."

It has been suggested that the members' emphasis on the hideous consequences of the Super had the opposite effect than they intended, for it sounded less like an argument for staying America's hand than a warning of the consequences of letting the Russians get the Super first. It is true that a lively fear of Russian intentions and capabilities had settled upon Washington, fueling the expectation that war with the Soviet Union was inevitable. By allowing their fearsome imagery to overwhelm their moral message, the GAC members may have generated more interest, not less, in pressing the program ahead.

Yet this view fails to reckon with the persuasive powers of Ernest Lawrence and Edward Teller, which played the crucial role in creating momentum for the H-bomb. Sensing that the GAC was poised to reject the Super, they worked to shore up support for the weapon well before the committee issued its report. One day after the GAC meeting, a disconsolate Lilienthal noted in his diary that briefings by members of the Joint Atomic Energy Committee who had been visiting Berkeley were "rather awful: the visiting firemen saw a group of scientists who can only be described as drooling with the prospect and 'bloodthirsty.' " Ernest Lawrence stood out as "quite bad," he reported, describing Lawrence's mind-set as: "There's nothing to think over; this calls for 'the spirit of Groves.' " Lilienthal concluded, "Things are certainly coming to the showdown stage and fast."

He was correct. The GAC report sharpened the divide between pro- and anti-Super camps on the AEC. Declaring themselves firmly opposed to the Super were Lilienthal, the Republican oil baron Sumner Pike, and Princeton physicist Henry DeWolf Smyth. On the other side, pressing for a development program at top speed, were Lewis Strauss and Gordon Dean, a veteran of the Roosevelt Justice Department now teaching law in California.

Strauss was the dominant personality of the two. A self-made banking and investment mogul who pronounced his surname "Straws," he bore down like a force of nature when in the grip of a personal obsession—and promoting the Super had become the defining obsession of his life. The matter brought out the most unpleasant aspects of a personality that was not endearing even at its best; Strauss would be described by the political observers Joseph and Stewart Alsop as a man with "a desperate need to condescend, to be always agreed with, to be endlessly approved and admired, to dominate and play the great man." They quoted one of his fellow AEC commissioners (anonymously) as observing, "If you disagree with Lewis about anything, he assumes you're just a fool at first. But if you go on disagreeing with him, he concludes you must be a traitor."

That characterization describes perfectly the course of the relationship between Strauss and Robert Oppenheimer. As Oppenheimer's position on the Super evolved from skepticism to outright opposition, Strauss's opinion of Oppie darkened steadily, like a cloud heralding a thunderstorm. The GAC report was a milestone in that process. It put Strauss in "an absolute dither," Rabi would recall: Strauss "went around and talked to newspapermen all over, and to the House and Senate, and whatnot." Meanwhile Lawrence, Teller, and Alvarez pitched in with appeals to the Pentagon and influential members of Congress. Oppenheimer stood fast as a counterweight to their campaign, but not an especially effective one, for his abstracted manner was not up to the task of communicating his over-optimistic argument that the Soviets would respond to an American offer to renounce the weapon by renouncing it themselves. Truman's secretary of state, Dean Acheson, declared himself flummoxed by Oppenheimer's words. "I listened as carefully as I knew how, but I don't understand what 'Oppie' is trying to say," he told his aide Gordon Arneson. "How can you persuade a paranoid adversary to disarm 'by example'?"

As the debate raged on, Truman made a show of keeping an open mind. On November 18 he appointed Lilienthal, Acheson, and Secretary of Defense Louis Johnson as a special committee to weigh the political, military, and technical aspects of the Super. Lilienthal, who had informed Truman of his intention to retire as AEC chairman early in the new year, now realized that his final weeks of service would be consumed in a fruitless battle to persuade his committee colleagues—and the president—of the folly of the hydrogen bomb. With Congress clamoring for approval, Truman's decision seemed preordained.

On January 31 the special committee gave Truman its recommendation that the United States proceed with the Super. The White House meeting that formally inaugurated the thermonuclear age took all of seven minutes, and it only lasted that long because Lilienthal asked for the time to express his dissenting view. He got out only a few words before the president cut him short. "What the hell are we waiting for?" Truman barked. "Let's get on with it."

Later the same day, Truman announced in a nationwide radio address that he had instructed the AEC to "continue its work on all forms of atomic weapons, including the so-called hydrogen or superbomb." That night, Lilienthal confided to his diary: "There is nothing but pain for the decision made today." The small personal satisfaction he felt came from his having passed one of the most grueling tests of his career by showing the courage "to 'stand up in meeting' and say 'No' to a steamroller . . . Whether time proves me right and the Pres.'s two Secretaries, and the Hill boys, and the E. O. Lawrences wrong, I suppose no one will ever know."

• • •

Truman's announcement came on Lewis Strauss's fifty-fourth birthday. The news transformed the cocktail reception he had scheduled for himself at a Washington hotel into a triumphal celebration of his quest for the Super. Oppenheimer, who had earlier tendered his acceptance, felt obliged to show up. Despondent and morose, he held himself apart from the festivities, seated alone with his back to the room. Not even when Strauss came by to introduce his daughter and her husband did Oppie turn around, instead merely offering a hand over his shoulder in silent greeting. Lewis Strauss's sensitivity to slights, real and perceived, was exquisite. This was one affront he would not forget.

Anti-Super physicists were crestfallen by the president's announcement. "I never forgave Truman for buckling under pressure," Rabi recalled.

But for Ernest Lawrence, opportunity beckoned.

Chapter Eighteen

Livermore

On a midsummer's day in 1950, Ernest Lawrence and Luis Alvarez stood on the tarmac of an abandoned naval air station on the outskirts of a hot, dusty farm town. The sole claim to fame of Livermore, California, was as the former home of a onetime heavyweight champ who had moved there in the 1920s. An arch spanning the main drag marked the distinction: *Home of Maxie Baer.*

They strolled about the base, taking in its cracked and overgrown runways, its derelict barracks, its empty gymnasium, and a drained swimming pool filled with debris.

"Well, Luie," Lawrence said, "this is it."

He sounded like a Moses sizing up the promised land, which was not too far from the truth. For months Ernest had been searching for a place to locate his newest project, one so big and ambitious that it would not fit on the Berkeley campus—not even on the hillside overlooking San Francisco Bay where he had built the 184-inch cyclotron. Now he had found it. Ernest Lawrence would bring the town of Livermore far more worldwide fame than Maxie Baer ever did. A hub of hydrogen bomb research, the Livermore National Laboratory would be Ernest Lawrence's final monument. To this day, it remains one of the United States government's largest and most secretive research institutions.

• • •

The road to Livermore started with a disappointment. Lawrence's campaign to add a nuclear reactor to Berkeley's arsenal of atom-smashing technologies dated back to his wartime effort to wrest Enrico Fermi's nuclear pile away from Arthur Compton and the University of Chicago. It would be Fermi who delivered the final blow to his dream, at the General Advisory Committee's meeting in October 1949—the same meeting that had produced the negative report on the Super. The GAC concurred that building a heavy-water reactor was a promising research project, but Fermi vetoed Berkeley as its home. Lawrence and the Rad Lab, he observed waspishly, had "absolutely no experience in reactor design or operation." Why should the Rad Lab get a reactor now when there were more experienced labs all over the country? Robert Serber, who had carried Berkeley's proposal to the GAC, had to acknowledge that Fermi had posed "the obvious question," and that it was unanswerable.

Exiting the meeting room, Serber delivered the bad news to Luis Alvarez, pacing the lobby downstairs. Alvarez had hoped to bring home to Ernest two pieces of good news: the GAC's approval of the reactor, and its assent to a crash program for the Super. But after Serber's report and his lunch with Oppenheimer, he knew he would be returning to Berkeley empty-handed. Without waiting for the GAC meeting to end, he packed up and left the capital. "I went back to doing physics," he would write. "But not for long." The GAC had closed the door to Berkeley's reactor. But another door was about to open wide.

After the GAC dashed his dream of a reactor, Ernest mobilized the Rad Lab to find another way of generating a heavy neutron flux. A new impetus for the project had emerged: fears of a shortage of uranium ore, the raw material for the production of plutonium bomb cores. The nation's entire nuclear arsenal was dependent on only two sources: a single rapidly depleting mine in the Belgian Congo and another in Canada, near the Arctic Circle.

Lawrence reasoned, however, that although domestic sources of uranium ore were practically nonexistent, uranium was plentiful in another form: tons of uranium tailings piled up as waste at Oak Ridge and Hanford. To Ernest, this detritus of U-235 separation and plutonium manufacture was untapped treasure; only neutrons were needed to unlock its value. "If you have neutrons, you can make any commodity," including plutonium, he told an enthralled Joint Atomic Energy Committee. Bombarding the waste uranium with neutrons would allow the United States to "break the bottleneck of this raw material problem," he promised. "We can make all the atomic bombs anyone could want. We could step up the atomic bomb production tenfold . . . We have got thousands of tons of [U-]238, and we can convert that into thousands of tons of plutonium."

But where to get the neutrons? The Rad Lab's idea was to develop a new type of accelerator. This involved a return to Ernest's old "brute force" method: when all else failed, dial up the energy, step up the current, and wait for something to happen. In this case, he reckoned, one could obtain the required profusion of neutrons by bombarding an appropriate target with a high-energy deuteron beam. The neutrons in turn would act upon a secondary target; obtaining the end product one sought was merely a matter of selecting the right secondary target. If tritium was the sought-after product, for example, you used lithium-6; if it was plutonium, the raw material would be uranium—namely, that waste U-238.

Lawrence settled on a linear accelerator as the most efficient neutron producer for his purpose. Essentially, he was uncoiling the cyclotron that years earlier he had conceived by twisting Rolf Wideröe's linear accelerator into a spiral. On a Saturday morning in late 1949, he called an urgent staff meeting to outline the concept, unnerving not a few staff members roused from their slumbers by the unexpected weekend alert. Staff physicist Don Gow, summoned groggily to the telephone with the message "Ernest wants to see you," thought instantly, "My God, what have I done?" Instead, he was being invited to witness Ernest's unveiling of his most audacious project yet.

Ernest conceived the accelerator on a stupendous scale, its prototype a tube 60 feet in diameter and 87 feet long, the first step in construction of a prodigious machine he envisioned as 1,500 feet in length. He estimated the power demand of the full-sized accelerator at 150,000 kilowatts, enough to light a city of three hundred thousand residents. "We just sat there with our mouths hanging open," Gow recalled. "Had he really gone mad at last? But he was serious."

Lawrence had returned to the wartime mind-set of applying unlimited resources to a problem. That approach had raised Oak Ridge from an unpopulated valley in remote Tennessee; when national security hung in the balance, as it did now, who would begrudge a few hundred thousand kilowatts? His machine would be capable of producing a half kilogram of plutonium—just over a pound—a day. "This isn't a large amount of power for the product one is getting," he would assure his rapt congressional audience.

The Lawrence magic was alive again. The more his staff talked out the concept that Saturday morning, the more it coalesced into a plausible scheme. Certainly if the theory of heavy neutron bombardment was sound, then the linear accelerator was the right type of machine. By the end of the meeting, design tasks were already being parceled out. Without fanfare, work on the Bevatron was halted and ramped up instead on the Materials Testing Accelerator, or MTA, as the new machine was inelegantly dubbed. Scientists who had put in their time with Ernest thought nothing of the hairpin turn in the lab's priorities. "The first priority was what he wanted done," shrugged Brobeck, laying aside his work on the Bevatron. The new machine, like so many of Ernest's ideas before it, was perched on the edge of technical plausibility, offering the prospect of limitless triumph or abject failure. The latter possibility was papered over, naturally, by the boss's vibrant optimism.

Five weeks after Truman's endorsement of the Super, with the nation committed to a hydrogen bomb program requiring immense quantities of fissionable material, the Atomic Energy Commission approved a $10 million grant for the prototype MTA, designated the Mark I. The full-size machine, Mark II, was costed out at $100 million. For that, the AEC gave only conditional approval, with the final nod to be based on the Mark I's performance.

It was enough for a start. The Mark I was what brought Lawrence and Alvarez to Livermore on their scouting visit, Ernest again embracing his wartime role as impresario of a partnership of government, academia, and industry. The University of California was loath to stand as the sole sponsor of what it viewed as a pure manufacturing scheme, but Jack Neylan, lubricating a project conceived by his protégé, persuaded his old friend Gwin Follis, the chairman of Standard Oil of California, to participate; the MTA would be managed by Follis's company through a newly organized subsidiary called California Research and Development Company.

Luis Alvarez, who was named the MTA's chief designer, accepted the job with a distinct sense of foreboding. He would be proven right. The MTA, he would recollect, "occupied most of my time for the next two years. It was not a happy time." Lawrence had built his career by overreaching and then making good on his promises. He was about to reach too far. On a strictly technical level, the MTA was a "tour de force," Alvarez acknowledged. "Ernest loved technical extrapolation, and we had fun drawing such an accelerator." But as for building the beastly thing, "we were working in absolutely unexplored territory." The sixty-foot main cylinder would contain the largest vacuum ever produced thus far, an enormous engineering challenge in itself. The giant metal "drift tubes" suspended inside the cylinder, which the deuteron beam would traverse, each weighed several tons. They were connected to one another by ramshackle wooden bridges over which construction men, technicians, and scientists clambered nervously.

The biggest challenge was posed by the immense energy coursing within the drift tubes. Stored electrical energy is discharged by sparking, a common bug in new accelerators that was customarily eliminated by polishing down any interior imperfections attracting the discharge. Inside the Mark I, these sparks packed the power of lightning bolts, ripping through the drift tubes and leaving behind what Pief Panofsky described as "spectacular stalagmites and stalactites of copper." Alvarez assumed personal responsibility for making sure the tanks were polished smooth. "Night after night for weeks on end, like a test pilot learning to fly a new plane," he recalled, "I sat at the control panel gradually running the drift tube voltage up to the point where the tank would spark." That point would be marked by a gigantic thunderclap, after which Alvarez would haul his gangly frame into the tube to strip away the stalagmites and stalactites and buff the surface smooth. Then he would crank up the voltage again until another sharp report summoned him back inside.

While the Mark I was still undergoing this shakedown procedure, Lawrence spent his time soothing official doubts about the MTA by citing the continuing threat of a shortage of fissile material. Well into 1951, this served as an effective all-purpose defense: asked by the Joint Atomic Energy Committee to predict when questions about the MTA's feasibility would be settled, he replied brusquely, "I am not waiting to see." Considering the urgency of finding a way "to free us from the raw material bottleneck," he said, "if we can do that for a few hundred million dollars, why, let's get going and do it."

Yet the bottleneck was already disappearing. In 1950, uranium production from mines in Colorado increased sharply, exceeding Canadian output for the first time. New ore deposits were discovered in western New Mexico, and procurement officials expressed high hopes for a scheme to extract uranium from phosphate processed by fertilizer manufacturers in Florida. The key to all this increased production turned out to be the magic of the free market: after the AEC announced a hike in the price it was willing to pay for ore, deposits that had seemed locked deep within Mother Earth suddenly started reaching the surface.

These developments eroded the rationale for the costly MTA. The United States was still dependent on South Africa and the Congo for most of its ore, but with the new sources coming on line, the panic that had driven approval of the Mark I only a year earlier dissipated. Lawrence suddenly found himself waging a fierce battle against rising complaints at the AEC about the MTA's high cost and indifferent performance. Interrogated at an AEC meeting by Commissioner Henry DeWolf Smyth, he repeated wanly, "MTA breaks the bottleneck in raw materials . . . It seems self-evident to me we should push the process development ahead just for this reason, even though we feel there is now only a remote possibility that our foreign supplies might be insufficient or cut off."

Smyth's query underscored that Ernest's government patrons were becoming less susceptible to his charm. In part this new skepticism reflected the superior technical knowledge of his official overseers, who now included trained physicists with knowledge of nuclear physics that equaled or even exceeded his own. Smyth was a former chairman of the Princeton physics department who had authored an official history of the bomb program released to the public following Hiroshima and Nagasaki. He was never going to be swayed by Ernest's enthusiasm and vague promises alone, as had Cottrell, Weaver, and Conant—scientists all, but relative strangers to the mysteries of the nucleus—much less Sproul, Crocker, or Neylan, to whom Lawrence's science was as indecipherable as a conjuror's repertoire.

Smyth was also unhappy with Lawrence's habit of proposing increasingly ambitious projects before his previous ones had borne fruit. So when Lawrence unveiled yet another audacious scheme in 1952—this one for a neutron-producing cyclotron code-named J-16 and bearing an astounding price tag of $30 million—Smyth was distinctly unmoved. "I don't understand clearly enough what it is you propose to do to see $30 million in the machine," he wrote Lawrence—a rare but telling rebuff to a man who had routinely emptied his backers' wallets through his sheer vision and optimism. Smyth observed that Congress had slashed the agency's research budget for 1953 from $43 million to $33 million, "as you no doubt have heard." Even if he were inclined to buy into Lawrence's vision, funds were scarce.

Lawrence's ability to maintain AEC backing for the MTA was fading fast. In July the agency voted to defer construction of the Mark II indefinitely. Standard Oil's California Research and Development subsidiary read the handwriting on the wall: the decision was tantamount to cancellation. The cost of the Mark I had already doubled from the original estimate of $10 million, and CRD's president, Fred Powell, was forced to question "the wisdom of continuing further research and development work." At Livermore, the project was hemorrhaging personnel. Some staff were laid off as the project approached what increasingly looked like a dead end, while others—mainly those whom Lawrence had reassigned from the Rad Lab—were returning to their former responsibilities at Berkeley. The project staggered on for eighteen months and was shuttered for good in December 1953.

"The prototype MTA attempted to carry a technology beyond its reasonable limits," Alvarez reflected. Lawrence delivered his own judgment nonverbally by becoming more detached from the project as its fortunes waned. As Lawrence machines went, this one was a fiasco, not even yielding substantial or unique knowledge to advance the science of accelerator design.

The MTA was a lesson in the management of Big Science in the new postwar landscape. Competition for the favors of academia, industry, and government was more intense, the qualifications of the judges higher, and the standards of success more exacting than before. Gone were the days when Ernest Lawrence's laboratory reigned as virtually the only bidder for millions of dollars in research support, his record and reputation sufficient to persuade his patrons to open their wallets; now there were rivals with records and reputations that matched his own. Nor could dispensers of government patronage merely weigh scientific projects against one another: the cost of high-energy physics had become so great that broader priorities were implicated. Officials and lawmakers now had to balance any spending on science against other burgeoning demands—for social programs, highways, school buildings, and other physical infrastructure. In the coming decades, these issues would only become more pressing.

Still, the MTA's failure marked only a temporary stumble in Lawrence's influence over domestic nuclear policy. The MTA might be dead, but the Super was still very much alive. Even before the first pieces of the Mark I were dismantled, Ernest was planning a new venture—one that would exploit the official mania for the hydrogen bomb to bring Livermore the prominence he originally expected to come from the MTA. Torrents of AEC cash would soon be on their way, transforming the old air station. The guiding principle was that a crash program to build the Super bomb required a Los Alamos of its own, and that place should be Livermore.

The idea that America required a second weapons laboratory to supplement Los Alamos originated with Edward Teller and was driven by two factors. One was the conviction of the stocky, beetle-browed Hungarian physicist that the lethargic pace of thermonuclear research had placed the United States at the mercy of the Soviet Union. "If the Russians demonstrate a Super before we possess one, our situation will be hopeless," he wrote in a frantic memo for a 1949 meeting on research priorities at Los Alamos.

The second factor was Teller's inability to get along with anyone who doubted either the need for a crash H-bomb program or the likelihood of its eventual success. Teller's obsessive personality created a managerial dilemma for Norris Bradbury, the able Berkeley-trained physicist who had succeeded Oppenheimer as Los Alamos director. Teller's intellect was indispensable at Los Alamos, but his presence was intolerable. The quandary deepened in mid-1950, when it became evident that the solution to a vexing technical issue in hydrogen bomb design rested on the collaborative work of Teller and the Polish mathematician Stanislaw Ulam.

Teller continually disturbed the peace at Los Alamos, where uncertainties about the lab's role in postwar research already had the staff on edge. His chief tactic was the threat of resignation—underscored now and then by an actual resignation. These outbursts forced Bradbury into a repetitive routine of pleading with Teller to stay and persuading him that he was loved. Meanwhile, Teller was constantly sharing his grievances with powerful friends and supporters in Washington, including Congressman Brien McMahon, chair of the Joint Atomic Energy Committee; William Borden, the committee's executive director and a fanatical hawk on the subject of the Super; and AEC commissioner Lewis Strauss, Washington's best-placed and most passionate supporter of the bomb program.

Notwithstanding Teller's grousing, Los Alamos actually was making significant progress on the Super. In the spring of 1951, the lab staged Project Greenhouse, a test of thermonuclear technology at the South Pacific coral atoll of Eniwetok. Greenhouse was a turning point in the American H-bomb program. Teller and AEC chairman Gordon Dean were on hand for the key trial on May 8, which produced an awe-inspiring fireball that melted a crater into the coral, obliterated the two-hundred-foot tower holding the device, and destroyed three hundred tons of equipment around it. The device was not a hydrogen bomb, exactly, but it was "the first thermonuclear test explosion on earth," in the words of one witness, the young Berkeley physicist Herbert York. The blast proved the feasibility of a controlled thermonuclear device so decisively that Teller claimed paternity immediately, wiring back to Los Alamos, "It's a boy."

But the success of Los Alamos seemed to make Teller even more insistent about a second laboratory. His attitude in turn intensified the debate over how to handle him. A few weeks before Greenhouse, Bradbury had tried to mollify the willful physicist by creating a semiautonomous thermonuclear division at Los Alamos, giving Teller authority over 25 scientists. Teller dismissed the offer out of hand, damning it as exactly the sort of half measure he had been fighting for years. He went over Bradbury's head to Gordon Dean to propose instead an independent new lab in Boulder, Colorado, where some 130 physicists would report to him alone.

Dean rejected this audacious plan, in part because he was unconvinced that a second lab was needed at all, much less on Teller's megalomaniacal scale. The AEC chairman could not imagine how a bifurcated weapons program would operate or how the reduction in responsibility for the

Super would affect morale on the New Mexico mesa. His most important question was whether the mercurial and messianic Teller could manage anybody. The list of prominent scientists who had refused to work with him was long and growing longer. A typical concern was voiced by Emilio Segrè, who had known Teller for decades and had turned aside his invitation to collaborate on the Super with the accurate if gracious judgment that he was a man "dominated by irresistible passions much stronger than even his powerful rational intellect."

Like Dean, Norris Bradbury could not conceive of Teller's helming a major lab of his own. Most division leaders at Los Alamos "wouldn't work with him," he recalled later. "I wouldn't put him in charge. Oppie hadn't put him in charge. Oppie knew him just as well as I did, perhaps better." Finally, tiring of his unending battles with Teller over research policy, in September 1951 Bradbury appointed Marshall G. Holloway, a Los Alamos veteran with a no-nonsense administrative style, as head of the thermonuclear division. Although Holloway's temperament was ideal for supervising a lengthy schedule of South Pacific tests stretching through the coming year, his appointment was also a calculated affront to Teller, who suspected Holloway of inadequate devotion to the Super. Informed by Bradbury that his role would be limited to "coordinating" with Holloway, an infuriated Teller again threatened to resign. His assistant, Frederic de Hoffmann, placed a panicky phone call to Gordon Dean to warn that Holloway's appointment was "like waving a red flag in front of a bull" and that "Teller would never stay under those circumstances." Dean refused to intercede, and within a week Teller made good on his threat. He returned to the University of Chicago to brood, declaring that he would continue working on the Super independently. He was there when Ernest Lawrence tracked him down and invited him to visit Livermore.

Teller's departure from Los Alamos failed to quell interest in a second laboratory. His unrelenting disparagement of Los Alamos had made inroads at the AEC, forcing Bradbury to shuttle to Washington to defend his team.

Communing with a sympathetic Colonel Kenneth Fields, the AEC's chief of military application, Bradbury pointed out "the somewhat ironic fact that every current weapon development has arisen out of the suggestion (and in many cases, the urging) of this laboratory." Had not the lab proven its worth to the AEC by meeting the agency's ever-greater demands for more thermonuclear tests? Yet instead of receiving recognition and gratitude, Los Alamos received "rather thinly veiled criticism" that it was not up to the job of weapons research and development.

The General Advisory Committee, still chaired by Oppenheimer, was also leaning toward a second lab. The GAC had reached the conclusion that the intensified test schedule really was placing insupportable pressure on Los Alamos; sooner or later, Oppenheimer feared, the workload would erode the quality of the lab's product. The GAC also was responding to a change in the wind in Washington resulting from the invasion of South Korea by North Korean troops in June 1950, followed by the intervention of Communist China on the North's side. As fears rose that the United States was fated to stand alone against a Sino-Soviet Communist alliance, it was becoming harder for the committee to resist the prevailing idea that everything possible should be done to move the Super along.

The question of a second lab, furthermore, had become a proxy for the question of what to do with Edward Teller. Everyone agreed that despite his vexing personality, Teller's intellect was a crucial asset for the bomb development program; the difficulty was finding a management formula that would satisfy both Teller and Bradbury, avoid exacerbating morale problems at Los Alamos, and maintain progress in thermonuclear research. None suggested itself, and 1951 ended with the issue still unresolved at the AEC.

On the West Coast, however, Ernest Lawrence was developing his own solution to the agency's dilemma. At a physics department reception on New Year's Day, he pulled Herb York aside. "Drop in on me this week," he said. "There's something I want to discuss with you."

A couple of days later, in Lawrence's office, York was startled when

Ernest asked him bluntly: "Does the United States need a second nuclear weapons laboratory?" York, who knew nothing of the high-level politics driving the Super debate, avoided giving a straight answer, but Lawrence's question was mostly rhetorical. He outlined a concept for a second lab as initially "a small group . . . in support of Los Alamos and controlled thermonuclear work . . . We'll start small and see what happens." He instructed York to pack his bags for a trip east to quietly sound out scientists and government officials about locating this new group at Livermore. Thirty years of age, the burly, crew-cut York found it "heady and exciting" to act as Ernest Lawrence's personal envoy in meetings with "all those other physicists whom I had been reading about in books and hearing about in lectures." Since the names on his itinerary all favored the second lab—they were AEC and air force officials who had been listening to Teller for months—it was unsurprising that York returned to Berkeley as a convert to the idea. "I reported to Lawrence that I, too, felt it would probably be useful to establish a second laboratory," he recalled. "The idea of doing so at the Livermore site was, for us, a natural one."

Lawrence drove Edward Teller out to Livermore during the first week of February, hoping that Teller would become the instrument of his hopes to create a permanent foothold for Livermore in the thermonuclear program. The hulking Mark I, not yet dismantled, still sat in its barnlike home: a corrugated metal building the length of a football field that hove into view while they were still miles away. Upon returning to Berkeley later that day, Ernest asked Teller to leave Chicago and help him establish the second laboratory. For this brief moment, the two physicists' ambitions and skills fell into perfect accord. Teller could have a place of his own to pursue the Super with a team of high-quality scientists; Lawrence, whose reputation continued to lure the world's best young physicists to Berkeley, would supply that elite manpower while expanding the boundaries of his research empire.

The AEC, Congress, and the Pentagon all recognized that Livermore, at a stroke, answered every objection raised against the second lab. "We debated this at some length," Gordon Dean recounted later. The second lab "had to be a place that was already established if you were going to save time. It had to be a place where you had . . . a man in there who commanded respect, that Teller would work for and work with, and be comfortable working with. There was only one place that I could finally fasten on that fitted this, and this was to work under Ernest Lawrence." The arrangement worked, Dean recalled a few years later, "because of Teller getting along very well with Dr. Lawrence."

Yet the assumption that Lawrence and Teller would work together in unalloyed harmony turned out to be an instance of hope triumphing over reality. Once the lab won the AEC's blessing, it became clear that the two scientists actually had very little in common. Their approaches to physics were divergent: Lawrence was an experimentalist nonpareil, Teller a profoundly intuitive theorist. They also held diametrically opposed views of how to organize and staff Livermore and how to structure the thermonuclear program. Teller saw the lab as a reflection of his own brimming self-esteem—as a huge laboratory staffed with name physicists working in lockstep according to his handcrafted research strategy. But "Ernest couldn't see that for beans," York recorded. "He was only willing to step in slow . . . no big names and no big plans." Lawrence conceived Livermore as a reflection of his own style, or more precisely as a replication of the original Rad Lab, staffed with talented but unsung young physicists eager to make their names at Berkeley, not coasting on reputations established elsewhere.

"You get a bunch of bright young fellows, and they'll learn it all," he advised York. "Those that have famous names—that's not because they're any better, it's only because they're a little older." In Lawrence's flat-management paradigm, there would be no titles among the scientists at Livermore, no pecking order among the PhDs on the laboratory floor. No titles were necessary, he declared proudly, because "there is no higher title than professor in the Radiation Laboratory."

No one who knew Teller would have expected him to submit meekly to Lawrence's authority. They both were fully alive to their own eminence, each backed by powerful and influential patrons. Lawrence's position as doyen of the Rad Lab and Teller's as the indispensable genius of thermonuclear technology made compromise impossible. Once the time came to organize the new lab, the only way to maintain friendly relations between the two was to keep them far apart. This task fell to the beleaguered York, who served as their go-between, shuttling between Berkeley and Chicago, where Teller had set up his Super lab in exile. York could see that fashioning a modus operandi for the new lab that met both men's specifications would be practically impossible, for "they were poles apart."

With Teller's dawning realization that he might not be slotted in for the role of supreme authority at Livermore, his old habit of undermine-and-conquer returned. A golden opportunity for his scorched-earth bluster arrived when the AEC released its official mission statement for Livermore in June 1952. The document set forth as the lab's objective the "development and experimentation of methods and equipment for securing diagnostic information on behavior of thermonuclear devices . . . in close collaboration with the Los Alamos Scientific Laboratory." (This was meant to recognize the particular expertise of York and other Berkeley scientists who had participated in Greenhouse—namely, the measurement and analysis of nuclear blasts.) The commission further expressed its "hopes that the group at UCRL"—that is, the University of California Radiation Laboratory at Livermore—"will eventually suggest broader programs of thermonuclear research to be carried out by UCRL or elsewhere."

Livermore's nebulous mandate was an ideal fit for Lawrence's scheme. For Teller, however, it replicated the unfocused dithering he had been grousing about. His response came at a reception in July at Berkeley's elegant Claremont Hotel, marking Livermore's launch as a national laboratory. Before a cheerful crowd that included Lawrence and Gordon Dean, the "well-lubricated" Teller dropped a bombshell, declaring that he would have nothing to do with the new lab.

This was tantamount to a threat to throttle Livermore at birth. Lawrence, for his part, was inclined to call Teller's bluff, for his putative partner was already proving entirely too fractious for his taste. "We would probably be better off without him," he muttered to York. To the AEC, however, the standoff signified that the Teller issue had not been solved by the creation of the second lab after all, which made Livermore a costly and time-consuming way of making no progress. At Dean's insistence, "intense negotiations were resumed among all concerned," York related. Within days, these yielded a commitment from the AEC that thermonuclear weapons development would play a part in Livermore's program from its inception. In Livermore's organizational chart, Teller would be designated as one of several members of the Scientific Steering Committee, but in recognition of his "obvious special status," he was awarded veto power over the committee's decisions. In other words, Teller was to have no formal authority at Livermore but was granted the implicit authority to direct its thermonuclear research program as he saw fit.

That still left open the question of who would manage Livermore day to day, in effect as Lawrence's viceroy. Ernest solved that problem by handing the reins to York. The offer surprised York almost as much as Lawrence's original question about the need for a second lab. He was struck by the nonchalance of Lawrence's decision making: there had been no search committee; no bureaucratic procedure or vetting of candidates. Lawrence had asked him if he thought he could "run it," and as soon as York assented, "he simply instructed me to do so . . . He gave me no new title, no immediate raise in salary, or any other change in status. He made no announcement about it, except an informal one to his immediate associates." York chalked up his new opportunity to Lawrence's visionary audacity: "It certainly was a matter of great guts," he reflected. "Who else would take the major responsibility for a new lab and then ask a thirty-year-old man with no experience to run it?"

He may have ascribed too much free will to his mentor. In fact, Lawrence had come under pressure from Oppenheimer and the General

Advisory Committee to establish a conventional management structure at Livermore, with the goal of creating a counterweight to Teller, who would be inclined to seize any unclaimed authority. York might be young and untested, but as Ernest's designated director, his managerial authority would be, at least in formal terms, indisputable. The arrangement with Teller, York reflected, was "peculiar," but it did the job: "None of the strained relationships that had surrounded Teller at Los Alamos developed at Livermore."

For all that, Lawrence remained unmistakably the animating force at Livermore, which soon acquired the operational features of a classic Lawrence laboratory. It was collegial and interdisciplinary, with only the blurriest lines distinguishing the individual researchers' responsibilities. As one Berkeley scientist related in contrasting the approaches of the two labs toward designing a bomb, at Los Alamos "some folks would work on one part . . . others on another part . . . They would write memos to each other." At Livermore "there were no fiefdoms. We were all working together."

But the AEC's dream of thermonuclear research blooming via a collegial relationship between Los Alamos and Livermore proved chimerical. The labs squabbled continually over credit for technical advances. The first friction point was the "Mike" test at Eniwetok in November 1952, involving a large thermonuclear device. The Mike device was based on the Teller-Ulam design and built by Los Alamos; Livermore, which at that time was fully engaged in "simply coming into being," played no role whatsoever. Yet following the government's announcement of the test, if not its spectacular results—a 10.4-megaton blast that completely obliterated Elugilab, an island in the Eniwetok Atoll—it was Livermore that received popular credit for the feat.*

The reasons, York surmised, were Teller's association with Livermore

---

*A megaton, then as now the conventional measurement of the energy of a nuclear blast, is defined as the equivalent energy of one million tons of TNT. The Hiroshima bomb has been estimated at about fifteen kilotons, or the equivalent of fifteen thousand tons of TNT.

and the AEC's "absurdly strict secrecy policy," which prevented the issuance of even a simple clarification of the roles of the two laboratories. An accurate picture emerged only in 1954, when Norris Bradbury was granted permission to counter Teller's public charges that Los Alamos had dragged its feet on the Super. The impetus was the publication of a highly sensationalized book, *The Hydrogen Bomb: The Men, The Menace, The Mechanism*, by journalists James Shepley and Clay Blair Jr., which accepted Teller's version of events as gospel. Bradbury, incensed at Teller's implication that Los Alamos scientists had shown disloyalty by holding back on the Super, observed at a press conference that his staff had "built a laboratory that developed every successful thermonuclear weapon that exists today." Three years had passed since Bradbury had offered an almost identical defense to the AEC's Colonel Fields, but Teller had never ceased carping about Los Alamos. A few months after the press conference, Teller, wounded at being shunned by former friends and colleagues over his continued bellyaching, finally attempted to set the record straight in an article for *Science*. Livermore's work thus far, he wrote, "has been mostly that of learning the difficult art of inventing and making nuclear weapons. All the magnificent achievements that have become in the meantime known to the world have been accomplished by Los Alamos."

In fact, within the thermonuclear research community, Livermore's reputation was marred by the almost comic fizzles of the first devices it readied for testing. As products of the lab's emerging interest in small-scale hydrogen bombs that could be delivered by aircraft or ballistic missile, these were designed to have modest yields—but not as modest as the embarrassing pops they emitted on the test range. At the first explosion of a Livermore device in Nevada in 1953, just six months after the lab's founding, the blast was not even powerful enough to incinerate the tower holding the device, a key metric for trials of this type. (For the follow-up trial two weeks later, the tower's height was halved to ensure that no trace of it would remain after the blast.) Livermore defended its flop as the not-unexpected product of vigorous innovation, on the theory that one can often learn more from mistakes than from easy successes. The new lab had inherited this tenet from its parent, Ernest Lawrence, who had experienced not a few such mishaps during his career. In any case, despite the anticlimactic results, the test results were interpreted by the AEC as validating Livermore's role as the developer of "new ideas" in thermonuclear technology.

But more embarrassment lay ahead. The occasion was Operation Castle, a test of Los Alamos and Livermore inventions at Bikini Atoll in March 1954. The first trial, of a Los Alamos device code-named Bravo, was a fiasco, though one that brought a sort of perverse credit to the Los Alamos lab. Its designers had miscalculated the physics of the fusion reaction and drastically underestimated its yield, with the result that an expected five-megaton device exploded with triple that energy. The enormous fireball showered dangerous radioactive fallout over task force personnel who were thought to be safely out of range. A radioactive plume of vaporized coral drifted east, covering seven thousand square miles of ocean. The cloud necessitated the hasty evacuation of nearby Marshall Islands residents to Kwajalein Island, four hundred miles distant, where they were treated for radiation sickness. Worse, from the standpoint of international relations, the fallout reached a hapless Japanese fishing boat, the *Daigo Fukuryu Maru* ("Lucky Dragon Five"), sickening its twenty-three crew members, one of them fatally. Lewis Strauss, who had succeeded Dean as AEC chairman, unjustly labeled the boat a likely "Red spy ship," intensifying the furor in Japan and creating a very public black eye for the American testing program.

Then came Livermore's turn, with a device code-named Morgenstern that was expected to yield one megaton. It was another dud, its 110-kiloton blast barely visible to observers watching from naval vessels on the fog-shrouded horizon. Mortified, the Livermore team canceled a second test. The Los Alamos devices had shown up their designers too, but at least their results truly were super; indeed, the Castle tests of the Los Alamos devices pointed toward what Rabi, who had succeeded Oppenheimer as chairman of the GAC, called a "complete revolution" in nuclear weaponry; a sudden maturing of the technology that opened the door to a vastly expanded role for thermonuclear bombs in America's strategic arsenal.

The Los Alamos staff returned from Operation Castle with a self-confidence they had not felt since the end of the war. By contrast, a pall of humiliation fell over Livermore, whose dismal record was now oh-for-three. "It was not surprising that some Los Alamos scientists filled the air with horse laughs," acknowledged York. More disturbingly, the serial fizzles prompted new questions about Livermore's management, its work product, and its cost. Bradbury renewed his attack on the very concept of a second lab. It had been "believed in some quarters that brilliant new ideas would flow from the establishment of competition," he wrote Colonel Fields at the AEC. "The brilliant new ideas have not appeared." Bradbury's complaint found an increasingly sympathetic audience at the General Advisory Committee. Livermore "had [not] been an effective organization in the two and a half years of its existence," Rabi observed at the committee's meeting in December 1954. He wondered aloud whether it might ever "really be an important laboratory."

Doubts about Livermore's future spread among the staff. They interpreted their slapdash working conditions, created by chronic underfunding by the AEC, as an ominous sign that the agency's commitment to Livermore remained conditional. Working plumbing was scarce, and the lack of air-conditioning tormented scientists and engineers trying to function in the baking heat of California's Central Valley. Livermore seemed to be an auxiliary outpost caught between two larger labs, Los Alamos and Berkeley, and it was hardly unreasonable to wonder, with Rabi, if it would ever amount to more.

As it turned out, the staff worried for naught, for they had not accounted for the insatiable demand for research and product created by competition among the military service branches, each striving to secure thermonuclear arsenals it could call its own. Notwithstanding the test fizzles, the AEC had committed fully to Livermore, its faith in Ernest Lawrence still strong. Livermore's first-year budget in fiscal 1953 had been $3.5 million and its staff contingent 698. By 1956, the end of York's five-year tenure as director, the lab boasted a $55 million budget supporting a staff of 3,000. One year later, the professional staff numbered 4,000 and Livermore ranked as the largest of the AEC's research labs. That year, its research in lightweight thermonuclear bombs came together with the navy's urgent desire for a nuclear ballistic missile that could be launched from a forthcoming class of submarines with unprecedented range. The result was Polaris, a weapons system that marked Livermore's "coming of age." Within five years, the lab's budget was $127 million, and the staff reached five thousand. Livermore would stay.

Chapter Nineteen

The Oppenheimer Affair

J. Robert Oppenheimer made many enemies with his persistent critique of US nuclear policy and the rush to the Super, but none as implacable as Lewis L. Strauss. Given Lawrence's relationship with both Oppenheimer and Strauss, it was inevitable that he would be drawn into their conflict.

Strauss's enmity for Oppenheimer deepened as the latter's opposition to the hydrogen bomb solidified. After the appearance of the General Advisory Committee's broadside against the Super in 1949, Strauss became convinced, in his customary fashion, that Oppie was not merely a fool but a traitor. As long as Strauss remained a minority voice on the Atomic Energy Commission, however, he could only chafe powerlessly at Oppenheimer's exalted role as the AEC's leading scientific advisor. His determination to remove Oppenheimer from the high councils of government went quiescent after his retirement from the AEC in February 1950, when he recognized that Truman's decision to pursue the Super marked the successful completion of his long campaign for this leap forward in national security.

But the interlude was brief. With the arrival of a Republican administration in 1953, Strauss returned to Washington endowed with extraordinary access to the White House. President Dwight Eisenhower installed him as his personal advisor on atomic energy in March and appointed him to the AEC chairmanship three months later. From his new perch, Strauss fixed his sights on Oppenheimer's last means of direct influence on commission policy, a consultantship he had been awarded by the previous chairman, Gordon Dean.

Even before Strauss's appointment was made public, he began moving against Oppenheimer, impelled by the physicist's increasingly outspoken campaign for a public debate on nuclear policy. Oppie's campaign took form in a speech at the Council on Foreign Relations in New York on February 17, 1953, before an elite audience of opinion makers and financial leaders, Strauss among them. His theme was the need for "candor" from political leaders about the dangers of nuclear proliferation and the necessity of international disarmament. "We do not operate well when the important facts, the essential conditions, which limit and determine our choices are unknown," Oppenheimer declared. "We do not operate well when they are known, in secrecy and in fear, only to a few men."

Strauss seethed as Oppenheimer outlined the state of the arms race between America and the Soviet Union. Oppie spoke in abstract terms because, as he acknowledged, the government's demand for nuclear secrecy required him to gloss over the horrific details. He warned that during the Cold War, the "atomic clock ticks faster and faster" and closed his speech with a vivid image of two great powers in mortal conflict: "two scorpions in a bottle, each capable of killing the other, but only at the risk of his own life."

Oppie's speech had been cleared by the White House and was said to have impressed Eisenhower himself. Strauss felt differently. Speaking of Oppenheimer's call for candor, he advised the president, "The campaign is dangerous and its proposals fatal." The facts Oppenheimer favored disclosing publicly "are of the greatest significance to the general staff of an enemy." To advocate placing such information in the public record, Strauss asserted, was perforce an act of disloyalty.

Over the following year, Strauss orchestrated a public campaign of vilification against Oppenheimer, as if to soften him up for the final blows, which would be the termination of his AEC consultantship and the revocation of his security clearance. He forged a close relationship with

J. Edgar Hoover, hectoring the FBI director to institute ever more intensive surveillance of Oppenheimer's movements and telephone calls. Hoover's fattened dossier on Oppenheimer eventually landed on Eisenhower's desk, with Strauss near at hand to interpret. His efforts succeeded: in December Eisenhower ordered that a "blank wall" be placed between the physicist and all classified or sensitive government information, pending further investigation. This was the first step in withdrawing Oppenheimer's security clearance. Meanwhile, through his network of well-connected friends sharing his fear of Communist world domination, Strauss placed articles attacking Oppenheimer in Henry Luce's *Time*, *Life*, and *Fortune*, and other popular magazines. The articles suited the tenor of the moment established by Senator Joseph McCarthy's febrile accusations of Communist influence in every corner of government. Robert Oppenheimer was destined to become the most prominent victim of the witch hunt.

Following Eisenhower's order, Strauss summoned Oppenheimer to his Washington office for what he expected to be a decisive encounter. He presented Oppenheimer with a lengthy accusation of communist fellow traveling and suspect loyalty, most of it based on warmed-over charges that had been traded wholesale among security officials since the launch of the Manhattan Project. Strauss informed Oppenheimer that his security clearance was suspended and urged him to resign quietly from his AEC position. To his astonishment, Oppenheimer balked. The following day, having weighed the demand with his Washington lawyers, Oppenheimer informed Strauss that he would fight the charges before an AEC review board. The decision, as Oppenheimer's biographers Kai Bird and Martin Sherwin would observe, "set in motion an extraordinary American inquisition."

Strauss stage-managed the AEC hearing into Oppenheimer's security clearance as he had the campaign of calumny that preceded it. He handpicked the review board's three members and its chief counsel, Roger Robb, a former federal prosecutor with a reputation for courtroom ferocity and political conservatism. He plied Robb with documents and notes to help him impeach Oppenheimer's character witnesses, many of whom were scientists who had worked with Oppie for a decade or more, and urged Robb to seek out witnesses who might be inclined to attack his quarry's trustworthiness from the stand. This quest led, inevitably, to Berkeley. There Robb found a hive of anti-Oppenheimer sentiment, presided over by Ernest Lawrence.

Robb's visit to the Rad Lab early in March 1954 could not have been better timed. Lawrence was incensed by an unsavory fact he had picked up at a recent cocktail party: Oppie had carried on an affair in 1947 with the wife of Caltech physicist Richard Tolman, a close friend of Ernest's who died of a heart attack only a few months after learning of the betrayal—in Ernest's view, the victim of a broken heart. The news crystallized all the doubts and resentments Ernest had suppressed about Oppenheimer in the years since they had first come together as youthful Berkeley faculty members—Oppie's "leftwandering," his arrogant individualism, his bohemian style, and of course his opposition to the Super and skepticism about the second laboratory. Under questioning by Robb's deputy, C. Arthur Rolander, these all came pouring out with an un-Lawrencian passion. In the heat of the moment, Ernest offered a shocking judgment of Oppenheimer that would come to haunt them both. Oppie, he said, "should never again have anything to do with the forming of policy." Even more momentously, he agreed to come to Washington to testify personally against his old friend.

Lawrence was not alone among the Berkeley crowd in disparaging Oppenheimer's character. Before they left town, Robb and Rolander added to their witness list Kenneth Pitzer, who had returned to Berkeley after his stint as the AEC's research director, and Luis Alvarez and Wendell Latimer. Despite the latter's objections to Neylan's security inquiries a few years earlier, Latimer would participate willingly in a process that was tilted even more harshly against its target. Berkeley would be well represented at the trial of J. Robert Oppenheimer's life and opinions, and entirely on the side of the prosecution.

• • •

The AEC Personnel Security Board hearing into J. Robert Oppenheimer's security clearance convened on April 12, 1954, in the agency's dilapidated headquarters, a temporary building on Washington's National Mall left over from wartime. The bill of particulars had been drafted and signed by the AEC's general manager, Kenneth D. Nichols, who had worked closely with Oppenheimer during the war as Groves's second in command but had emerged from the experience regarding him as a "slippery sonofabitch." The document, which read like a criminal indictment, covered Oppenheimer's associations with liberals, leftists, and supposed Communist front groups, as well as his purported efforts to stifle the Super program.

Over the next three and a half weeks, almost every aspect of Oppenheimer's life and career would be picked apart by the AEC's legal team, acting as prosecutors in a proceeding that lacked even the most rudimentary evidentiary standards. From some of the nation's leading scientists would come denunciations of Oppenheimer, veiled and overt, among them Edward Teller's assertion that "I personally would feel more secure if public matters would rest in other hands." From witnesses on the other side came denunciations of the very effort to humiliate a man who had served his country with distinction. None was as biting as I. I. Rabi's. Thanks to Oppie, he snapped, "we have an A-bomb . . . and what more do you want, mermaids?"

Among the hearing's more sinister features was its preoccupation with an encounter between Oppenheimer and his friend Haakon Chevalier just before work began at Los Alamos. Chevalier was a Berkeley French professor and, along with his wife, a member of the Oppenheimers' social and intellectual circle. One evening in early 1943—the precise date was never established —the Chevaliers joined Robert and Kitty for dinner at the Oppenheimer home in Berkeley. While Oppie was alone mixing martinis in the kitchen, Chevalier walked in on him to communicate an extraordinary offer. He stated that George Eltenton, a British physicist with leftist sympathies working in the Bay Area for Shell Oil Company, wished to know if Oppenheimer would be willing to pass information about his research to a contact of Eltenton's at the Soviet Consulate in San Francisco. Later accounts by Oppenheimer, Chevalier, and Eltenton leave no doubt that Oppenheimer instantly and heatedly turned away the offer. "I thought I said, 'But that is treason,' but I am not sure," Oppenheimer would testify at his hearing. "I said, anyway, something, 'This is a terrible thing to do.' . . . That was the end of it. It was a very brief conversation."

But Oppenheimer's maladroit handling of the conversation's aftermath elevated what had seemed at first to be a momentary exchange between friends into something that became known, ominously, as the "Chevalier affair"—an affair in which Lawrence was a peripheral and unwitting figure. As Oppenheimer acknowledged, he should have reported the approach promptly to Manhattan Project security officials. But since Chevalier had agreed on the spot that the offer was inappropriate, Oppie had put it out of his head. As it later transpired, Eltenton had been asked by Peter Ivanov, a Soviet intelligence officer working undercover as a consular official in San Francisco, to develop a pipeline to "three scientists" associated with the Rad Lab. Eltenton identified his targets as Oppenheimer, Ernest Lawrence, and a third physicist whose name he did not recall but thought might have been Luis Alvarez. Eltenton was not sufficiently close to any of them to make contact, but he did know Chevalier and asked him to undertake the initial approach to Oppenheimer.

Nothing ever came of the offer. Oppenheimer plainly did not agree to it, and there is no evidence that Lawrence or Alvarez even knew they had been mentioned until years later. Oppenheimer's greater problem was that over the course of many years, he had given security officials several variant versions of his talk with "Hoke" Chevalier, all aimed ineptly at shielding his friend from the security apparatus. Oppenheimer's dissembling would be exploited by Strauss and Robb to discredit virtually his every word, becoming a linchpin of the campaign to destroy him.

The hearing progressed inexorably to the point when Ernest Lawrence would be called to testify. Ernest anticipated the moment grimly, wracked with second thoughts about his commitment to take the stand. Finally, he changed his mind. But quailing at the thought of Strauss's certain apoplexy at his withdrawal, he chose to put off the uncomfortable conversation until the last minute. As late as Friday, April 23, he assured Nichols by phone that he would be in Washington no later than the following Tuesday, with his appearance at the witness table scheduled for a day or two later.

Lawrence spent the intervening weekend at Oak Ridge, attending a conference of AEC laboratory directors. If he had had any doubts about how his willingness to testify against Oppenheimer played in the physicists' fraternity, these were dispelled by the "barely civil" reception he got from his colleagues, among them Henry DeWolf Smyth, one of the AEC commissioners who ultimately would rule on Oppenheimer's fate, and Rabi, who had delivered his uncompromising defense of Oppenheimer just a few days earlier. As Ernest could not fail to recognize, much of the physics community was lined up behind Oppenheimer. The Rad Lab stood alone.

From Oak Ridge, Lawrence telephoned Berkeley to pour out his misgivings to Alvarez. As he explained over the long-distance line, he dreaded that his testimony, combined with the distinctly anti-Oppenheimer mindset of the other Berkeley witnesses and of the Livermore-based Edward Teller, "would seem to reflect a sort of cabal." There could be no upside for the Rad Lab if it got dragged into the controversy over Oppenheimer's past and behavior. The lab's experience with the loyalty oath had been bruising enough; its involvement in this even more explosive battle would inextricably identify the lab with one side of the most fraught political issue of the day.

Alvarez tried to strengthen Lawrence's spine. "I sort of thought Ernest buckled under to pressure that he shouldn't have," he recounted later. As the emotional conversation drew to a close, Lawrence pleaded with Alvarez to follow his lead for the good of the lab and refuse to testify. Alvarez had spent nearly his entire professional career in Lawrence's thrall, following his orders without question. “I wasn’t about to change now,” he reflected later, and reluctantly agreed to his boss’s request.

On Monday morning, Lawrence telephoned Strauss from Oak Ridge. By then, his body had given him a painful yet plausible pretext for backing off: a severe attack of ulcerative colitis, a medical condition from which he had suffered for years—though perhaps brought on at this moment by emotional tension. But the infuriated Strauss dismissed Lawrence’s plea of illness and delivered a vicious tongue-lashing, capping it bluntly with an accusation of cowardice. Lawrence rang off. Visibly shaken, he summoned his fellow Oak Ridge guests to bear witness that he was not feigning illness by showing them his toilet, filled with bright red blood. The next day he flew home.

Alvarez, disturbed by the abject misery he had heard in Lawrence’s voice—“I had never seen Ernest intimidated before,” he reflected—telephoned Nichols that same day to withdraw. A few hours later, Strauss called him back, determined to prevent Lawrence’s “illness” from spreading. Despite his loyalty to Lawrence, Alvarez proved to be more pliable to the AEC chairman’s haranguing. “If you don’t come to Washington and testify, you won’t be able to look yourself in the mirror for the rest of your life,” Strauss growled. Torn between his incompatible loyalties to Lawrence and to Strauss, Alvarez finally opted to serve the latter, and booked a seat on the red-eye flight to Washington.

Alvarez would write later of his uneasiness about giving testimony “that might hurt a friend.” From the witness stand, he attested to his own “admiration and respect for Robert,” as well as his certitude that while Oppenheimer’s judgment about the Super was “faulty,” it was “in no way related to his loyalty to the country, of which I had no doubt.”

The record, however, discloses Alvarez’s testimony to be an ill-disguised act of score settling. Contemptuous of Oppenheimer’s arguments against the Super, he depicted Oppenheimer as a Svengali ruthlessly hypnotizing some of the world’s most sophisticated scientists into joining his opposition campaign. "Every time I have found a person who felt this way," he testified, "I have seen Dr. Oppenheimer's influence on that person's mind."

Yet some of his examples plainly arose from his own misconceptions. He testified that Fermi's opposition to the Super had surprised him, because he knew Fermi was "one of two men who had signed an appendix to the [October 30, 1949, General Advisory Committee] report expressing views somewhat different from those of the majority group led by Dr. Oppenheimer." He seemed to think that Fermi had favored the Super. But this was exactly wrong, for the appendix by Fermi (and Rabi) had expressed the same iron opposition to the Super as the majority statement.

Alvarez also spoke of his perplexity that Rabi had "changed his mind so drastically after talking with Dr. Oppenheimer." Rabi's initial position regarding the Super, Alvarez testified, "was one of enthusiasm." This was flatly untrue, as Rabi had already informed the hearing board at considerable length. To the contrary, he had tried to bring the maniacally optimistic Alvarez and Lawrence down to earth from the moment they first approached him about joining the H-bomb program. His "enthusiasm" for the weapon was a figment of Alvarez's imagination.

Alvarez did deliver one genuine bombshell. He swore that one day in October 1949, while driving from Stanford University back to Berkeley in company with Lawrence and Vannevar Bush, he heard Bush state that President Truman had asked him to chair a committee to assess the evidence that Russia's Joe-1 explosion had been an atomic bomb. Bush said he thought it odd that he should be chairman, since he was not a physicist; Oppenheimer, the best choice, already had been named to the committee. As Alvarez related Bush's explanation, "I think the reason the president chose me is that he does not trust Dr. Oppenheimer." Alvarez claimed that "this was the first time I had ever heard anyone in my life say that Dr. Oppenheimer was not to be trusted."

Yet Bush had mentioned no such conversation during his testimony only six days earlier. (Instead, he had crisply reproached the board for attacking Oppenheimer "because he expressed strong opinions," adding, "When a man is pilloried for doing that, this country is in a severe state.") Summoned back to the stand by Oppenheimer's lawyer, Lloyd K. Garrison, to rebut Alvarez's revelation, Bush firmly denied having said any such thing. "I am quite sure I didn't say to him that the president had doubts about Dr. Oppenheimer simply because it was not true," he testified.

It fell to Ernest Lawrence, purportedly the only other witness to the conversation, to break the deadlock. He did so via an affidavit notarized in Berkeley on May 4, the day of Bush's return appearance. Lawrence's statement had the curious effect of undermining both the other witnesses, while placing his own credibility at the service of a smear of Oppenheimer. "I remember driving up from Palo Alto to San Francisco with L. W. Alvarez and Dr. Vannevar Bush when we discussed Oppenheimer's activities in the nuclear weapons program," he stated. "In the course of the conversation, [Bush] mentioned that Gen. Hoyt Vandenberg had insisted that Dr. Bush serve as chairman of a committee to evaluate the evidence for the first Russian atomic explosion, as General Vandenberg did not trust Dr. Oppenheimer."

Attributing the supposed mistrust of Oppenheimer to Vandenberg rather than Truman made superficial sense, for it was Vandenberg, commander of the air force, and not Truman who had assembled the committee and appointed both Bush and Oppenheimer. But Lawrence's recollection only added more fog to the picture, for Bush also had been asked if it might have been Vandenberg who had doubted Oppenheimer's trustworthiness—and he had denied that with the same vehemence. As he observed, if Vandenberg truly harbored doubts about Oppenheimer, why would he have appointed him to the committee in the first place? A clear version of the mysterious conversation never did emerge.

Despite his failure to testify in person, Lawrence's voice did not otherwise go unheard at the hearing. Before closing the hearing record, Robb inserted the transcript of the interview Ernest had given Rolander in Berkeley some two months earlier. Lawrence's words, which remained safely immune to cross-examination by Lloyd Garrison through this tactic, were withering. He described Oppie as arrogant, naïve, and suspiciously hostile to a thermonuclear program that was manifestly in the nation's best interests. His conclusion that J. Robert Oppenheimer, a man with whom he had shared the triumphs and setbacks of conjoined personal and professional lives for twenty-five years, "should never again have anything to do with the forming of policy" reverberated in the record with devastating finality.

Ernest Lawrence's enmity left Oppenheimer deeply perplexed. In 1963, four years before his death, Oppie still found the subject difficult to understand and painful to discuss. "I think there was probably warmth between us at all times, but there was bitterness, which I think became very acute in '49 and which was never resolved before his death," he told Herbert Childs, who had been commissioned by John and Molly Lawrence to write Ernest's official biography. "He disapproved of my leftward course and told me so. There was never anything there that would have led to any great bitterness."

Oppenheimer recognized that Lawrence's politics had evolved to reflect those of the people who had become his patrons and friends, men like Alfred Loomis and Jack Neylan. They were not individuals inclined to soft-pedal their views about national politics and national security or, in Neylan's case, about Robert Oppenheimer, whom Neylan described as "a man so conceited he just shoved God over." Continuing in this vein, Neylan conjectured, "Ernest's very modesty angered him . . . I think he hated Ernest because Ernest was so kind to him."

That judgment may say more about Neylan than about Oppenheimer. Friends and colleagues of Ernest and Oppie typically sought an explanation of their break in their divergent personalities, rather than the character of either man. But they found it no easier to pinpoint the cause. James Brady, who had worked with Lawrence and Oppenheimer almost from the time of their arrival at Berkeley, tried to get to the bottom of it one day while he and Ernest were together in the Rad Lab during the Oppenheimer hearings.

"How does this happen, that all the other physicists except the group here in Berkeley are defending Oppie?" he asked.

"There's a very good reason for it," Ernest snapped. "We're the only ones around here who really know that man." Brady, shocked at Ernest's vehemence, tried to delve deeper. Ernest seemed to be especially scandalized that Oppenheimer had lied to the Manhattan Project's security officials about Haakon Chevalier. "I got Oppenheimer that job in the first place," he complained, as though Oppie's blunder reflected poorly on his own judgment. "I could excuse almost anything except lying to the security people. This I can't believe, I can't understand. A man can't even be a good physicist, if you'll lie that way." Even that answer seemed to leave too much unsaid, Brady thought. "It seemed to me to be personal," he recollected.

Yet Oppenheimer almost always gave Ernest the benefit of the doubt. He refused to accept that Ernest might have participated in Strauss's campaign against him or engaged in any of the rumormongering about his Communist tendencies that periodically swept the Berkeley campus. "I never heard anyone attribute that to Ernest," he said.

There was one thing that hurt, however. After the Soviet Union's first reported H-bomb test in August 1953—a relative fizzle known to US intelligence as "Joe-4"—"Ernest said, I think to DuBridge, 'Well, it's sure lucky that some people's advice wasn't taken.' " That thinly veiled reference to himself, Oppenheimer said, was the "harshest thing that I heard."

By then, their break was complete. After the loyalty hearing, Oppenheimer said, "we saw almost nothing of each other." All that would survive of their friendship was the scientific legacy they had built together, and the very distinct legacies that would outlive them both.

Chapter Twenty

The Return of Small Science

On June 29, 1954, the Atomic Energy Commission voted 4 to 1 to revoke J. Robert Oppenheimer's security clearance, one day before it would have expired anyway. The majority opinion was written by Lewis Strauss, who spared no effort to paint Oppenheimer as a weak-minded perjurer whose behavior had materially undermined national security. "The work of Military Intelligence, the Federal Bureau of Investigation, and the Atomic Energy Commission—all, at one time or another have felt the effect of his falsehoods, evasions, and misrepresentations," Strauss wrote. He cited "the proof of fundamental defects in [Oppenheimer's] 'character' " developed during the hearing and declared that "his associations with persons known to him to be Communists"—a category that encompassed at one time or another not only Haakon Chevalier but also Oppenheimer's wife, Kitty, and his brother, Frank—"have extended far beyond the tolerable limits of prudence and self-restraint which are to be expected of one holding the high position that the Government has continually entrusted to him since 1942."

The lone holdout on the commission was Henry DeWolf Smyth, who had telegraphed his distaste for the inquiry to Ernest Lawrence at Oak Ridge. In his dissent, he observed that the "massive dossier" assembled by the personnel board had failed to unearth a single indication "that Dr. Oppenheimer has ever divulged any secret information . . . In spite of all this, the majority of the Commission now concludes that Dr. Oppenheimer is a security risk. I cannot accept this conclusion or the fear behind it."

Smyth's concern about the "fear" underlying the Oppenheimer verdict was widely shared in the research community. It had become disturbingly clear that in the supercharged political atmosphere of the time, scientists could be placed on trial for their personal views, with their careers and reputations hanging in the balance. The government's immense financial sway over research funding gave its interests—including its political interests—immense weight. Thus did the economics of Big Science create a double-edged sword.

Under the circumstances, scientists' fealty to military and political orthodoxy trumped their honest scientific judgment. The result was a sea change in the relationship between scientists and the Atomic Energy Commission. At the agency's inception, scientists had hoped it would serve as a civilian bulwark against military monopolization of nuclear research. Now, under Lewis Strauss, it had become even more security-obsessed than the colonels and generals the scientists had worked with during the war, in purposeful if somewhat prickly companionability. In the words of the AEC's official historians, "It was not likely that an agency that had destroyed the career of a leader like Oppenheimer could ever again enjoy the full confidence of the nation's scientists."

The Oppenheimer case did more than heighten the conflict between scientists and bureaucracy; it also caused a deep rift within the scientific community that would not be healed for years. A great deal of obloquy fell on Lawrence and his colleagues at the Rad Lab, who had stood united, and virtually alone, against Oppenheimer. Compounded by Lawrence's campaign for the Super, a project that many physicists considered technically and morally dubious, this record undermined his reputation as a scientist's scientist driven by the quest for knowledge alone. But Ernest was not the only physicist whose judgment was clouded by politics; almost everybody's was, for it had become almost impossible to take a stand for or against the pursuit of thermonuclear technology on purely technical grounds. James

Franck had been correct when he wrote after Hiroshima and Nagasaki that scientists could no longer "disclaim direct responsibility for the use to which mankind had put their disinterested discoveries."

The Oppenheimer case left Lawrence with a horror of overt dabbling in the security politics of the time. Among his longtime friends and colleagues, he had seen those who supported Oppenheimer get ritually humiliated by the personnel board, while those who doubted Oppie, like himself, were stigmatized by the scientific community. Now every security inquiry posed similar risks. These were nothing like the Berkeley security cases of 1948, which Ernest could manage smoothly via his friendship with Jack Neylan; they were ferocious inquisitions conducted by ruthless anticommunists, before whom a man's life and reputation counted for little.

One such case involved the Rad Lab's former staff chemist Martin Kamen, who had been fired by Ernest during the war in response to an unproved—and groundless—charge of disloyalty. Kamen had struggled to remake his career, survived a suicide attempt, and finally succeeded in gaining a high-level faculty appointment at Washington University in St. Louis, thanks to Arthur Compton, its chancellor. In 1954 he brought a libel suit against the arch-conservative *Chicago Tribune*, which had identified him as a Russian spy on the strength of a vague accusation made by the right-wing Iowa senator Bourke Hickenlooper. Kamen reached out to Lawrence for a deposition attesting to his loyalty, of which Lawrence in fact had no doubt. To Kamen's surprise, Lawrence agreed—but on condition that he not be subjected to cross-examination. Kamen could not fathom how Lawrence expected to be immune from courtroom questioning; he supposed, undoubtedly correctly, that Ernest was "not prepared to undergo the trauma of an adversary confrontation." Ernest's response contrasted sharply with that of Compton, who sat for a blistering daylong deposition by the *Tribune*'s lawyers without wavering for a moment in his defense of Kamen. In the end, Kamen won his lawsuit and $7,500 in damages.

• • •

Ernest's health had deteriorated toward the end of the war and grown even worse after the armistice. Now, as he felt weighed down by his failure to keep the Rad Lab above the turmoil of the Oppenheimer case, illness began to get the better of him. He was gray, tired, and distracted, to the point that he received a rare tongue-lashing from Luis Alvarez for his inattention to Rad Lab business. His most troubling conditions were a strangely recurrent viral pneumonia, chronic sinusitis, and the ulcerative colitis that had laid him low the weekend before his scheduled testimony against Oppenheimer. In the view of his brother, John, who had long played the role of his medical advisor, Ernest's basic problem was overwork. Through the thirties and into early wartime, his natural vigor had kept these maladies at bay, but the sheer magnitude of the demands on his time and energy finally had caught up with him. There was more to do and less time for relaxation—not that Ernest could ever keep still for long—and a weaker constitution to hold it all together.

The end of the war brought an unceasing parade of high-profile visitors to Berkeley—military men, foreign dignitaries, important scientists from abroad—and a stream of invitations to join government commissions and corporate boards. Jack Neylan and Rowan Gaither steered him away from most of these, including a lucrative invitation to join the Monsanto Company board. Neylan's concern was that the offer was too rich: "I knew that any time Ernest took that amount of money, he would break his neck to earn it," he recalled. But Neylan and Gaither found it harder to stop him from accepting invitations to join public committees and boards, with the result that he was constantly shuttling between Berkeley and Washington to attend one government function or another.

John and Molly tried to steer Ernest toward rest and relaxation, but he had a habit of transforming even innocuous holiday pursuits into physical challenges. During a 1946 visit to his in-laws on Balboa Island, a resort spot south of Los Angeles, he conceived the idea of renting a home on the island for the summer. Learning that the house he coveted was unavailable for rent but up for sale, he bought it on the spot. Molly's dream of a quiet summer evaporated, for the house was desperately in need of rehabilitation—"an old wreck, one of the oldest around, and built without benefit of architect" she recollected. Worse, Ernest decided that he should manage the renovation work personally. "The remodeling was a true Ernest venture," recalled his daughter Margaret, who turned ten that summer watching the builders tear off the front of the house and rebuild the structure room by room. Yet once the project was finished, the press of business kept him from visiting his refurbished vacation home for more than a few weekend days at a time.

What consumed him in this period mostly was something new: his first purely commercial venture, built around his first invention totally unconnected with the work of the Rad Lab. He was creating a tube for color television.

Ernest's interest in color TV dated back to 1949, when he and Luis Alvarez were invited to a demonstration of a rudimentary tube. The very idea of a color broadcast was so implausible that Alvarez secreted a small magnet in his pocket to verify that the picture was actually produced by electrons striking a phosphor screen. So it was, though the picture was hopelessly blurry. The device did, however, start Ernest thinking about how to produce a higher-quality display.

The technical challenge was right up his alley: it involved the electromagnetic focusing of a stream of charged particles and its synchronization with an oscillating electric current—in other words, the basic elements of the cyclotron. The project returned him to the milieu of small science he had left behind after his first few handmade cyclotrons. He now was fashioning a device by himself, without an army of technicians and engineers, working on a laboratory bench at home. Within a few months, he had rigged up an improved tube, which he showed to Alfred Loomis, who contributed a few minor technical suggestions, and to Alvarez and McMillan, who reacted with unease to Ernest's enthusiasm for what they considered a most unprepossessing project. "Ed and I were embarrassed by the tube's poor picture quality and its even poorer commercial potential," Alvarez recalled. But Ernest "dismissed our doubts with a wave of his hand. There were technical problems that had to be solved, he said, but the tube was certain to do the job."

Greater encouragement came from Rowan Gaither, whose businessman's soul was stirred by a device he thought might have great commercial possibilities. An impeccably groomed and proper San Francisco lawyer with extensive connections on Wall Street and in Washington, Gaither was a habitué of the worlds of finance and of technology. After the war, he had helped the Pentagon reorganize its research and development arm, known as Project RAND, into the independent RAND Corporation. Then he assisted Henry Ford II with restructuring the Ford Foundation, of which he would later serve as president. Gaither's relationship with Ernest had broadened from that of financial and legal counselor to one of personal friendship. After the new project had been effectively blessed by Alfred Loomis, the two Loomis protégés formed a partnership to transform Lawrence's tinkering on color TV into a business. Their only disagreement involved the division of ownership: Gaither proposed an eighty-twenty split in Lawrence's favor, and Lawrence insisted on fifty-fifty. They sought a tie-breaking ruling from Loomis, who supported Lawrence. On those lines, Chromatic Television Laboratories was incorporated on March 31, 1950.

Color TV was an embryonic technology, but the size of its perceived market already had triggered a furious rush to develop a consumer product. The contestants included RCA, the Columbia Broadcasting System, and General Electric, all seeking approval from the Federal Communications Commission for mutually incompatible technologies. Hollywood was also interested, and Gaither soon lined up financial backing from Paramount Pictures. Over the next six years, its investment would run to millions of dollars.

In the thrall of his usual optimism, Ernest Lawrence began recruiting workers for the venture, installing them in a garage laboratory in a vacation home he had acquired in Diablo, a resort community less than an hour's drive from Berkeley. The Diablo house had been another failed effort at a relaxation hideaway. Ernest had bought it on the spur of the moment in 1950 as "a place where he could get away from the telephone and the pressures of his regular work," Molly recalled. It was a smallish bungalow, not really spacious enough for their family of eight, though it was located near a country club with a swimming pool for the children. The house never appealed much to Molly, for the inland summers were beastly hot, and Ernest himself never seemed to be around—to him, the inactivity defined by "rest and relaxation" meant boredom. But now the house acquired a new purpose. Ernest equipped the garage with bunk beds and a kitchenette for Chromatic technicians, and soon doubled its size to accommodate a full-scale electronics shop.

The staff swelled with scientists and technicians recruited from Livermore and the Rad Lab. One of the first was Don Gow, a former military engineer who had worked with Alvarez on the linear accelerator and the MTA. Gow was swept into Lawrence's magnetic field and held there by the boss's "sense of urgency, his willingness to try ideas rapidly and drop them when a better idea came along." He and his fellow Chromatic employees became accustomed to receiving calls from Lawrence at dinnertime with the message, "I've got a new idea—let's stop what we've been doing and look at it tonight." They would all speed out to Diablo as if in a caravan, not to return home to Berkeley until dawn.

Lawrence worked on the tube design with an intensity he had not shown since the Alpha racetrack project. The blue notebooks he always carried filled up with notes and designs scribbled at all hours of day and night and all circumstances: on the drive from Balboa to Berkeley, on the train to Chicago, during a Bohemian Grove retreat, on a flight to New York. In mid-1951 he developed a radically new design based on placing microscopically thin wires just behind the tube's glass viewing surface. These were to focus the electron beam onto colored phosphors behind the screen. The Diablo shop built a prototype, and the result was stunning: a much brighter picture than the RCA and CBS tubes, with cheaper components. On the strength of a demonstration, Paramount agreed to build a manufacturing plant to turn out prototype tubes in Oakland.

Paramount also stepped up its involvement in Chromatic's management. Suddenly there were production managers and financial executives looking over Ernest's shoulders. The essential contradiction between the operational methods of research laboratories and industrial factories became excruciatingly clear. Production engineering, Gow reflected, "is a very expensive game and none of us, including Ernest, had the faintest notion what it meant." Lawrence had run the Rad Lab frugally—for a laboratory. But the cost of experimental equipment had always been secondary to the goal of the research, especially during the war, when the urgency of the Manhattan Project made profit-and-loss calculations irrelevant. "We were all used to building one of something, and when it contributed to a million-dollar experiment, you hardly noticed the cost," Gow said. Lawrence's efforts at Chromatic similarly were aimed at producing a single working prototype, but Paramount's goal was to mass-produce tubes to be marketed at a nominal price of $50 or $75. The demands of mass manufacturing and mass marketing began to overwhelm Ernest's resolutely sunny disposition.

Adding to the pressure were the many other demands on his time. The debate over the Super had not let up, resulting in frequent summonses to Washington. Livermore was undergoing its transition from the MTA project to its role as the second bomb laboratory. And of course the Rad Lab was running at full bore, with the Bevatron back on track. "Ernest was probably the busiest man in the country," Gow observed. "There were very important people with respect to defense matters and very important people with respect to basic science and basic research, and very important people with respect to color television, and very important people with respect to God knows what else, in and out of the office in an absolutely steady string." Lawrence's hands were on everything, and everything required decisions, sometimes split-second decisions.

Finally, in the spring of 1952, Ernest broke down, landing back in the hospital with another attack of colitis. The doctors counseled rest and solitude, but it was a prescription he could not tolerate. A week at Balboa was interrupted by daily phone calls from Washington, New York, Los Alamos, and Berkeley. He tried to maintain his usual breezy mien, but it often cracked, producing outbursts of temper over even minor problems. On one occasion, he slapped a Livermore staff member who had questioned his instructions, an unprecedented breach of the Lawrence style and Radiation Laboratory protocol. The victim gave notice but was persuaded to stay after Ernest assembled the entire staff to witness his personal apology.

Plainly, a way had to be found to enforce a rest cure. The ingenious solution came from Jack Neylan: a round-the-world voyage on tankers owned by Standard Oil of California. As company chairman Gwin Follis was a personal friend of Neylan's—and the company itself had been Berkeley's subcontractor on the MTA—arrangements were concluded rapidly. Ernest and Molly would be accompanied by sixteen-year-old Margaret, whose teenage suitors' attentions made Ernest uneasy; and by Dr. John Sherrick, a stodgy old family friend who had delivered five of the Lawrence children and was slotted in as Ernest's traveling medic.

The group departed from the Port of New York on January 24 aboard the oil company tanker *Paul Pigott*, the manifest of which identified them as members of the crew (Molly and Margaret listed as stewardesses and Ernest as ship's doctor), their wages set at one dollar each for the voyage. More surprising were the spacious and luxurious quarters available for VIP travelers on the seagoing vessels of a great American corporation. For two months they traveled in style—from New York to Beirut by sea; thence by car to Amman, Jordan, and onward by air to Bahrain, Karachi, and Ceylon, where they spent an enchanting two weeks as Paramount's guests, watching Vivien Leigh and Peter Finch struggle to film a blockbuster called *Elephant Walk*. (Leigh later suffered a nervous breakdown and was replaced by Elizabeth Taylor.) They were feted by premiers and emirs and entertained at embassies and consulates as they made their way back west to Palermo, Sicily, where they boarded another tanker for the homeward passage. Ernest disembarked in New York seeming more hale than he had in years, delighting Alfred and Manette Loomis, who greeted the voyagers at dockside.

Ernest's condition was not to last. There were issues to manage at the fledgling Livermore lab, thermonuclear tests to attend in the Nevada desert and plans to be made for new tests at Eniwetok. International duties also beckoned. In 1954 he visited Geneva to consult on the establishment of an international high-energy physics lab under the aegis of CERN, the European Organization for Nuclear Research; in time, this facility would host two linear accelerators, three synchrotrons, and the Large Hadron Collider, all of them technological offspring of that original, wax-slathered accelerator Ernest had held in his hand nearly a quarter century earlier. Later in the spring—on the very day the Oppenheimer hearing ended—Ernest departed for a speaking tour of Japan sponsored by the AEC's Atoms for Peace program, which aimed to relieve nuclear research of its martial coloration by emphasizing its potential for power generation and other drivers of peacetime prosperity.

Back home, a national debate was raging over H-bomb testing, provoked by the fallout fiasco of the Bravo tests. Lewis Strauss, determined to keep President Eisenhower on the path to thermonuclear supremacy, would demand more from Lawrence: more counsel, more public support, more time and effort. Soon after taking office as AEC chairman in 1953, Strauss expanded Project Sherwood, an Atoms for Peace program aimed at developing electric power reactors based on nuclear fusion. Livermore leaped onto the fusion bandwagon, following Lawrence's practice of adjusting his research priorities in step with where the money was; Strauss had expanded the budget for Sherwood from $1 million to $10 million, with Livermore in line for a third of the total. But ramping up Livermore to embark on a new program meant more pressure on Ernest and more disappointment, for after a burst of enthusiasm for controlled fusion among nuclear experts, the idea proved to be unworkable.

Then there was Chromatic. A few months after Lawrence's return from his round-the-world tour, success appeared to be nearly at hand: Lawrence color tubes manufactured in Oakland played a role in a landmark of television history: the televised coronation of Queen Elizabeth II. On June 2, 1953, while a worldwide audience estimated at 150 million watched the coronation procession in black and white, one hundred young patients at London's Great Ormond Street Hospital, a children's institution, viewed it live in color, on two twenty-inch Lawrence tubes displaying a closed-circuit image from three color cameras stationed along the route. But technical achievements were one thing; the commercial potential of color TV was another. Doubts were growing about the public demand for color sets. At Chromatic, the ebbing interest was sensed acutely. "Instead of having people clamoring at us for sample tubes, you couldn't give them away," Gow recalled.

By 1955, Paramount would be desperate to exit the field; Ernest, feeling duty-bound to help his financial partner reduce its exposure to losses, went on the road to help Paramount find a buyer. He met with executives from Columbia Broadcasting System and Philco in the United States and flew to Holland to make a pitch, fruitlessly, to the Dutch technology conglomerate Philips. In the end, Paramount transferred Chromatic to its DuMont Laboratories, a TV manufacturer affiliated with the DuMont television network, an early network that would fail in 1956 despite launching the careers of Jackie Gleason and other future television stars. Lawrence's technology did not quite disappear; in 1961 Paramount licensed what was left of it to Sony. A few years later the Japanese company incorporated elements of Lawrence's design into one of the most successful color TV technologies in history, a product it named Trinitron.

Chapter Twenty-one

The "Clean Bomb"

Ernest's colitis recurred in 1956, a flare-up he attributed largely to stress over the unresolved fate of Chromatic. But he had equally pressing concerns about Livermore. The lab's future was still uncertain, despite a series of successful trials dubbed Operation Teapot at the Nevada Test Site in February and March 1955. After one of the blasts, the lab's business manager ran down the hall to York's office with the news, shouting, "We're still in business!"

Teapot did not end the debate in Washington over the country's two duplicative bomb labs. Only two months later, Congress's Joint Atomic Energy Committee convened yet another in its seemingly endless string of hearings on the topic. The agenda was framed by the committee's staff as a series of ominous questions: "Exactly what is the relationship between Los Alamos and Livermore? . . . In the event that the two laboratories come up with similar proposals . . . how does the [AEC] decide who will do the job?" Moreover, Teapot's success was incomplete. The operation had failed to achieve its most important goal, which was to show that radioactive fallout from weapons tests could be confined to the test site. Within days of the blasts, fallout was detected as far east as New York, New Jersey, and South Carolina. The health risks, to be sure, were negligible. But the political implications were enormous.

Americans made anxious by reports of the dangers of fallout were not comforted by Lewis Strauss's maladroit performance at a press conference at the White House on March 31, shortly after the Bravo tests in the Pacific. In Eisenhower's presence, Strauss spoke haltingly from a written statement, assuring his audience that radiation reaching the United States was "far below the levels which could be harmful in any way to human beings, animals, or crops." Unwisely, he then took questions. Asked by a reporter to describe how powerful this strange new weapon known as the hydrogen bomb might be, he replied: "As large as you wish . . . that is to say, an H-bomb can be made large enough to take out a city."

"Any city? New York?" he was asked.

"The metropolitan area, yes."

Eisenhower accompanied him out of the room. "Lewis, I wouldn't have answered that one that way," the president said levelly. "I would have said, 'Wait for the movie.' "

It was too late. Strauss's words created alarm over the prospect of nuclear holocaust without laying to rest the public's anxiety about fallout. For Livermore, this was the worst of all possible worlds, for it gave new life to a campaign to end nuclear testing.

The president himself did not need the Bravo results to make him sensitive to the international drive for a test ban and nuclear disarmament. As a military man, he was skeptical about explosive devices that were too powerful to serve as practical weapons, and less impressed than Truman with the idea that bigger was invariably better. Herb York had learned this in 1954, soon after Lawrence bestowed on him the formal title of Livermore director. His authority bolstered by a title on the door and personalized office stationery, York set forth a "working philosophy" for Livermore calling for "always pushing at the technological boundaries." In practice, that meant developing powerful but compact thermonuclear devices, with a high proportion of yield to weight. But when York sent a proposal to test a twenty-megaton bomb to the White House for Eisenhower's approval, he was crisply rebuffed. Upon being informed that the bomb would be much larger than any ever built, the president barked: "Absolutely not! They are already too big!" As York learned later from General Andrew Goodpaster,

Eisenhower's military attaché, his request had prompted the president to reflect that "the whole thing is crazy; something simply must be done about it."

Eisenhower soon launched an initiative to do that "something," appointing Harold Stassen as his special assistant for disarmament in March 1955. Stassen had been labeled the "boy wonder" of Republican politics after his 1938 election as governor of Minnesota at the age of thirty-one. His amiable North Country personality cloaked a sharp legal mind and soaring political ambition. The appointment carried cabinet rank, which Stassen was determined to exploit fully; seizing on a newspaper editorialist's calling him Eisenhower's "Secretary of Peace," he audaciously advised the president that if he was asked about the nickname at a press conference, Eisenhower should reply, "That certainly expresses it." Stassen leaped into his job with characteristic energy, declaring that he would assemble a supporting staff of "experienced men with brilliant analytical minds" to review American disarmament strategy from top to bottom. A few weeks later, he established eight task forces to take on the job. To chair the panel devoted to the crucial issues of international inspection and control of nuclear materials, he named Ernest Lawrence.

Strauss made the most of his influence over Lawrence by urging him to appoint Teller to the task force and keep thinkers of the Oppenheimer stripe off it. His advice may have been superfluous, for when it was fully assembled, nine of the Lawrence panel's twelve scientists were from Livermore. Besides Teller, they included Mark Mills, a bespectacled thirty-eight-year-old theoretician from Caltech who had joined Livermore the previous year and risen quickly to become one of Ernest's most trusted aides. The remaining task force members hailed from the RAND Corporation, Rowan Gaither's old fiefdom. This composition ensured that the panel would lean toward opposing a test ban based on doubts about the feasibility of inspections. That would create difficulty for Stassen, for international pressure to suspend tests was growing—indeed, the Soviets had proposed a test moratorium in May, just as Stassen's task forces were coming together.

Among the members of the inspection task force, Lawrence and Mills were willing to entertain the idea of a test ban in principle but were pessimistic that it could be adequately monitored. The hardliner, naturally, was Teller, who reminded his Livermore colleagues on the panel that a test ban struck at the heart of the lab's very purpose. "Edward was always unabashedly hostile to the whole idea [of a test ban]," York recalled. "If we were behind, we had to test to catch up, he said; and if we were ahead, we had to test to stay there. There was no circumstance under which a test ban could be in our interest."

Teller's viewpoint was shared by Lewis Strauss, who was struggling at the White House to hold the line against Stassen's lobbying for an international agreement on nuclear weapons. On the eve of a joint meeting of all the task forces in October 1955, Strauss cautioned Lawrence against any attempt by Stassen to co-opt his reputation for the cause of disarmament. "All that the man in the street will realize," he wrote Lawrence, "is that a great scientist, inventor of the cyclotron, has accepted this assignment and, because of the stature of his scientific ability, will pull the rabbit out of the hat." He need not have worried. The report from Lawrence's panel proved suitably skeptical about monitoring. The main problem, Ernest concluded, was the difficulty of tracking fissionable material in the Soviet Union to ensure that it was not finding its way into weapons. Reducing diversions even to 10 percent of total production, he calculated, would require a force of tens of thousands of inspectors fanned out across Russia's vast territory.

On this occasion, however, Lawrence's immense credibility as the biggest of the big scientists failed to carry the day. His estimate came in for ridicule when Stassen presented the task force findings to the National Security Council on December 22. President Eisenhower told Stassen he was "quite sure the Soviets have never given any thought to any inspection plan which involved the presence . . . of anything like twenty to thirty thousand foreign inspectors." Secretary of State John Foster Dulles complained that the figure would make the United States "a laughing stock"— not the best position for the United States to be in during the run-up to a summit meeting with the Russians, scheduled for the coming summer in Geneva. Stassen was instructed to "revise" his estimates and report back in mid-February.

The prospects for a test ban agreement faded, but only temporarily. As the presidential election year of 1956 dawned, international pressure on the White House intensified. During a state visit to Washington early in the new year, British prime minister Anthony Eden proposed that the United States and Great Britain unilaterally suspend testing. The move, he explained, would give him political cover at home, where public sentiment against the bomb was steeply on the rise. The proposal won him only a stern lecture from Lewis Strauss, who maintained that concerns over fallout were overblown, and therefore a ban was unnecessary.

Strauss already was orchestrating a propaganda campaign against fallout panic, featuring scientists on the AEC payroll. The campaign had been launched in December 1955 with a paper in the *Scientific Monthly* by Gordon Dunning, a "health physicist" at the AEC who assured readers that the technology for confining radioactivity to the site of a blast was constantly improving. Dunning's paper, which declared nuclear testing to be "mandatory to the defense of our country," was couched in the sober language of an academic treatise. It bristled with statistics that sounded comforting but were impossible for the average layperson to fully comprehend. ("Accepting the foregoing estimates, if several large thermonuclear detonations occurred every year for 30,000 years, the near equilibrium amount of carbon-14 thus created in the world would be about twenty times greater than the amount now present.") Dunning's bottom line was that "there is essentially no risk of hazardous amounts of fallout outside the control areas in the Pacific and continental United States."

Then came a widely reported lecture at Northwestern University by AEC commissioner Willard Libby, a former Berkeley chemistry professor. "It is possible to say unequivocally," he stated, "that nuclear weapons tests carried out at the present time do not constitute a health hazard to the human population."

But the pressure for an end to testing continued to mount. Adlai Stevenson II, the former governor of Illinois, established leadership on the nuclear issue early in his campaign against Senator Estes Kefauver of Tennessee for the 1956 Democratic presidential nomination by calling for a test moratorium; Kefauver was forced reluctantly to concur. Another source of pressure was nestled within the commission itself, in the person of AEC commissioner Thomas Murray, who originally had joined with Strauss to favor the Super project but had become unnerved by the intensity of the arms race. A millionaire businessman and a prominent Catholic layman, Murray framed his position in stark moral terms. Testifying before the Joint Committee on Atomic Energy, he called for a unilateral test moratorium on large devices and limitations on the size of weapons for the US stockpile, although he stopped short of an outright ban on all nuclear weapons. "God in His almighty power and goodness has given us the secret of atomic energy for purposes of peace and human well-being, and not for purposes of war and destruction," he told the committee in February 1956.

Murray also advocated development of a "clean bomb"—a device that could somehow deliver the destructive force of a nuclear blast minus the fallout. The rather chimerical notion of a nuclear bomb without radioactive side effects had emerged in late 1954 and had been embraced by Livermore, where it fit nicely with the lab's brief to explore new ideas. The concept appealed to both Teller and Lawrence, who understood that public alarm about fallout could threaten the lab's existence. In the spring of 1956, "clean" devices were placed on the program for Operation Redwing, a series of tests in the South Seas. Eisenhower mentioned the clean bomb obliquely at a press conference shortly before the operation began, when he fended off a question about Stevenson's call for a moratorium by declaring that America's goal was "not to make a bigger bang, not to cause more destruction [but] to find out ways and means in which you can . . . reduce fallout, to make it more of a military weapon and less one just of mass destruction.

"We know we can make them big," he said. "We are not interested in that anymore."

The program for Redwing included the first airborne drop of a thermonuclear device, which created an enormous fireball four miles in diameter when it detonated fifteen thousand feet over Bikini. Wind blew the fallout plume safely away from inhabited islands, which allowed Strauss, in his official statement after the tests were concluded, to crow about having achieved "maximum effect . . . with minimum widespread fall out hazard. Then, clumsily, he said too much: "Thus the current series of tests has produced much of importance not only from a military point of view but from a humanitarian aspect."

Strauss's allusion to a "humanitarian" H-bomb drew derision from critics of the arms race. The very notion of "clean" thermonuclear weapons was mercilessly demolished in the *Bulletin of the Atomic Scientists* by Ralph Lapp, a distinguished antinuclear physicist. By lucidly describing the process that produced an H-bomb blast, Lapp showed that a clean bomb was a fantasy. Because an H-bomb encompassed fission, which invariably produced radioactive emissions (dirty), and fusion, which did not (clean), one could reduce the *relative* radioactivity produced by the blast by increasing the ratio of fusion to fission in the bomb. But that increased the bomb's overall power, which created an *absolute* increase in dirtiness. "The superbomb can be designed to be either relatively clean or very dirty," Lapp wrote. "Part of the madness of our time is that adult men can use a word like *humanitarian* to describe an H-bomb."

Strauss plainly had overreached, but to what end? His goal may have been to dilute the impact of a fallout study released by the National Academy of Sciences. The study concluded that the effect on Americans of radioactive emissions from nuclear tests was vastly outweighed by their exposure to diagnostic X-rays and other common sources. (Lawrence himself had launched a personal campaign in Berkeley to eradicate the casual use of fluoroscopic X-ray machines on children's feet, a ubiquitous promotional practice that was eventually outlawed nationwide.) What drew headline attention, however, was the academy's finding that even low-level radiation exposure carried potential genetic consequences. "The concept of a safe rate of radiation," the study declared starkly, "simply does not make sense if one is concerned with genetic damage to future generations." This conclusion added to public alarm not merely about test fallout but also about the global arms race generally.

Far from giving up on the "humanitarian" H-bomb, Strauss redoubled his efforts on its behalf in his role of Eisenhower administration spokesman on the technicalities of nuclear policy. The idea he put forth was that a testing moratorium would halt research into making H-bombs safer, thereby leaving the United States with an obsolete "dirty" nuclear arsenal. While Adlai Stevenson assembled PhDs to talk about the risk of elevated cancer rates and "uncontrollable forces that can annihilate us," Strauss deployed his own cadre of scientists to dismiss these statements as lurid propaganda and depict the threat of radiation damage as minor.

Among them were Lawrence and Teller. Shortly before the election, Strauss prevailed on them to issue a statement opposing Stevenson's test ban. Teller produced the first draft and brought it to Lawrence's home on the Sunday before Election Day, just as Ernest was bidding farewell to guests with whom he had shared a liquor-fueled country club outing. When he finally was able to scrutinize the draft, he found that it bore all the flaws of Teller's oratorical style: it was too long, too argumentative, and too emotional. Ernest summoned Daniel Wilkes, the university's public information officer, to his house for an instant redrafting. Wilkes, perplexed at the sudden urgency for a statement about a policy debate that had been going on all year, warned that it would appear overtly partisan when it appeared in newspapers on election morning. But Ernest, who was "feeling no pain" from his sociable afternoon, as Wilkes recounted later, countered that the statement was a favor for Strauss and needed to be ready that night.

Wilkes's version, issued over the signatures of Lawrence and Teller, sounded familiar themes: the country had "no sure method of detecting nuclear weapons tests" by the Soviets and continued US testing was necessary to maintain "a fast-moving scientific technical nuclear weapons program . . . We are never sure a device will work until it is tested; and we cannot know that our last idea works." The text assured Americans that "the radioactivity produced by the testing program is insignificant" and that, no matter the outcome of the election, "tests will continue to be carried out with scrupulous regard to public health." The statement received prominent play in newspapers around the country on the morning of the election, which Eisenhower, as expected, won in a landslide.

In defeat, Adlai Stevenson had succeeded in placing nuclear testing policy solidly on the public agenda. Inside the White House, however, the standing of Harold Stassen, the most vociferous advocate of the test ban, had slipped. Before the Republican National Convention that summer, he had irked Eisenhower by leading the "Dump Nixon" movement, which aimed to place Christian Herter, the patrician governor of Massachusetts, on the ticket as vice president in Richard Nixon's stead. At disarmament talks, moreover, he had developed the disturbing habit of offering the Russians concessions that had not been approved by the president or Secretary of State John Foster Dulles. In March Stassen was stripped of his cabinet rank as a rebuke. Under normal circumstances, his demotion would have given Strauss an opening to lobby against the test ban; but Eisenhower was losing patience with Strauss, too, largely because of his intransigence on that very topic. The policy stalemate within the administration continued.

It was an uncomfortable time for Livermore, which was suffering through another in a long sequence of existential crises. This one dated back to the previous fall, when the AEC and the Pentagon had worked out the military program for both weapons laboratories. During that process, five projects had been canceled or suspended—all of them Livermore's. The new schedule idled "half of our potential capability," York grumbled to Norris Bradbury. AEC officials tried to ease Livermore's concerns by advising Lawrence and York to keep moving ahead on advanced research—because of the lab's role as the developer of new technologies, its products would always run ahead of the military's known needs at any given point; sooner or later, the Pentagon's requirements would catch up with the lab's innovations. This would turn out to be true, but in the fallow days of late 1956 and early 1957, Livermore's future looked bleak.

Earlier in the year, a new provocation had revived international protests against the bomb. This was Great Britain's announcement that it would test its first thermonuclear device over Christmas Island that spring. The prospect prompted indignation in Hawaii and Japan and a statement opposing further tests from the revered humanitarian and Nobel Peace Prize laureate Albert Schweitzer, issued from his home in the equatorial African nation of Gabon. Schweitzer's warning of "a catastrophe that must be prevented under every circumstance" was read by the president of the Nobel Prize committee in Oslo to a vast radio audience around the world—except in the United States. There his words went unbroadcasted and almost unnoticed, until AEC commissioner Willard Libby issued a public riposte that served chiefly to bring Schweitzer's statement to Americans' attention despite the blackout.

Meanwhile, the scientific community was stirring again. The animating figure was Caltech chemist Linus Pauling, a Nobel laureate known for his strong leftist views and stem-winding oratorical flair. After receiving a standing ovation for a speech at Washington University in St. Louis advocating a test ban, Pauling launched a petition drive among scientists calling for an immediate international halt to all nuclear testing. He started with twenty-seven signatories, among them Harold Urey, Merle Tuve, and the former Rad Lab scientists Martin Kamen and Edward U. Condon. Within two weeks, there were two thousand signatures, at which point Pauling delivered the petition to the White House and released it to the press.

The petition served to draw Lawrence and Teller back into the public debate. Their opportunity had been developing since Memorial Day, when Senator Henry Jackson of Washington called at Livermore. A member of the Joint Committee on Atomic Energy, Jackson's goal was to enlist Lawrence and Teller in his lobbying for increased production at Hanford, the plutonium plant located in his home state; but the conversation inevitably strayed onto the topic of the need for continued testing to preserve America's nuclear superiority. Encouraged by the scientists' firm opposition to a test ban, Jackson invited them to address the Joint Committee's Subcommittee on Military Applications, of which he was chairman.

Lawrence, Teller, and Mark Mills appeared before the subcommittee on Thursday, June 20. Jackson introduced them by alluding to their goal of making atom bombs cleaner, reporting "the gleam in the scientists' eye of making them almost like Ivory Soap"—then adding wryly: "but not quite."

The scientists urged the committee not to short-circuit testing that could make nuclear weapons more moral. "If we stop testing," Lawrence declared, "well, God forbid . . . we will have to use weapons that will kill fifty million people that need not have been killed." Seen in that light, he said, a test ban would be "a crime against the people." Teller's contribution was a warning that the United States could never devise a foolproof monitoring regime to guard against clandestine testing by the Soviets. The committee members, impressed by the scientists' presentations, wondered aloud if the White House was aware of their views. W. Sterling Cole, a Republican from New York, took the initiative of arranging a meeting for the three physicists with President Eisenhower for the following Monday. They holed up for the weekend in a Washington hotel suite, where they were drilled carefully by Lewis Strauss.

At nine o'clock in the morning on June 24, Strauss and his charges were ushered into Eisenhower's presence inside the Oval Office. Ernest Lawrence, a Nobel laureate and the intimate friend of statesmen and millionaires, suffered a sudden bout of stage fright at his first meeting with a US president. His discomposure astonished Edward Teller. "This awe was something I just could not imagine," he recalled. "He could not bring up a word. I mean he was all tight and excited."

Lawrence eventually found his voice and launched into a speech very much like the one he had delivered to Jackson's subcommittee. The minutes of the White House meeting quote him as stating: "If we know how to make clean weapons, but fail to do so and to convert existing weapons into clean ones, then the use of dirty weapons in war would truly be a crime against humanity." Eisenhower listened in fascination, but tactfully schooled his visitors in the realities of international arms policy. He reminded them that the United States was "up against an extremely difficult world opinion situation" and declared that he did not want the nation to be "crucified on a cross of atoms." But he assured them that no test ban would be accepted without a comprehensive agreement on disarmament. "We have not thought of stopping tests without some kind of package deal," he said.

After they left the Oval Office, Strauss paraded the scientists before the White House press corps. Lawrence assured the reporters that it would be possible "to produce nuclear weapons that are in a sense just like TNT [in other words, fallout-free], except tremendously more powerful."

A reporter asked how clean they could be. Strauss replied that fallout had already been reduced "between nine-tenths and ten-tenths, almost half the way."

The news conference ended with a question for Lawrence: "Do you think the tests should be continued?"

"Of course I do," he replied.

The next day, the national press parroted the figures it had been fed by Strauss: "U.S. Eliminates 95% of Fall-Out from the H-Bomb," the *New York Times* reported on its front page.

Reading those words in his Manhattan apartment, David Lilienthal found them repellent. "The irony of this is so grotesque it is rather charming," he wrote in his personal journal. "Ernest Lawrence and Edward Teller, with Strauss, were the ones who were so sure that the *super* H-bomb, big as all hell, would be the saving of the country; that those, who like myself, entertained strong doubts about this—well, there must be something queer or unpatriotic about us . . . But now it appears that the big H-bomb doesn't seem so clearly the answer to all the world's problems of security."

Lilienthal had made a long, dispiriting personal journey since that day in 1946 when he had met Ernest Lawrence for the first time and recorded his impressions of "a rather fabulous figure . . . full of vitality and enthusiasm." Now he regarded Lawrence as a scientist on the make for government largess and personal glory. "How greedy for headline fame can you get?" he wrote. His contempt for the "salesman type" of scientist was unbounded. "E. O. Lawrence, Luis Alvarez, Edward Teller—Madison Avenue–type scientists. Scientists in gray flannel suits."

Ernest and his colleagues left a profound impression on President Eisenhower, though not enough to prompt him to rule out a test ban entirely. Goaded by a reporter the day after their visit to give "an unequivocal yes or no on this business of immediate suspension of nuclear testing," Eisenhower continued to equivocate, indicating that while he now had second thoughts about a test ban, he was still determined to negotiate one. The president explained that he had just been "visited by people that certainly, by reputation and common knowledge, are among the most eminent scientists in this field, among them Dr. Lawrence and Dr. Teller . . . They tell me that already they are producing bombs that have ninety-six percent less fallout than was the case in our original ones, or what they call dirty bombs . . . They say, 'Give us four or five years to test each step of our development, and we will produce an absolutely clean bomb.' . . . It does show, as you so aptly say, the question is not black and white."

Eisenhower had considerably exaggerated the scientists' purported accomplishment. They told him they needed at least six or seven years for a cleaner bomb; he quoted them as offering an "absolutely clean" device in four or five. His version left the scientific community dumbfounded, for by general agreement, there was no way to create "an absolutely clean bomb." Even relative freedom from fallout, well short of the 95 percent cleanliness Strauss and the scientists had mentioned, could be achieved only from reducing yields. That was well understood at Livermore, where Herb York informed Brigadier General Alfred D. Starbird, the AEC's director of military application, that developing a clean tactical weapon by the early part of the following decade could not be done except with some "very lucky breaks." And that time frame was for weapons of immense size and doubtful transportability by aircraft or missile. Smaller weapons would take "several more years" at least, he said.

Several news publications looked into the predictions of powerful yet clean nuclear weapons made by Lawrence, Teller, and the president and determined they were talking fantasy. "What an 'absolutely clean' H-bomb might be remained a mystery to most scientists and congressmen, as well as to the public," observed *Newsweek*. Eisenhower's remarks revived the contempt for the very concept of a "safe" thermonuclear weapon that had been raised by Strauss's invention of the "humanitarian H-bomb." Asked *The New Republic*: "Is it 'cleaner' to be vaporized by H-bomb blast than to be poisoned by H-bomb fallout? Apparently so, if we correctly read the President's words . . . When Admiral Lewis L. Strauss and his technicians"—a scornful demotion for Lawrence and Teller, two of the nation's most eminent scientists—"plead for five more years of tests, they are asking, it seems to us, *not* for time to produce a 'humanitarian' bomb, but for a continuation of the arms race."

Outside the White House, Lawrence and Teller barely moved the needle of debate. Those who were opposed to testing found the scientists' promises absurd and immoral; those who favored continued tests adopted their argument that testing was the best way to make nuclear weapons less horrible and more practical. Strauss, for his part, was delighted by the performance of his trusted scientists, for plainly they had succeeded in buying time for continued testing, if not for sweeping a ban off the negotiating table entirely. When Lawrence and Teller arrived back at Livermore, they were greeted by letters bearing his congratulations. "Everything has worked out as we had hoped it would," Strauss wrote. His military aide, Navy Captain John H. Morse Jr., seconded his opinion: "You may detect some feeling that clean weapon potentialities have been over-stated and over-simplified recently," he wrote Lawrence. "I think you did exactly the right thing in statements to the President and the Press. The situation called for over-selling rather than under-selling."

Strauss had another reason to feel optimistic about a continuation of testing: he had seen off his most determined adversary on the Atomic Energy Commission, Thomas Murray. At Strauss's urging, Eisenhower declined to reappoint Murray when his term as commissioner expired on June 30. ("I mark off the days on my calendar" before Murray's departure, Strauss told the president.) Murray would continue his campaign for a test ban in public appearances and congressional testimony, but now he was on the outside looking in. With the death in February of the distinguished mathematician John von Neumann, who had served on the AEC since 1955, there were now two vacancies for Eisenhower to fill—with solid supporters of continued testing, Strauss hoped. But the Joint Committee on Atomic Energy held jurisdiction over AEC appointments, and to appease its Democratic majority, Eisenhower named two former Truman aides, John S. Graham and John F. Floberg, to the open seats. Neither was vulnerable to Strauss's sway.

The appointments were a further manifestation of Strauss's fading influence, but a greater change was to come. It was prompted less by developments on the disarmament front than by a very different event in the technological contest between the United States and Russia.

On October 5, 1957, America awoke to the news that a man-made object launched the previous day from a Soviet space complex was orbiting overhead. It was a 184-pound aluminum alloy sphere twenty-three inches in diameter which the Russians called *Sputnik*—meaning "traveling companion" or "fellow traveler." In the hysteria that followed over Soviet scientists' apparent outdistancing their American counterparts, a new group of technical advisors was brought into the White House. Lewis Strauss's monopoly on the scientific information reaching Dwight Eisenhower was about to end, taking with it Ernest Lawrence's influence over nuclear policy.

Chapter Twenty-two

Element 103

The Eisenhower White House moved swiftly to soothe Americans' nerves over *Sputnik*, dispatching spokesmen to dismiss the satellite orbiting in the sky overhead as "a silly bauble" and its launch as merely "a neat scientific trick." Keeping the administration's more independent outside advisors in line proved more of a struggle. When Edward Teller declared on Edward R. Murrow's popular television program that the launch was a loss for the United States "greater and more important than Pearl Harbor," Eisenhower upbraided Lewis Strauss for his protégé's loose talk. But the ominous implications of the Soviet achievement were hard to overlook, especially after the launch of *Sputnik 2*, a capsule carrying a live dog named Laika ("Barky"), one month later; the Soviet Union appeared to be outpacing the United States in the development of long-range ballistic missiles capable of carrying a large payload—a nuclear warhead, for example. *Sputnik 2* weighed 1,100 pounds; the heaviest American satellite, Vanguard, weighed 3.5 pounds and had yet to make it off the launch pad. All the pride Americans had felt in the Big Science that had won World War II was draining away at the spectacle of multimillion-dollar rockets exploding in flight.

Eisenhower understood that allaying the public's concerns required concrete action. For advice, he turned to I. I. Rabi, whom he had come to know during his five-year stint as president of Columbia University. Born in Polish-Ukrainian Galicia, Rabi was the levelheaded physicist who had tartly denounced Strauss's vendetta against Oppenheimer and had attempted, if fruitlessly, to talk Lawrence and Alvarez down from their starry-eyed pursuit of the Super. He also chaired the Science Advisory Committee of the Office of Defense Mobilization, a successor to the technology committees organized by Vannevar Bush during the war. In that role, Rabi strove to present a reasoned counterbalance to Strauss's unrelenting opposition to a test ban. At a White House meeting three weeks after the *Sputnik* launch, he joined with Hans Bethe, the Cornell physicist who also had flatly refused Teller's invitation to join in research on the Super, in arguing that a test ban would preserve America's nuclear advantages over the Soviet Union and that therefore the United States should accept a ban "as a matter of self-interest." Strauss simmered through most of the presentation but erupted when Rabi took a potshot at the illusory concept of a "clean" bomb. Strauss fired back furiously that *his* science advisors, Professors Lawrence and Teller, thought the assumptions underlying Rabi's findings baseless. The bewildered president shut down the bickering and later confided to his diary: "I learned that some of the mutual antagonisms among the scientists are so bitter as to make their working together almost an impossibility . . . Dr. Rabi and some of his group are so antagonistic to Drs. Lawrence and Teller that communication between them is practically nil."

It was Rabi who retained Eisenhower's confidence in this lofty scientific dispute, however. Asked for advice about a response to *Sputnik*, he replied that what Eisenhower needed above all was a sound, independent science advisor without ties to the political camps pitched on opposite sides of the disarmament issue. The best man for the job, he said, was a member of his Science Advisory Committee named James R. Killian.

Killian, a soft-spoken South Carolinian of fifty-three who had succeeded Karl Compton as president of MIT, was a management expert rather than a trained scientist. But he had proven himself as a superb academic administrator, earning his spurs honing consensus among the fractious faculty cliques on the MIT campus. In his new post as chair of what would be christened the President's Science Advisory Committee, or

PSAC, he would earn the trust of both the White House and the scientific community while keeping the president "enlightened"—Eisenhower's word—about the innumerable technological challenges facing his government. (Eisenhower called him "my 'wizard.' ") In his three years chairing the PSAC, Killian also would succeed in restoring much of the confidence of scientists in their government patrons that had been lost by the persecution of Oppenheimer.

Killian's committee contained a diversity of scientific opinion that had been missing from Eisenhower's scientific councils under Strauss. As members, he appointed Rabi and Bethe as well as Herbert York, who had grown bored with the routine of lab management after five years at the helm of Livermore and would presently resign to become the Pentagon's chief scientist. For the first time since Eisenhower took office, the views of Strauss, Lawrence, and Teller were balanced by learned voices on the opposite side of the nuclear debate.

Killian's rise at the White House coincided with the waning influence of Eisenhower's former chief advisors on disarmament, Strauss and Harold Stassen. The last hurrah for the two aging warriors was sounded during a marathon session of the National Security Council on January 6, 1958, when Stassen proposed a new framework for talks with the Soviet Union. His idea was to offer the Russians a two-year moratorium on nuclear testing, with compliance to be verified by eight to twelve monitoring stations on each country's territory. What made the scheme novel was that for the first time it decoupled a test moratorium from all other disarmament issues, thereby removing the major cause of the deadlock between the two countries.

Although Stassen's idea resembled one that Eisenhower was quietly beginning to favor, it inspired an extended row at the NSC meeting. Secretary of State Dulles fretted that it would prompt objections from Great Britain and France. Strauss reiterated his familiar theme that the cessation of testing would have "severe repercussions" for the clean weapons program and that Los Alamos and Livermore would "lose momentum" during the suspension. He cited the conclusion of Lawrence and Teller that "several score of inspection stations would be required to monitor testing in the Soviet Union," not the dozen or so proposed by Stassen. Eisenhower, now fully alive to the limitations of the scientific advice Strauss had been feeding him, disingenuously asked him to explain the discord in the scientific community about the efficacy of monitoring. How, he asked, could the White House reconcile the views of Edward Teller, who had just published an article doubting the effectiveness of any monitoring, and I. I. Rabi, who maintained that the necessary oversight was well within US capabilities? "Apparently Governor Stassen believes in the opinion of one group of scientists, and Admiral Strauss follows the views of another group," Eisenhower observed. In light of subsequent events, he might have been hinting that it was time for both men to leave the field. Killian, attending his first NSC meeting, smoothly reset the debate by promising to place a definitive, neutral study of the monitoring issue by "the most highly qualified US scientific and technical personnel" on the president's desk within a few weeks. The meeting ended with Eisenhower tabling Stassen's plan, ostensibly for the moment.

Stassen saw the handwriting on the wall: his tenure as Eisenhower's "secretary of peace" effectively had come to an end. Eisenhower made it official on February 7, when he offered Stassen a consolation prize of a position elsewhere in the administration. Stassen declined, having already turned his attention to a prospective run for governor of Pennsylvania. Like Adlai Stevenson, he had achieved much in defeat. His imaginative, if often overreaching, diplomacy kept disarmament policy alive at the White House, and his final plan for decoupling a test ban from disarmament was indeed the key to breaking the logjam of policy that had made both goals so hard to achieve.

At a White House meeting two weeks before Stassen's departure, Strauss had reminded Eisenhower that his term as chairman of the Atomic Energy Commission was to expire on June 30. Eisenhower offered to reappoint him for a new term, but Strauss had grown weary of dodging brickbats in Washington. He asked the president to "weigh very carefully" the question of his renomination, as he had "accumulated a number of liabilities, including [the enmity of] most of the columnists in the Washington press." Eisenhower replied levelly that he shared the same liabilities. But when the time came, he let Strauss go.

In the early months of 1958, as the prospects for a moratorium brightened, the United States and the Soviet Union each staged a massive series of tests with the expectation that the test window would soon be closed. The US tests, known as Operation Hardtack, had been planned for nearly three years and were to encompass surface, underwater, and high-altitude trials. The latter were a particular interest of Ernest Lawrence, who was concerned that the Russians might attempt testing at one hundred thousand feet or higher in an effort to evade American atmospheric detection systems; his goal was to gather data to help counteract any such evasion. Hardtack would also test Livermore bombs of both "clean" and dirty varieties, and payloads of differing shapes and sizes. Not the least of them was a prototype of Polaris, the submarine-based warhead for the navy, Livermore's important new government patron. By late February, fourteen thousand military and civilian technicians were on their way to the South Pacific to prepare for Hardtack, which continued to expand as project leaders tried to cram their devices into the schedule; at a White House meeting with Strauss and Dulles on March 24, Eisenhower noticed that the tests, which originally were to run from April through July, were now set to continue into September. He asked for an explanation from Strauss, who blamed the expectation of bad weather.

At the same meeting, John Foster Dulles observed that the Russians had stepped up their own missile and bomb testing sharply—there had been eleven blasts in the space of less than three weeks—and that intelligence reports were pointing toward a Soviet announcement of a unilateral suspension of testing by the end of the month. Any such initiative, he said, would place the United States in "an extremely difficult position throughout the world." Having completed their own tests, the Soviets would have America boxed in a corner over Hardtack. If the United States canceled the tests, it would lose technological ground to the Soviets; if it staged them, the nation would "lose the confidence of the Free World as the champion of peace." The prospect of being outmaneuvered had converted Dulles to fervent advocacy of a test ban treaty. On the spot, he proposed that Eisenhower announce that Hardtack would be the last test series he would order during his term in office, and tie the announcement to a redoubled push for comprehensive disarmament talks.

Despite his lame-duck status, Strauss opposed Dulles's proposal with his customary ferocity, reiterating that testing did not present a significant health hazard. The meeting broke up dispiritedly, its attendees braced to sustain a propaganda blow from the Soviet Union, as Eisenhower pleaded wanly for his staff to "think about what could be done to get rid of the terrible impasse in which we now find ourselves with regard to disarmament."

The intelligence reports were correct. Under its new premier, Nikita Khrushchev, the Soviet Union announced its unilateral suspension of nuclear bomb tests on March 31, called on the United States and Great Britain to do the same, and warned that if they continued their tests, the Russians would feel free to resume their own. As White House advisors had anticipated, the Russian announcement, for all its cynicism, was an enormous propaganda victory. Calls for America to match the Soviet initiative streamed in from overseas—Albert Schweitzer renewing his appeal for an end to the tests with the words, "Mankind insists that they stop, and has every right to do so"—and from Eleanor Roosevelt and other domestic luminaries.

As it happened, not everyone in the nuclear research and development community thought a test moratorium would be a bad thing. Livermore, to be sure, continually pressed for more testing and a larger budget—"There is far more useful work than a laboratory the present size of [Livermore] can possibly do in the immediate future," argued a staff report of this period. "We feel that . . . limitations of funds should not be a determining factor to pursue some of this work." At Los Alamos, however, Norris Bradbury relished getting a breather from the relentless demand for more designs and more tests. "The blunt fact appears to be that, while the country is spending more on atomic weapon research and development than it ever has before, it is almost certainly getting less return per dollar spent than it was getting in 1947–50 or in 1952–54," he informed Admiral Starbird. Government spending on nuclear weapons research was running about $150 million a year, Bradbury estimated, about three times its rate prior to the establishment of Livermore. The existence of the second lab, moreover, made even unpromising weapons programs impossible to kill: programs questioned by Los Alamos would simply be taken up by Livermore, infused as it was with Ernest Lawrence's bottomless optimism and the natural bureaucratic competitiveness for funding. Los Alamos constantly worried that if it rejected a project, it would be seen, to its discredit, as "less 'enthusiastic' " than its rival, Livermore. "All of these things," Bradbury wrote, "leads [sic] to a weapon program which is too long, too detailed, involves too much testing, and too much work for too little real improvement."

Preparations for Hardtack proceeded in a frenzy made more difficult by the unpredictable spring weather at Eniwetok. On the night of April 7, Mark Mills boarded a helicopter with a diagnostic crew to inspect preparations for one of the first tests when they ran into a sudden squall. The chopper crashed in the sea, and Mills drowned.

When the news reached Berkeley, Ernest was writing a letter to his brother, John, who was visiting Geneva. He reported that he had recovered from his recent bouts of illness and now felt "fit as a fiddle." Then the blow landed, devastating on both personal and professional levels. In the four years since Mills had joined the Rad Lab, he had emerged as Ernest's heir apparent. After York's departure for the Pentagon, Lawrence had agreed to name Teller as lab director for a single year, with the proviso that Mills would then take over permanently. The word of his death at sea provoked a colitis attack that kept Ernest bedridden for four days, until he roused himself and hobbled, ashen, into an East Bay church for Mills's memorial service.

As the work on Hardtack proceeded in the South Pacific, progress on a test ban started to be made in Washington.

In early April, Killian reported to Eisenhower and Dulles the conclusion of his experts that test ban violations could be detected with existing technology. Given the superiority of US knowledge about nuclear weapons over that of the Soviets, Killian confirmed, a bilateral agreement for the suspension of bomb tests following Hardtack "would be greatly to the advantage of the United States." Eisenhower seemed to take Killian's report as a comfort. A few days after receiving it, he confided to Killian that he "had never been too much impressed or completely convinced by the views expressed by Drs. Teller, Lawrence and Mills that we must continue testing of nuclear weapons." This was revisionist history, of course; the representations by the three scientists during their White House meeting the previous June were what had prompted Eisenhower to question publicly the wisdom of a test ban.

But the ground had shifted since then. Killian's report helped build momentum toward an overture to the Soviets. This took concrete form on April 28, when Dulles drafted a letter to Khrushchev proposing a "technical conference" in Geneva on test ban inspections. At long last, the United States had formally divorced a test ban from the broader question of disarmament. As Khrushchev recognized, the foundation was laid for a breakthrough. On May 9 he accepted.

Two weeks later, Eisenhower announced a preliminary agreement to the press, along with the names of the three American delegates to the technical conference: former AEC research director James B. Fisk, Caltech physicist Robert Bacher, and Ernest Lawrence. Of the three delegates, Bacher favored a test ban, Fisk was neutral, and Lawrence was opposed but flexible.

Ernest had been at the Balboa house trying to steal a few days of rest when the invitation to Geneva arrived from Washington. Molly objected to the assignment, not least because preparation for the conference would involve weeks of intensive briefings in Washington. But as Lawrence would tell Tuve before his departure for Europe, "We helped start this and have to do what we can about it. The president asked, so I must go."

The conference would mean his fourth visit to Geneva in four years. Each trip had carried its own burden of nostalgia, starting with the 1954 visit to help launch CERN's physics lab, based so heavily on offspring of Ernest's original accelerator. In 1955, during an Atoms for Peace conference connected to that year's US-Soviet summit meeting, Ernest had delivered a presentation on accelerators jointly with Vladimir Veksler, the Russian physicist who had discovered phase stability in 1944 simultaneously with Ed McMillan and had been irked by McMillan's mistaken claim to primacy. At Geneva, there were no residual hard feelings; Lawrence and Veksler shared a friendly dinner at the city's finest French restaurant, comparing notes on the technologies that had made their careers. A year later, at a Geneva symposium sponsored by CERN, Ernest ran into Stan Livingston. They spent a long afternoon in reminiscences of a distinctly autumnal hue, marveling at how far they had come since those early experiments with the quirky eleven-inch accelerator—it was not even called the cyclotron then—and how greatly their work had influenced the course of science during the quarter century that followed.

The new assignment presaged another reunion. Bob Bacher and Ernest Lawrence had first met in 1930, when Bacher visited Berkeley to examine one of Edlefsen's early four-inch accelerators. Before the war, Lawrence recruited him for the radar lab at MIT, where he forged a lasting friendship with Lawrence's brother-in-law Ed McMillan; later he collaborated with Ernest on the electromagnetic separation of uranium.

But there would be little time for retrospection over the next few weeks, for the briefing schedule for the three delegates was punishing. The delegates "practically lived together," Bacher recalled, cramming on the technical aspects of test monitoring like grad students preparing for their orals. Watched over by armed soldiers during the week, they left the briefing room only for meals and sleep. After the first Friday, a bone-weary Ernest Lawrence boarded a red-eye flight west for a few days at Balboa and was back in the air two days later for another grueling week. He returned to Berkeley the following Friday, though he arrived too late to make his son Robert's high school graduation. Meeting him at the airport, Molly judged him "as near exhaustion as I had ever seen him" and lobbied him again to withdraw from the Geneva conference. Instead, he returned to Washington for two more weeks of briefings.

That final stage included sessions with State Department diplomats, who fretted that the scientists, rank novices at diplomacy, would be dragged into political discussions by their wily Soviet counterparts; at one briefing, Secretary Dulles himself emphasized that the delegates were to do "a purely technical scientific job"—no politics whatsoever. To maintain the aura of technical consultation, Dulles decreed that no diplomatic staff would be formally assigned to the delegation.

The stress on Ernest did not let up. The State Department botched the travel arrangements for him and Molly, placing them on a plane without sleeper berths, which Ernest desperately needed for the long first leg from New York to Lisbon, Portugal; he arrived in Europe prostrate. It did not help that the Soviets kept the Americans guessing until the last moment about whether they would even appear. Two weeks before the scheduled July 1 start of the talks, the Russians suddenly had insisted that the agenda be broadened to encompass negotiation of the test ban itself. It remained unclear whether they would relent in the face of firm American opposition to the idea. "We were going, and hoped they'd show," recalled Bacher.

The Soviets did show, though it became evident with their first appearance on the tarmac at Geneva that the State Department's concerns about their political intentions were well taken. Among the delegates was a "shaggy man with an unprepossessing manner and a crooked smile," whom American Kremlinologists identified promptly as Semyon Tsarapkin, one of Moscow's craftiest diplomatic negotiators. The Americans came to call him "Old Scratchy." Tsarapkin and Fisk, the American delegation's chair, would wrestle during the succeeding weeks over the course of the talks—politics versus science.

The meetings were wedged in among constant banquets and receptions, at which Ernest considered his attendance essential to forging a personal connection with his Russian counterparts. At one garden party, he had an unexpected encounter with Robert Oppenheimer, who was in Geneva on other business. It was their first and only meeting since the security hearing, and it would be the last time they saw each other. They exchanged a few anodyne words—"it certainly was not unpleasant," recalled Oppenheimer, who stayed only briefly and left first.

There were sightseeing trips up the Swiss Alps designed as getaways for relaxation, but these excursions only taxed Ernest's weakened constitution. After two weeks in Switzerland, he developed a hacking cough and a persistent fever. Molly, fearing the reappearance of his viral pneumonia, summoned a doctor. He detected no respiratory infection but prescribed shots of liver and placenta extract, a novel therapy for weakened immune systems. After that, Ernest was seldom out of bed except for brief outings for fresh air and visits to the doctor. "He just didn't seem to get well," recalled Bacher. Finally, in the fourth week of the ordeal, he decided to go home.

John met Ernest at the Berkeley house and the next day whisked him to Oakland's Peralta Hospital, where his gray pallor shocked the attending physicians. But two days later, he was strong enough to call for his painting gear, intent on pursuing a hobby he had taken up at the urging of Manette Loomis. He promised to join Molly and the children in Balboa within the week. The next day, however, his condition worsened again. The paints were left untouched, the resolve to visit Balboa gone. It was clear to the doctors that Ernest's chronic colitis had severely compromised his digestive system. The prospect of surgery was raised, which would mean the removal of much of his colon and a permanent change in his quality of life. At John's insistence, Ernest was transferred across the bay to Palo Alto Hospital, the better to consult with its Dr. Albert Snell, one of the nation's preeminent specialists in colitis.

On the Berkeley campus, Cooksey engaged in hushed discussions with Clark Kerr, the new president of the University of California, about transferring the directorship of the Radiation Laboratory to Ed McMillan, whose appointment Ernest endorsed. Molly, summoned by her sister, Elsie McMillan, returned north to sit with Ernest full-time. After resisting the idea of abdominal surgery for years, he had finally relented after consulting with Snell, but now he seemed to be looking ahead to a more distant prospect. "You know," he told Molly, "I wish I'd taken more time off. I would have liked to, but my conscience wouldn't let me."

The surgery was scheduled for August 27. Molly leaned over Ernest as he was being wheeled into the operating room and whispered, "I'll be here when you wake up." The operation lasted five hours. After it was over, she saw him for only a moment. This time, when she leaned close, she thought she heard him say, "Molly, I'm ready to give up now." It was ten o'clock at night. The doctors encouraged her to take a rest—Ernest would be slow coming out of sedation, they said—so she repaired to a late-night coffee shop. But the wait was agonizing, and she rushed back to the hospital. When she came out of the elevators, she knew at once from the look on the nurses' faces that Ernest was gone, at the age of fifty-seven.

Ernest's condition had astonished the surgeons in the operating room. Infection and colitis ulcers had so compromised his system that they were amazed he had been able to play tennis just a few months earlier. The chances that he could have had a normal recovery, much less a normal lifestyle, were virtually nonexistent. Ernest's friends and family members contemplated with mixed feelings of dread and relief what his existence would have been like: "A man of Ernest's disposition," Cooksey told a condoling colleague, "would have been in complete misery as an invalid." But now there was other work to do, he wrote. "Ernest left such a heritage that for those of us here, we have a fine challenge to carry on."

Many of Ernest's friends and colleagues pondered that heritage after his death, but perhaps the oddest judgment came from Robert Oppenheimer. On a wintry day in late January 1959, he stopped at David Lilienthal's house in Princeton. Oppie's hair had turned silvery; he seemed ruddier and less gaunt than Lilienthal remembered from the last time they had seen each other. They had a long talk about Lilienthal's efforts to promote disarmament through the nonprofit Twentieth Century Fund. Toward the end, Lilienthal elicited "the one touch of anything resembling bitter feeling in this man, who was treated so badly when he should have been knighted." It came when Lilienthal mentioned Ernest Lawrence.

As Lilienthal recorded the conversation in his journal: "Lawrence died of frustration, Robert as much as said, because of the long strain of an over-reaching ambition, culminating in his efforts to torpedo the talks in Geneva concerning the ending of the bomb tests." Oppie's startled host replied that Ernest "had always seemed to my observation to be a very picture of the extrovert, the satisfied man, the man of bounce and buoyancy." No, Oppenheimer said: "I have known him longer and closer than you; his fears that he was being, or might be, undermined in his position were a terror for him."

It was indeed, as Lilienthal reflected, a "strange sidelight." It was also at odds with the facts, and with Oppenheimer's own judgment as he relayed it a few years later to Lawrence's biographer, Herbert Childs. Basing his words on the description of Lawrence's role at the Geneva talks that he had received from his own close friend Robert Bacher, Oppenheimer acknowledged that although Lawrence "had very grave doubts" about the possibility and even the wisdom of a test ban, "apparently he was quite willing to subordinate his doubts to the mission that he was contracted to make." That was much closer to the truth: Lawrence had gone to Geneva out of a sense of duty and a commitment to seeing the talks through, and the effort shortened his life.

The Geneva conference continued for three more weeks after Lawrence's return home. James Fisk, who supervised the American side of the talks with a diplomatic skill unexpected of a scientific appointee, brought home an agreement designating the detection of nuclear bomb tests as "technically feasible." It was the indispensable prelude to talks on a test ban. The very day after the Geneva meeting adjourned, Eisenhower proposed that negotiations begin on October 31. To set the stage for talks, he announced that the United States would observe a one-year moratorium on testing, renewable annually on condition that the Russians also refrained from testing during the period and that "satisfactory progress" continued to be made toward a disarmament agreement. Within a few days, Khrushchev agreed.

Since Hiroshima, there had been more than 190 atomic and thermonuclear tests in the atmosphere, underground, and at sea. The United States accounted for 125 of these; the Russians, 44; and Great Britain, 21. Now, for the first time since 1945, the man-made thunder and lightning at nuclear test ranges around the world ceased.

The talks continued for more than two years, in peaks and valleys of hope and gloom, while the test ranges remained silent. The silence ended during the early months of John F. Kennedy's presidential term. The Soviet Union announced on August 31, 1961, that it would resume atmospheric testing and made good on the threat the very next day. Kennedy held out until March 2, 1962, when he announced in a televised address that the United States would resume nuclear tests in the atmosphere. The new round of testing ended in the summer of 1963, when America and the Soviet Union reached a limited test ban agreement, outlawing tests in the atmosphere, at sea, and in space—everywhere, in short, except underground.

The three-year test moratorium had been a difficult period for Livermore, now under the leadership of Edward Teller. Its devices typically were more sophisticated and complicated than those of Los Alamos, and therefore harder to perfect via theoretical modeling rather than testing. But the lab, which was rechristened Lawrence Livermore National Laboratory in 1980, thrived and expanded by exploiting its productive relationship with the Air Force and the Navy. Livermore continued to work on Polaris during the moratorium, then followed with the next-generation submarine-launched nuclear missile, Poseidon. In the 1970s, its work yielded MIRVs, multiple independently targeted reentry vehicles. These were warheads loaded with eight to fourteen nuclear payloads, each one up to twenty-five times more powerful than the Hiroshima bomb. MIRVs, which encompassed the land-based MX missile program, were the ultimate product of Livermore's long quest for a high-yield, low-weight nuclear device.

The Radiation Laboratory lived up to its heritage and its challenge as a centerpiece of Big Science. Seven of Ernest's students and colleagues there would win Nobel Prizes in physics or chemistry after his death. Four more members of what was presently rechristened Lawrence Berkeley National Laboratory would join them in the pantheon—scientists who had not worked directly with Ernest but continued the traditions he established at the Rad Lab as his heirs.

Many others earned the esteem of their fellow chemists and physicists even if they fell short of the ultimate honor—a Nobel Prize. One was Albert Ghiorso, a squat, stubborn physicist who had arrived at Berkeley just before the war and became known equally for his wizardry with electronics and his unimpeachably liberal views—"the only person I know who describes Franklin Roosevelt as a conservative," Seaborg would write of Ghiorso. He also became known as the most assiduous and creative hunter of new elements in history, credited with the discovery or codiscovery of twelve transuranics.

Among them was element 103, which Ghiorso and colleagues at Berkeley believed in 1961 that they had found via a process known as "atom-at-a-time" chemistry—literally detecting an average of a single atom in each of hundreds of trials. Continuing a tradition in which they had honored the lab with the christening of element 97 as berkelium in 1949 and element 98 as californium in 1950, they named this one—an element that to this day has not been seen by the human eye—lawrencium.

Epilogue

The Twilight of Big Science?

Molly Lawrence, resistant to spectacle as ever, rejected a public funeral for her husband. Instead, at ten o'clock on the third day after Ernest's death, a few hundred invited friends and colleagues filled the pews of Berkeley's First Congregational Church, two blocks from the campus and a short stroll from the site of the old Rad Lab.

Presiding was Clark Kerr, the balding labor economist who had succeeded Bob Sproul as Berkeley's chancellor soon after the loyalty oath conflict. (Sproul was elevated to the presidency of the rapidly expanding University of California system, a post in which Kerr succeeded him in 1958.)

Ernest surely would have been gratified to hear Kerr place his life and career within the continuum of humanity's unending quest for knowledge:

> *Manlike creatures have lived on this planet for at least a million years. Throughout those million years they have constantly groped to understand more about and to control better the world about them. A few of them have shot some ray of light into the great unknown darkness of ignorance and illuminated a new era for all future generations. One of the strongest of these beams of light was created by Ernest Lawrence, and men forever after will see farther and understand more because of it. Each of us and each of our children owe to Ernest Lawrence a debt beyond price. For by his expansion of our understanding, by his*

> *reduction of our ignorance, he has added a little to the human dignity of each of us and something more to the meaning of life.*

Those who knew Ernest best might have heard echoes in these words of Raymond Fosdick's testimonial two decades earlier, when he marked the Rockefeller Foundation's huge $1.15 million grant for the 184-inch cyclotron by calling it "a mighty symbol, a token of man's hunger for knowledge, an emblem of the undiscourageable search for truth which is the noblest expression of the human spirit." Those of a more practical cast of mind pondered Ernest's role as an impresario of research administration. "He will always be remembered as the inventor of the cyclotron," wrote Luis Alvarez in a memorial for the National Academy of Sciences. "But more importantly, he should be remembered as the inventor of the modern way of doing science."

Over the decades that followed, Ernest Lawrence's way of doing science remained the model. The creation of the large-scale interdisciplinary laboratory might have happened without him, for discoveries in high-energy physics created their own demand for ever-larger, more complex, and more costly accelerators. But the development of Lawrence-style labs seemed inevitable only in retrospect. To Ernest's contemporaries and the first generations of scientists who succeeded him, there were many other paths Big Science might have followed. To them, his invention of the big laboratory "where physicists, engineers and technicians could work together symbiotically to construct ever larger and more complicated particle accelerators" (the words are those of the British physicist John Bertram Adams, who served as CERN's director from 1975 to 1980) was uniquely American, and uniquely Lawrencian. It was Ernest Lawrence, for instance, who elevated engineers to coequal status in the accelerator lab—physicists in Europe, by contrast, "tended to shun the 'dirty' details of engineering," which surely accounted for Europe's lagging behind the United States in accelerator technology, though it may also have encouraged a freer, less machine-bound approach to physics on the continent.

The momentum created by Lawrence's leadership of the Rad Lab carried physics forward into the 1970s. Steven Weinberg, a future Nobel laureate (in 1979, for his contributions to the theory of the electroweak force), arrived at the Rad Lab as a postdoc in 1959, when the Bevatron that Bill Brobeck built for Ernest reigned on the hilltop. The Bevatron, Weinberg recalled, "had been built specifically to accelerate protons to energies high enough to create antiprotons [protons with a negative charge], and to no one's surprise antiprotons were created." But so were many other types of particles, which demanded the construction of yet another generation of accelerators, more energetic and of course more expensive, to penetrate the new mysteries. Just as the first glass and wax accelerators begat the 27-inch and 37-inch cyclotrons, and they begat the 60-inch and the 184-inch, the Bevatron pointed the way to accelerators too big to fit in the ravine and too costly to be built by a single university. So the new machines were built by academic consortiums and university-government collaborations like the ones underlying Fermilab outside Chicago, and CERN. The new generation of scientific instruments were beyond monumental in scale; on the Illinois prairie and in the pastoral belt on the border between France and Switzerland they became, in Weinberg's words, "features of the landscape."

But within a few short years of Ernest Lawrence's death, skeptics were questioning the scale and expense of the enterprises his methods had fostered. Among the doubters was physicist Alvin M. Weinberg, who in 1961 had coined the term "Big Science." Weinberg, as it happens, was then the director of the Oak Ridge National Laboratory—the laboratory founded by Ernest Lawrence and General Groves to enrich uranium using electromagnetic separation. But that gave him a unique vantage point from which to survey the research harvest Ernest Lawrence had sowed.

Alvin Weinberg posed three fundamental questions about Big Science: Is it ruining science? Is it ruining the nation financially? Should the money it commands be redirected—spent on eradicating disease and other efforts aimed directly at "human well-being," for example, rather than on "spectaculars" like space travel and particle physics?

Merely to ask the questions was to hint that the answers must be affirmative. Big Science thrived on publicity, Weinberg observed, which reduced discussions of technical merit to debates about which projects might make the biggest splash in the press. The surfeit of money for big projects encouraged more building and less thinking: "There is a natural rush to spend dollars rather than thought—to order a $\$10^7$ [$10 million] nuclear reactor instead of devising a crucial experiment with the reactors at hand." Weinberg illuminated the uneasiness already emerging about the impact of Big Science on research and the university. "I suspect that most Americans would prefer to belong to the society which first gave the world a cure for cancer," he wrote, "than to the society which put the first astronaut on Mars."

Others expressed concerns about the impact of Big Science on the traditional structure of academia, which melded basic research, applied research, and teaching into a unified yet multifaceted whole. Once physicists' machines burst the confines of the university campus this relationship began to break down; it was fragmented even further by the dominance of military funding during World War II, the Korean War, and the Cold War. "When the machines outgrew their university environment," John Bertram Adams told his audience at the Rad Lab's fiftieth anniversary symposium, "the place where experiments were carried out became separated from the place where students were taught physics . . . It takes a particularly robust personality to teach academic courses at, say, Harvard University and carry the heavy responsibility of a major experiment at, say, Fermilab or CERN, especially when the experiment lasts several years."

Big Science was no longer part of the academic institution, but an institution unto itself. Experiments run on billion-dollar machines had to be approved by committees, which based their decisions not only on the objective merits of the proposals but on their subjective judgments of the applicants' reputations and standing in their fields. The widening gulf between this style of science and the serendipitous research of the past showed how much had been lost with the passing of Ernest Lawrence, indeed of his entire generation. While he was alive and presiding over the Rad Lab, Ernest served as his own experiment committee, guiding research directly through his own demands and indirectly through his choice of top associates—at the Rad Lab, important research often was defined as whatever Ernest Lawrence, Luis Alvarez, or Ed McMillan wished to pursue.

Ernest Lawrence's generation comprised scientific statesmen who drew their authority in peacetime from the roles they had played during World War II. By the end of the third decade after the war, many had passed on: J. Robert Oppenheimer in 1967, Arthur Holly Compton in 1962, Vannevar Bush in 1974. No one in the succeeding generation commanded the respect of members of Congress or residents of the White House as they had; none could claim to represent the unified interests of the scientific community as they could; none had the fund-raising skills of an Ernest Lawrence.

The passing of American science's greatest and most influential cohort occurred just as demand burgeoned for new accelerators, and as doubts about their necessity began to be heard even among physicists. To the particle physicists who had dominated the science during the cyclotron era, the need for bigger and more energetic machines was an article of faith. "We simply do not know how to obtain information on the most minute structure of matter (high-energy physics) or on the grandest scale of the universe . . . without large efforts and large tools," wrote Pief Panofsky, the Rad Lab veteran who had become head of Stanford University's competing high-energy accelerator program, in 1992. The projects, moreover, were all-or-nothing. "Big Science has the special problem that it can't easily be scaled down," Steven Weinberg observed. Writing of the new multibillion-dollar accelerators that were designed to send beams of particles in opposite directions around giant underground tunnels and smash them into one another, he stated, "It does no good to build an accelerator tunnel that only goes halfway around the circle." But not all science was physics, and not all physics was high-energy physics.

As the Manhattan Project's scientific leaders began to depart the battlefield in the mid-1960s, doubts about scientists' role in setting national priorities also became stronger. "A 20-year honeymoon for science is drawing to a close," wrote *Science Magazine*'s editor, Phil Abelson—former Rad Lab researcher and developer of the thermal diffusion uranium separation process for the atomic bomb—in 1966.

A grand honeymoon it had been. During those twenty years, a period that started with Hiroshima and received a powerful boost from *Sputnik*, scientists rose to figures of great consequence in all aspects of political life. Coming out of the war, Vannevar Bush, Ernest Lawrence, and their colleagues were able to persuade Congress that "basic science was worth supporting for its own sake—or at any rate without inquiring too closely about its connection with practical results," observed Don K. Price, an expert in public administration at Harvard. The Eminent Scientist became a "political animal," wrote Ralph Sanders, a political scientist at the government's Industrial College of the Armed Forces, in the pages of the *Bulletin of the Atomic Scientists*. "The President's Science Advisor now comments upon issues which forty years ago were the exclusive preserve of politicians . . . Hordes of scientists today scamper about the landscape of public affairs, eager to grapple with an increasing range of questions," given free reign by politicians "dazzled by the brilliant achievements of science, disturbed by the often esoteric nature of science, and bothered by the Soviet challenge in science." The process had started with Ernest Lawrence, but it had grown immeasurably after his death.

By then, however, signs of the waning influence of scientists were already emerging. In absolute terms, to be sure, science still commanded an enormous share of national resources: federal government spending on research and development had grown from $74 million in 1940 to $15 billion in 1965, an increase averaging nearly 20 percent a year. But the growth rate had fallen sharply. From 1950 to 1955, the annual growth had been 28 percent; from 1961 to 1965, it was 15 percent.

This trend surely reflected the sheer impossibility of sustaining the growth rate of the war years and immediate postwar period. But there was more to it. Big Science had allowed its past achievements to be oversold, and its promoters overpromised gains for the future. By the mid-1960s, the successes of wartime were receding into the mists of memory, and the expenses of competing with Russia in the post-*Sputnik* era began to seem staggering.

Then came Vietnam. The war placed a heavy strain on government resources; the debate it engendered over America's role in the world placed a spotlight on the nation's social priorities; meanwhile the participation of the academic and scientific establishment in the war machine led to new questions about military funding of research. Congress moved to wean academia from the mother's milk of Pentagon funding. In 1969 the Mansfield Amendment, named for Senate Majority Leader Mike Mansfield, Democrat of Montana, barred the Pentagon from spending money on any research not directly related to the military. The change struck at a host of Big Science university projects funded by the Defense Department's *Sputnik*-era Advanced Research Projects Agency, or ARPA—not least among them a network linking university research computers known as the ARPANET, the grandfather of today's Internet. (In recognition of the change in its mission, ARPA would be renamed the *Defense* Advanced Research Projects Agency, or DARPA.) And it was especially hard on physicists, many of whom had based their career aspirations on expectations of continued government funding for Big Science. At MIT, which suffered a 30 percent drop in its government support since 1968, physics chairman Victor Weisskopf lamented in 1972 the declining prospects of "a generation of people who studied physics under the stimulus of *Sputnik*. As kids in school they were told this was a great national emergency, that we needed scientists. So they worked hard." Now, he said, "they are out on the street and naturally they feel cheated."

Big Scientists tried to push back against the skepticism about their work inspired by the Vietnam debate and their rising budget demands. They resorted to several old justifications for big spending on basic research. They claimed that, given enough money, practical applications from basic science were just around the corner: the conquest of cancer "or heart disease, or stroke, or mental illness, or whatever," as *Harper's* editor John Fischer repeated dismissively—a line, he remarked, overused "to justify research with only remote relevance to the conquest of anything." Or they projected a military breakthrough, or world domination by the Russians if the US effort in Big Science faltered.

In this atmosphere of national stringency and reconsidered priorities, Big Science faced challenges unknown in Lawrence's days. In that era and well into the first postwar decade it had been widely accepted that national pride demanded government funding of the quest for knowledge for its own sake. "The great ideas arise when you give freedom," declared the Rockefeller Foundation's Warren Weaver—"freedom to think, freedom from other pressures—to individuals of great intellectual capacity . . . and let them be motivated primarily by their curiosity to find out how nature operates."

Now that notion sounded hopelessly elitist. Multimillion-dollar projects aimed at knowledge for its own sake came under increasing practical scrutiny. That was the fate, for instance, of Project Mohole, launched in 1958 as an audacious plan to drill into the Earth's mantle through the deep ocean floor. Mohole was conceived as a geological analogue to the space race and the exploration of subatomic physics. But it could not survive the mushrooming of its cost estimate to $127 million from the original $15 million. Its promoters had tried to save it by depicting Mohole as "a panacea for virtually everything but poison ivy," scoffed *Science* magazine in July 1966, after the project was canceled by Congress.

No event brought the limits of Big Science into sharp relief in the

United States like the bitter debate over the Superconducting Super Collider, the SSC, in the 1980s and early 1990s. An ultramodern descendant of the glass-and-sealing-wax accelerators of Ernest Lawrence and Niels Edlefsen, the SSC was projected to cost $6 billion over ten years. Its supporters' sales pitch to Congress came straight from Lawrence's playbook. They evoked national pride, the prospect of discoveries with life-saving applications, the glory of mankind's search for the fundamental truths of nature. Sheldon Glashow and Leon Lederman, the SSC's most energetic promoters, wrote that the project would spin off new knowledge about superconducting magnets (useful for "super-rapid transit," batteries, and electrical transmission), in construction techniques, in computer science. But their bedrock case was a warning that American science risked being outrun by Europe. If America rejected the SSC, they wrote, "the loss will not only be to our science but also to the broader issue of national pride and technological self-confidence. When we were children, America did most things best. So it should again."

Yet the SSC camp lacked Lawrence's ability to hold a lay audience in thrall, and to hold the scientific community together. As early as 1967, the *New York Times* had decried the "expensive irrelevance" of high-energy accelerators given the pressing social problems of the era (its specific target was a planned new machine at Fermilab). As the SSC campaign progressed, budgetary considerations came to trump the promise of technological spinoffs, national pride, and human aspiration. Steven Weinberg came face-to-face with the challenge during an appearance on the Larry King radio show with an anti-SSC congressman. "He said that he wasn't against spending on science, but that we had to set priorities," Weinberg recollected. "I explained that the SSC was going to help us learn the laws of nature, and I asked if that didn't deserve a high priority. I remember every word of his answer. It was 'No.' " No mere congressman would have dared deliver such a blunt rebuff to Ernest Lawrence in his day.

The SSC was further undermined by an open split within the physics community over whether it was necessary at all. High-energy physicists like Panofsky and Weinberg said yes, but they were contradicted by those in other branches of the science who had long felt shortchanged by the high-energy group's voracious appetite for research funding. Finally, in 1993, amid continuing discord over the Superconducting Super Collider's cost, necessity, and utility, and a deepening economic slump, Congress killed the project.

Was it a death knell for Big Science? That remains unclear at this writing, decades later. After the SSC's cancellation, the center of gravity of high-energy physics shifted to CERN and its Large Hadron Collider, which became the world's most powerful accelerator by default. The LHC keeps thousands of physicists employed, many of them Americans, and in 2013 achieved the signal success of identifying the elusive Higgs boson. But as has been the pattern in physics for about a century, the discovery only pointed the way to more questions about the fundamental particles and forces of nature—questions that might require yet bigger and more powerful machines to answer. "In the next decade," Steven Weinberg predicted, "physicists are probably going to ask their governments for whatever new and more powerful accelerator we then think will be needed. That is going to be a very hard sell."

In the years since the cancellation of the SSC, government's role in funding Big Science has continued to wane. In the first decades of the twenty-first century, the dominant patron is business, which contributes two-thirds of the funds spent on research and development in the United States. Of that, nearly two-thirds is "development" spending—that is, efforts to bring the results of applied research to market. Only about one of every six dollars spent on R&D goes to basic research. Business was the source of almost all of the increase in funding detected by the National Science Foundation from 2003 through 2008. The future of Big Science appears to depend on industry, whose R&D priorities are very different from those of universities, research foundations, and government.

• • •

It was not only the cost and relevance of Ernest Lawrence's Big Science that came under attack in the years following his death. There was also the issue of the ends that had been served by his style of raising money and deploying scientific talent. The focus of that question was Livermore.

By the 1980s, Livermore's increasingly prominent role in the international arms race began to raise disturbing doubts in a home on Tamalpais Road in the hills north of the Berkeley campus. There Molly Lawrence pondered whether her husband's legacy was being properly served at Livermore. Her conclusion was that Ernest would have disapproved, and she became determined to tell the world.

Watching the news about a controversial ICBM project one day in 1982, she told a reporter from her local newspaper, "I heard that 'Lawrence Livermore' would be designing parts for the MX. Suddenly it hit me how dreadful it was that Ernest's name was associated with this, was lending legitimacy and respectability to it." She was convinced that Ernest would be as appalled as she was at how the effort to develop the Super as a matter of national security had become transformed into an escalating race to produce destructive power without limits. He would have been angered, she said, by "the middle-class idiots who refuse to face the horror we've brought upon ourselves, refuse to try and stop the madness when we have ten to twenty times what we need for deterrence."

That spring she wrote to the University of California regents, expressing her "shame and remorse" at Ernest's identification with Livermore and asking that his name be removed from the lab. The regents demurred, telling her that since Livermore was a federal government laboratory, its name was a federal issue. It may not have helped that one of the regents was her brother-in-law, John Lawrence, who sharply disagreed with her position. She then took her battle to Congress, asking for the assistance of California's US senator, Alan Cranston. She never got it, and Ernest's name remains on Livermore to this day.

But whether he would truly have objected to its later role in nuclear weapons development is by no means certain. The possibility that developing the Super would unleash a permanent arms race between the United States and the Soviet Union was well understood in 1952; indeed, it was part of the standard brief against the Super offered by Oppenheimer, Fermi, Rabi, and other opponents. Livermore campaigned energetically for an ever-larger role in weapons development during Lawrence's life and continued to do so under Edward Teller, his designated successor as director. Lawrence himself was the promoter of weapons schemes as elaborate for their time as anything Livermore developed later.

Molly Lawrence's purpose was to safeguard a legacy that her husband had developed over the course of a half century. Her goal was just. To repeat Robert Oppenheimer's observation: through his genius, Ernest Lawrence had not merely illuminated some of the darkest mysteries held by nature but also invented a new approach to "the problem of studying nature." Although Lawrence's approach facilitated to a new degree the alliance between science and the military, it also enriched science, and indeed enriched our understanding of the natural world. Until nearly the end of his life, Ernest was able to keep the excesses that big money would bring to science in check. Even he would succumb, in time, to the conviction created by human nature that one's own aim is invariably true. But that should not obscure his real achievement in bringing a new level of knowledge to science.

The year before Molly started campaigning to cleanse Ernest's name of its association with militarism and mass destruction, she evoked more uplifting memories in a talk at Berkeley. The occasion was the fiftieth anniversary of the Rad Lab's founding. Citing John Greenleaf Whittier's lines "Of all sad words of tongue or pen, the saddest are these: It might have been," she pondered the unique combination of fortune, drive, and serendipity that had enabled Ernest to make his mark in the world of science—"a whole series of 'what if's.' "

What if Rolf Wideröe had not published an article on the acceleration of potassium ions in 1928? What if Ernest had not come across it in the library one day and managed to understand the general principles even though he couldn't read German very well? What if Niels Edlefsen had not been persuaded to build the first Berkeley accelerator, that messy little glob of glass and sealing wax? . . . What if Stan Livingston had not undertaken the task of building larger accelerators and come up with some very ingenious solutions to some of the knotty problems that arose? What if that wonderfully inspired, dedicated, hard-working, long-suffering bunch of young people had not gravitated to Berkeley to work night and day, Sundays and holidays, for their demanding maestro? What if Robert Gordon Sproul had been an old fuddy-duddy of a university president instead of a young, dynamic one? . . .

What if any of these substantial elements in the success of the Laboratory had been lacking? What if the right people had not had the right ideas at the right time, the right degree of enthusiasm and persistence, at the right time and the right place?

She concluded: "Surely the Radiation Laboratory would not have been founded in 1931 at Berkeley, and we wouldn't be here tonight celebrating the golden anniversary of those auspicious circumstances. But they did, and it was, so here we are."

Acknowledgments

Almost all the men and women who lived the formative years of high-energy physics, the development of the atomic bomb, and the birth of the thermonuclear age with Ernest O. Lawrence have passed on now. But the voices of many of them live on in the remarkable collection of interviews conducted by Herbert Childs for his 1968 authorized biography of Lawrence, *An American Genius*, and deposited at the Bancroft Library of the University of California, Berkeley. Childs's work and materials, along with *Lawrence and His Laboratory*, the history of the Radiation Laboratory published by J. L. Heilbron and Robert W. Seidel in 1969, are among the indispensable starting points for any examination of Lawrence and his times.

So too are the Ernest O. Lawrence Papers at the Bancroft, where Susan Snyder, head of public services, and the rest of the Bancroft staff were unfailingly courteous and helpful during the long hours I spent with these archival materials in the library's reading room. I am also indebted to Pamela Patterson of the Lawrence Berkeley Laboratory for her assistance with my request for archival materials at the lab and her hospitality during my visit. The Library of Congress, the Niels Bohr Library and Archives of the American Institute of Physics, and the U.S. Military Academy Library at West Point also afforded me access to important historical materials. Robert Lawrence, Ernest and Molly's second son, generously provided me with copies of photographs in the family's possession.

My agent, Sandra Dijkstra, contributed the enthusiasm, advocacy, and faith in this project on which I long ago came to depend. My editor at Simon & Schuster, Thomas LeBien, provided invaluable guidance on structuring the manuscript and shaping the final product.

Finally but most importantly, this book could not have been researched and written without the love, forbearance, collaboration, and support of my wife, Deborah, or the inspiration of my sons, David and Andrew.

Bibliography

Alvarez, Luis W. *Alvarez: Adventures of a Physicist*. New York: Basic Books, 1987.

Appleby, Charles A. *Eisenhower and Arms Control, 1953–1961*, vol. 1: *A Balance of Risks*. Baltimore: Johns Hopkins University, 1983.

Barrett, Edward L., Jr. *The Tenney Committee*. Ithaca, NY: Cornell University Press, 1951.

Beisner, Robert L. *Dean Acheson: A Life in the Cold War*. New York: Oxford University Press, 2006.

Bernstein, Jeremy. *Plutonium: A History of the World's Most Dangerous Element*. Washington, DC: Joseph Henry Press, 2007.

Bird, Kai, and Martin J. Sherwin. *American Prometheus: The Triumph and Tragedy of J. Robert Oppenheimer*. New York: Alfred A. Knopf, 2005.

Brands, H. W. *Traitor to His Class: The Privileged Life and Radical Presidency of Franklin Delano Roosevelt*. New York: Doubleday, 2008.

Brown, Laurie Mark, Max Dresden, and Lillian Hoddeson, eds. *Pions to Quarks: Particle Physics in the 1950s*. Cambridge: Cambridge University Press, 2009.

Buderi, Robert. *The Invention That Changed the World: How a Small Group of Radar Pioneers Won the Second World War and Launched a Technological Revolution*. New York: Simon & Schuster, 1996.

Bush, Vannevar. *Pieces of the Action*. New York: William Morrow, 1970.

Byrnes, James F. *All in One Lifetime*. New York: Harper & Bros., 1958.

Cantelon, Philip L., Richard G. Hewlett, and Robert C. Williams, eds. *The American Atom: A Documentary History of Nuclear Policies from the Discovery of Fission to the Present*. Philadelphia: University of Pennsylvania Press, 1992.

Carroll, Sean. *The Particle at the End of the Universe: How the Hunt for the Higgs Boson Leads Us to the Edge of a New World*. New York: Dutton, 2012.

Carson, Cathryn, and David A. Hollinger, eds. *Reappraising Oppenheimer: Centennial Studies and Reflections*. Berkeley: Office for History of Science and Technology, University of California, 2005.

Childs, Herbert. *An American Genius: The Life of Ernest Orlando Lawrence*. New York: E. P. Dutton, 1968.

Clark, Ronald W. *Einstein: The Life and Times*. New York: World Publishing, 1971.

Crelinsten, Jeffrey. *Einstein's Jury: The Race to Test Relativity*. Princeton, NJ: Princeton University Press, 2006.

Cole, K. C. *Something Incredibly Wonderful Happens: Frank Oppenheimer and the World He Made Up*. Orlando, FL: Houghton Mifflin Harcourt, 2009.

Compton, Arthur Holly. *Atomic Quest: A Personal Narrative*. New York: Oxford University Press, 1956.

Conant, James B. *My Several Lives: Memoirs of a Social Inventor*. New York: Harper & Row, 1970.

Conant, Jennet. *Tuxedo Park: A Wall Street Tycoon and the Secret Palace of Science That Changed the Course of World War II*. New York: Simon & Schuster, 2002.

Davis, Nuel Pharr. *Lawrence & Oppenheimer*. New York: Simon and Schuster, 1968.

Dean, Gordon E. *Forging the Atomic Shield: Excerpts from the Office Diary of Gordon E. Dean*. Edited by Roger M. Anders. Chapel Hill: University of North Carolina Press, 1987.

Divine, Robert A. *Blowing on the Wind: The Nuclear Test Ban Debate, 1954–1960*. New York: Oxford University Press, 1978.

Eisenhower, Dwight D. *The White House Years: Waging Peace, 1956–1961*. Garden City, NY: Doubleday, 1965.

Eve, Arthur S. *Rutherford: Being the Life and Letters of the Rt. Hon. Lord Rutherford, O.M.* Cambridge: Cambridge University Press, 1939.

Fosdick, Raymond Blaine. *The Story of the Rockefeller Foundation*. New York: Harper and Brothers, 1952.

Galbraith, John Kenneth. *The Great Crash, 1929*. Boston: Houghton Mifflin, 1988.

Galison, Peter, and Bruce Hevly, eds. *Big Science: The Growth of Large-scale Research*. Stanford, CA: Stanford University Press, 1992.

Gardner, David P. *The California Oath Controversy*. Berkeley: University of California Press, 1967.

Goldsmith, Maurice. *Frederic Joliot-Curie*. London: Lawrence & Wishart, 1977.

Grant, James. *Bernard M. Baruch: The Adventures of a Wall Street Legend*. New York: Simon and Schuster, 1983.

Greenberg, Daniel S. *The Politics of Pure Science*. Chicago: University of Chicago Press, 1999.

Groves, Leslie R. *Now It Can Be Told: The Story of the Manhattan Project*. New York: Harper & Row, 1962.

Guerlac, Henry E. *Radar in World War II*. Los Angeles: American Institute of Physics, 1987.

Hagerty, James C. *The Diary of James C. Hagerty: Eisenhower in Mid-Course, 1954–1955*. Edited by Robert H. Ferrell. Bloomington: Indiana University Press, 1983.

Hansen, Chuck. *The Swords of Armageddon*, vol. 4: *The Development of U.S. Nuclear Weapons*. Sunnyvale, CA: Chukelea Publications, 1995.

Heilbron, J. L., and Robert W. Seidel. *Lawrence and His Laboratory: A History of the Lawrence Berkeley Laboratory*, vol. 1. Berkeley: University of California Press, 1989.

Hendry, John, ed. *Cambridge Physics in the Thirties*. Bristol, UK: Adam Hilger, 1984.

Herken, Gregg. *Brotherhood of the Bomb: The Tangled Lives and Loyalties of Robert Oppenheimer, Ernest Lawrence, and Edward Teller*. New York: Henry Holt, 2002.

Hershberg, James G. *James B. Conant: Harvard to Hiroshima and the Making of the Nuclear Age*. New York: Alfred A. Knopf, 1993.

Hewlett, Richard G., and Oscar E. Anderson Jr. *The New World: A History of the United States Atomic Energy Commission*, vol. 1: *1939/1946*. Washington, DC: U.S. Atomic Energy Commission, 1962.

Hewlett, Richard G., and Francis Duncan. *Atomic Shield: A History of the United States Atomic Energy Commission*, vol. 2: *1947/1952*. Washington, DC: U.S. Atomic Energy Commission, 1972.

Hewlett, Richard G. *Nuclear Navy, 1946–1962*. Chicago: University of Chicago Press, 1974.

Hewlett, Richard G., and Jack M. Holl. *Atoms for Peace and War, 1953–1961: Eisenhower and the Atomic Energy Commission*. Berkeley: University of California Press, 1989.

Hoffman, Darleane C., Albert Ghiorso, and Glenn T. Seaborg. *The Transuranium People: The Inside Story*. London: Imperial College Press, 2000.

Holton, Gerald, ed. *The Twentieth-Century Sciences: Studies in the Biography of Ideas*. New York: W. W. Norton, 1970.

Josephson, Paul R. *Physics and Politics in Revolutionary Russia*. Berkeley: University of California Press, 1991.

Kamen, Martin D. *Radiant Science, Dark Politics: A Memoir of the Nuclear Age*. Berkeley: University of California Press, 1985.

Kelly, Cynthia C., ed. *Oppenheimer and the Manhattan Project: Insights into J. Robert Oppenheimer, "Father of the Atomic Bomb."* (Record of a symposium on Oppenheimer and the Manhattan Project, June 26, 2004, Los Alamos, New Mexico, Atomic Heritage Foundation.) Hackensack, NJ: World Scientific Publishing, 2006.

Kennedy, David M. *Freedom from Fear: The American People in Depression and War, 1929–1945*. New York: Oxford University Press, 2005.

Kevles, Daniel J. *The Physicists: The History of a Scientific Community in Modern America*. Cambridge, MA: Harvard University Press, 1977.

Lilienthal, David E. *The Journals of David E. Lilienthal*, vol. 1: *The TVA Years 1939–1945*. New York: Harper & Row, 1964.

———. *The Journals of David E. Lilienthal*, vol. 2: *The Atomic Energy Years 1945–1950*. New York: Harper & Row, 1964.

———. *The Journals of David E. Lilienthal*, vol. 3: *The Road to Change, 1955–1959*. New York: Harper & Row, 1969.

Livingston, M. Stanley. *Particle Accelerators: A Brief History*. Cambridge, MA: Harvard University Press, 1969.

———, ed. *The Development of High-Energy Accelerators*. New York: Dover Publications, 1966.

Manchester, William. *The Glory and the Dream: A Narrative History of America 1932–1972.* Boston: Little, Brown, 1974.

Nichols, K. D. *The Road to Trinity: A Personal Account of How America's Nuclear Policies Were Made.* New York: William Morrow, 1987.

Oliphant, Mark. *Rutherford: Recollections of the Cambridge Days.* Amsterdam: Elsevier Publishing, 1972.

Pais, Abraham, and Robert P. Crease. *J. Robert Oppenheimer: A Life.* New York: Oxford University Press, 2006.

Panofsky, Wolfgang K. H. *Panofsky on Physics, Politics, and Peace: Pief Remembers.* New York: Springer, 2007.

Pfau, Richard. *No Sacrifice Too Great: The Life of Lewis L. Strauss.* Charlottesville: University Press of Virginia, 1984.

Regis, Ed. *Who Got Einstein's Office? Eccentricity and Genius at the Institute for Advanced Study.* New York: Perseus Books, 1987.

Rhodes, Richard. *The Making of the Atomic Bomb.* New York: Simon & Schuster, 1986.

———. *Dark Sun: The Making of the Hydrogen Bomb.* New York: Simon & Schuster, 1995.

Rigden, John S. *Rabi: Scientist and Citizen.* Cambridge, MA: Harvard University Press, 1987.

Roosevelt, Franklin Delano. *F.D.R.: His Personal Letters*, vol. 3: *1928–1945.* Edited by Elliott Roosevelt. New York: Duell, Sloan and Pearce, 1950.

Seaborg, Glenn T. *Nuclear Milestones: A Collection of Speeches.* San Francisco: W. H. Freeman, 1972.

———. *The Plutonium Story: The Journals of Professor Glenn T. Seaborg, 1939–1946.* Edited and annotated by Ronald L. Kathren, Jerry B. Gough, and Gary T. Benefiel. Columbus, OH: Battelle Press, 1994.

———, with Eric Seaborg. *Adventures in the Atomic Age: From Watts to Washington.* New York: Farrar, Straus and Giroux, 2001.

Seabrook, William B. *Dr. Wood, Modern Wizard of the Laboratory: The Story of an American Small Boy Who Became the Most Daring and Original Experimental Physicist of Our Day—But Never Grew Up.* New York: Harcourt, Brace, 1941.

Segrè, Emilio. *Enrico Fermi, Physicist.* Chicago: University of Chicago Press, 1972.

———. *A Mind Always in Motion: The Autobiography of Emilio Segrè.* Berkeley: University of California Press, 1993.

Serber, Robert. *The Los Alamos Primer: The First Lectures on How to Build an Atomic Bomb.* Berkeley: University of California Press, 1992.

———, with Robert P. Crease. *Peace & War: Reminiscences of a Life on the Frontiers of Science.* New York: Columbia University Press, 1998.

Sherwin, Martin J. *A World Destroyed: Hiroshima and Its Legacies.* 3rd ed. Stanford, CA: Stanford University Press, 2003.

Smith, Alice Kimball. *A Peril and a Hope: The Scientists' Movement in America, 1945–1947.* Rev. ed. Cambridge, MA: MIT Press, 1971.

———, and Charles Weiner, eds. *Robert Oppenheimer: Letters and Recollections.* Cambridge, MA: Harvard University Press, 1980.

Smith, Richard Norton. *The Colonel: The Life and Legend of Robert R. McCormick.* Boston: Houghton Mifflin, 1997.

Snow, C. P. *Variety of Men.* New York: Scribner, 1967.

Steinberger, Jack, *Learning About Particles: 50 Privileged Years.* New York: Springer, 2005.

Strauss, Lewis L. *Men and Decisions.* Garden City: Doubleday, 1962.

Stuewer, Roger H., ed. *Nuclear Physics in Retrospect: Proceedings of a Symposium on the 1930's.* Minneapolis: University of Minnesota Press, 1979.

Truman, Harry S. *1945: Year of Decisions.* New York: New American Library, 1955.

Weart, Spencer R., and Melba Phillips, eds. *History of Physics: Readings from Physics Today.* New York: American Institute of Physics, 1985.

Weiner, Charles, ed. *Exploring the History of Nuclear Physics: Proceedings of the American Institute of Physics on the History of Nuclear Physics, 1967 and 1969.* New York: American Institute of Physics, 1972.

York, Herbert F. *The Advisors: Oppenheimer, Teller, and the Superbomb.* San Francisco: W. H. Freeman, 1976.

———. *Making Weapons, Talking Peace: A Physicist's Odyssey from Hiroshima to Geneva.* New York: Basic Books, 1987.

Notes

Abbreviations

AIP Niels Bohr Library & Archives, American Institute of Physics, College Park, MD

BANC Bancroft Library Collections, University of California, Berkeley

EOLP Ernest O. Lawrence Papers, Bancroft Library, University of California, Berkeley

FRUS Foreign Relations of the United States

HCP Materials Assembled for a Biography of Ernest O. Lawrence (Herbert Childs Papers), Bancroft Library, University of California, Berkeley

IMJRO *In the Matter of J. Robert Oppenheimer*, Transcript of Hearing before Personnel Security Board: US Government Printing Office, Washington, DC, 1954

LBNL Lawrence Berkeley National Laboratory

LOC Library of Congress

NYT New York Times

OH Oral history

RF Rockefeller Foundation Archives

Introduction: Creation and Destruction

1 *When the presentation ended*: Carroll, *The Particle at the End of the Universe:* (New York: Dutton, 2012), p. 187.

2 *"It is almost as hard"*: Robert R. Wilson, "My Fight Against Team Research," in *Twentieth-Century Sciences*, Holton (New York: W. W. Norton, 1972), p. 468.

5 *"The first to disintegrate"*: Stuewer, *Nuclear Physics in Retrospect*, p. 107.

6 *"haggard, nervous"*: "Science: Einstein's Field Theory," *Time*, February 18, 1929.

6 *"Mathematician Einstein"*: "Science: He Is Worth It," *Time*, December 22, 1930.

6 *"easy to talk to"*: Bruce Bliven, "New Miracles of Atomic Research," *New Republic*, June 16, 1941, pp. 818–20.

7 *The term* Big Science *was coined*: Alvin M. Weinberg, "Impact of Large-Scale Science on the United States," *Science* 134, no. 3473 (July 21, 1961): pp. 161–64.

7 *"The logistics of keeping"*: Alvin M. Weinberg, "Scientific Teams and Scientific Laboratories," in *Twentieth-Century Sciences*, Holton, p. 430.

7 *"We simply do not know"*: Peter Galison, "The Many Faces of Big Science," in Galison and Hevly, *Big Science*, p. 7.

8 *"physicists, mathematicians, chemists"*: Weinberg, "Scientific Teams and Scientific Laboratories," p. 427.

8 *"abnormal competitive element"*: Karl T. Compton to M.C. Winternitz, November 24, 1941, EOLP. Winternitz was then a member of the government's wartime Committee on Medical Research, and Compton's letter was in the nature of an appeal for a federal contract.

9 *"Could the theory"*: Wigner, E., "The Limits of Science," *Proceedings of the American Philosophical Society* 94, no. 5, October 1950.

9 *"spending their time"*: Galison and Hevly, *Big Science*, p. 4.

10 *"The new cyclotron"*: 1940 Annual Report, The Rockefeller Foundation, p. 43.

Chapter One: A Heroic Time

15 *"a big, rather clumsy"*: Snow, *Variety of Men*, p. 2.

15 *"Rutherford was an artist"*: Quoted in Rhodes, *Making of the Atomic Bomb*, p. 45.

17 *"uranium radiation"*: Rutherford, "Uranium Radiation and the Electrical Conduction Produced by It," *Philosophical Magazine* 40, no. 109 (January 1899): pp. 109-163.

17 *"I was brought up"*: Rhodes, *Making of the Atomic Bomb*, p. 47.

18 *But a tiny number*: Eve, Arthur S., "Modern Views on the Constitution of the Atom," *Science 40*, no. 1021 (July 24, 1914): pp. 115–21.

19 *"uncarpeted floor boards"*: Oliphant, *Rutherford*, p. 19, cited in Rhodes, *Making of the Atomic Bomb*, p. 134.

19 *"bad taste"*: Weiner, *Exploring the History of Nuclear Physics*, p. 177.

19 *Its entire annual budget*: The figure is from Chadwick, James, "Some Personal Notes on the Search for the Neutron," 10th International Congress of History of Science, Cornell University, 1962.

20 *"We must conclude"*: Rutherford, "Collisions of Alpha Particles with Light Atoms," *The Philosophical Magazine and Journal of Science*, vol. 37, no. 22 (1919): p. 537.

21 *"a heroic time"*: Quoted in Kevles, *The Physicists*, p. 163.

21 *"the whole system"*: See Forman, Paul, "The Doublet Riddle and Atomic Physics circa 1924," *Isis* 59, no. 2 (Summer 1968): pp. 156–74.

21 *"decidedly confused"*: Kevles, *The Physicists*, p. 162.

21 *"so desperate"*: Chadwick, "Some Personal Notes."

21 *"Revolution in Science"*: For Eddington and his campaign, see Sponsel, Alistair, "Constructing a 'Revolution in Science'," *The British Journal for the History of Science* 35, no. 4 (December 2002): pp. 439–67.

22 *"Mme. Curie Plans"*: *New York Times*, May 12, 1921; for retraction, see *New York Times*, May 13, 1921.

22 *"endeavoring, with some initial success"*: Eve, Arthur S., "Modern Views."

23 *"It is my opinion"*: Clark, *Einstein: Life and Times*, p. 163.

23 *"The very strange situation"*: interview of Werner Heisenberg by Thomas S. Kuhn, February 25, 1963, AIP.

24 *By Rutherford's reckoning*: Rutherford, "Nuclear Constitution of Atoms," Bakerian Lecture, June 3, 1920.

25 *No one could explain*: See Kevles, *The Physicists*, p. 224.

25 *"much deformed"*: Rutherford, "Nuclear Constitution of Atoms."

25 *"He expounded"*: Chadwick, "Some Personal Notes."

25 *"What we require"*: Rutherford, February 28, 1930, cited in Eve, *Rutherford*, p. 338.

26 *"There appears to be"*: "Address of the president, Sir Ernest Rutherford, O.M., at the Anniversary Meeting," November 30, 1927, *Proceedings of the Royal Society of London* 117, no. 777 (January 2, 1928): pp. 300-316.

26 *When it arrived*: Heilbron and Seidel, *Lawrence and His Laboratory*, p. 50.

27 *rather than fitting*: C. C. Lauritsen and R.D. Bennett, "A New High Potential X-Ray Tube," *Physical Review* 32 (November 30, 1928): p. 850.

27 *"long sinuous snarling arc"*: McMillan, "Early History of Particle Accelerators" in Stuewer, *Nuclear Physics in Retrospect*, p. 115.

27 *"All of us youngsters"*: Tuve in Weiner, *Exploring the History of Nuclear Physics*, p. 27.

27 *"albatross"*: Tuve to McMillan, April 21, 1977, ibid., p. 135.

Chapter Two: South Dakota Boy

29 *"Most of us"*: Livingston in Weiner, *Exploring the History of Nuclear Physics*, p. 33.

30 *Carl taught*: Childs, *American Genius*, p. 26

30 *"If a man doesn't"*: John Lawrence OH.

30 *"born grown up"*: Margaret (Lawrence) Casady to Herbert Childs, April 15, 1963, HCP.

30 *A family yarn*: Mabel Blumer recollection, HCP; see also Childs, *American Genius*, pp. 28–29.

31 *one day he took*: Childs, *American Genius*, p. 34.

31 *Merle's through diligent*: Interview of Merle Tuve by Charles Weiner, March 30, 1967, AIP.

32 *Ernest and Merle dug*: Tuve recollections, HCP; Tuve, AIP.

32 *"When the president says"*: Tuve recollections, HCP.

32 *"the wickedness"*: Childs, *American Genius*, p. 46.

32 *D in electricity and magnetism*: See John Lawrence OH, BANC.

33 *"There's a fellow"*: John Lawrence OH, BANC.

33 *theory of relativity*: On the Swann-Einstein letters, see Hagar, Amit, "Length Matters: The Einstein-Swann Correspondence," *Studies in History and Philosophy of Modern Physics* 39, no. 3 (2008): pp. 532–56.

33 *Swann had crossed*: See "W. F. G. Swann" (obituary), *Physics Today* 15, no. 4 (April 1962): p. 106.

34 *"I like to think"*: W. F. G. Swann, "The Teaching of Physics," *American Journal of Physics* 19 (March 1951): pp. 182–87.

35 *"Swann was unhappy"*: Loeb to Childs, November 15, 1960, HCP.

35 *"Every two years"*: Childs, *American Genius*, p. 65.

36 *Swann delivered*: Ernest Lawrence, "The charging effect produced by the rotation of a prolate iron spheroid in a uniform magnetic field," *Philosophical Magazine* 47 (May 1924): pp. 842–47.

36 *"He was different"*: Childs, *American Genius*, p. 89.

37 *"We decided"*: Beams recollections, HCP.

37 *"There is no definite"*: Ernest Lawrence and Jesse Beams, "On the Nature of Light," *Proceedings of the National Academy of Sciences* 13 (April 15, 1927): pp. 207–12.

37 *"He worked me"*: Beams recollections, HCP.

38 *"a sort of roving"*: Ibid.

38 *"I felt out"*: Leonard Loeb to Elmer E. Hall, May 8, 1926, reprinted in Childs, *American Genius*, pp. 98–99.

40 *"the modern developments"*: Crelinsten, *Einstein's Jury*, p. 21.

41 *In vain, the university*: Heilbron and Seidel, *Lawrence and His Laboratory*, p. 19. For Compton offer, see Birge OH, BANC.

41 *"I got the following"*: Birge to EOL, May 23, 1927, EOLP. Birge's reference was to Floyd K. Richtmeyer, an X-ray expert who would in fact spend his entire career at Cornell, becoming dean of its graduate school in 1933. Lawrence indeed had received an offer from Cornell, which he turned down.

42 *"not behind"*: Childs, *American Genius*, p. 114.

42 *"the teaching schedules"*: Birge to Lawrence, February 23, 1928, EOLP.

43 *the university had granted him*: Childs, *American Genius*, p. 108.

43 *"the younger men"*: Birge to EOL, March 5, 1928, EOLP.

43 *"the biggest mistake"*: Beams recollections, HCP.

44 *"He responded"*: Tuve to Edwin McMillan, April 21, 1977, in Stuewer, *Nuclear Physics in Retrospect*, p. 135.

Chapter Three: "I'm Going to Be Famous"

46 *"I merely looked"*: Lawrence, "The Evolution of the Cyclotron" (Nobel Prize lecture delivered December 11, 1951), in Livingston, *Development of High-Energy Accelerators*, p. 137.

47 *"The merit," he told a friend, "lies"*: Szilard to Otto Stern, undat., cited in Heilbron and Seidel, *Lawrence and His Laboratory*, p. 82.

47 *Tom Johnson*: Johnson to Lawrence, September 15, 1931, EOLP.

47 *"But what are you"*: Childs, *American Genius*, p. 139.

48 *"But what can they do"*: Brady to E. M. McMillan, April 21, 1977, in McMillan, "Early History of Particle Accelerators," Steuwer, *Nuclear Physics in Retrospect*, pp. 131–32.

48 *Everyone seemed to have a different reason:* See Heilbron and Seidel, *Lawrence and His Laboratory*, p. 86.

49 *"Why don't you"*: See McMillan, "Early History," p. 126. The German quote appears in John H. Lawrence, MD, oral history, interview by Sally Smith Hughes, 1979–1980, BANC.

49 *"any serious difficulty"*: M. A. Tuve, G. Breit, and L. R. Hafstad, "The Application of High Potentials to Vacuum-Tubes, *Physical Review* 35 Jan. 19, 1930): p. 66.

50 *"Mind if I work"*: Childs, *American Genius*, p. 134.

50 *"I've got a crazy idea"*: Ibid., p. 146.

51 *"If the work"*: Lawrence to parents, February 23, 1930, EOLP.

51 *"Preliminary experiments"*: E. O. Lawrence and N. E. Edlefsen, "On the Production of High Speed Protons," *Science* 72 (October 10, 1930): pp. 376–77.

51 *Livingston acknowledged*: Interview of M. Stanley Livingston by Charles Weiner and Neil Goldman, August 21, 1967, AIP.

52 *"He didn't think"*: Ibid.

53 *"all the basic features"*: Livingston, *Development of High-Energy Accelerators*, p. 117.

53 *"At last we seem"*: Quoted in Heilbron and Seidel, *Lawrence and His Laboratory*, p. 95.

53 *"We are having a bit"*: Lawrence to Swann, January 24. 1931, EOLP.

54 *"Lawrence was really"*: Livingston, AIP.

54 *"the best experimental"*: Childs, *American Genius*, p. 155.

55 *"If there is one chance"*: Birge, *History of the Physics Department, University of California, Berkeley*, p. 388.

55 *"angling for funds"*: Livingston, AIP.

56 *"asked me the outright"*: Ibid. The book was the comprehensive text "Radiations from Radioactive Substances" (1930) by Rutherford, Chadwick, and Charles D. Ellis.

56 *Two weeks later*: The estimate of 750,000 volts is in Lawrence to Cottrell, July 17, 1931; that to 900,000 volts is in a letter the same day to Donald Cooksey. Both in EOLP.

57 *"I am hastening"*: Lawrence to Cottrell, July 17, 1931, EOLP.

57 *"I could never work"*: Gray, George W., "Science and Profits," *Harper's Monthly*, April 1936.

58 *"[T]he moment that"*: Abraham Flexner, "University Patents," *Science* 77, no. 1996 (March 31, 1933): p. 325.

58 *"the dean of one"*: Gregg, Alan, "University Patents," *Science* 77, no. 1993 (March 10, 1933): pp. 257–59.

58 *"a vicious influence"*: Gray, "Science and Profits."

58 *"Science is dependent"*: Ibid.

59 *"as to the action"*: Quoted in Archie MacInnes Palmer, "University Patent Policies," *Journal of the Patent and Trademark Office Society* 16 (February 1934): pp. 96–131.

59 *"endowment for scientific work"*: Cameron, Frank, *Cottrell: Samaritan of Science*, p. 151.

60 *But its grant portfolio*: The figures on assets and grants are from Research Corporation for Scientific Advancement, *100 Years of Supporting Science Innovation*, pp. 8–26.

60 *"full freedom"*: Palmer, "University Patent Policies."

61 *"may prove to be"*: Cottrell to Poillon, July 7, 1931, cited in Cameron, pp. 288–89.

62 *"the advantage of brevity"*: Lawrence to Sproul, January 6, 1932, EOLP.

62 *"If he is"*: Poillon to Cottrell, August 6, 1931, cited in Heilbron and Seidel, *Lawrence and His Laboratory*, p. 111.

63 *"your proton merry-go-round"*: Slater to Lawrence, September 4, 1931, EOLP.

63 *"it never occurred"*: Lawrence to Slater, September 8, 1931, EOLP.

63 *"his method for spiraling"*: Johnson "to whom it may concern," September 15, 1931, EOLP.

63 *"during my stay"*: Stern to Lawrence, November 2, 1931. The German original and a contemporaneous English translation, from which this quotation is taken, are in EOLP.

63 *"It is apparent"*: Lawrence to Arthur P. Knight, December 2, 1932, EOLP. The cyclotron patent is 1,948,384, "Method and Apparatus for the Acceleration of Ions." Based on Livingston's eleven-inch design, it was issued on February 20, 1934, and assigned to the Research Corporation, with Lawrence identified as the sole inventor.

64 *"It is entirely"*: Lawrence to Poillon, October 19,1935, EOLP.

64 *master's degree*: Childs, *American Genius*, p. 160.

64 *"I am beginning"*: Lawrence to Cooksey, July 17, 1931, EOLP.

65 *"more or less intuitively"*: Livingston, AIP.

65 *"Dr. Livingston has asked"*: Rebekah Young to Lawrence, August 2, 1931, EOLP.

Chapter Four: Shims and Sealing Wax

67 *"We were heading"*: Livingston, AIP.

67 *"It was Lawrence's genius"*: Livingston, AIP.

68 *But they spent*: The process of testing shims is documented in Lawrence's workbooks for 1932, EOLP.

68 *Eventually they discovered*: See Ernest O. Lawrence and M. Stanley Livingston, "The Production of High Speed Light Ions Without the Use of High Voltages," *Physical Review* 38 (April 1, 1932): p. 834.

68 *"Lawrence literally danced"*: Livingston, *Particle Accelerators*, p. 28.

69 *"The place on the coast"*: Boyce to Cockcroft, January 8, 1932, cited in Weiner, Charles, "1932—Moving into the new physics," *Physics Today* 25, no. 5 (May 1972): p. 40.

70 *"To a superficial"*: P. V. Danckwerts, "From the Cavendish to Harwell," *New Scientist* 101, no. 1404 (April 5, 1984): pp. 24–25.

71 *"We weren't ready"*: Livingston, AIP.

71 *the pages of the journal* Nature: J. D. Cockcroft and E. T. S. Walton, "Disintegration of Lithium by Swift Protons" (letter dated April 16, 1932), in *Nature* 129, no. 242 (April 30, 1932): p. 649.

71 *he wired his graduate student Jim Brady*: Brady recollections, HCP.

72 *the newlyweds made their circuitous way*: The honeymoon itinerary is described in Childs, *American Genius*, pp. 186–87.

72 *"Everyone has to have"*: Molly Lawrence recollections, HCP. John Lawrence recalled a very similar observation from his father (John Lawrence OH, BANC).

72 *"The place was beginning"*: Interview of Milton White by Charles Weiner, May 11, 1972, AIP.

73 *The Rad Lab's letter*: E. O. Lawrence, M. S. Livingston, and M. G. White, "The Disintegration of Lithium by Swiftly-Moving Protons," *Physical Review* 42 (October 1, 1932): pp. 150–51.

73 *"I don't know what"*: Malcolm Henderson recollections, HCP.

73 *The department's budget*: See Birge, p. 460.

73 *By contrast, the Rad Lab's spending*: See Heilbron and Seidel, *Lawrence and His Laboratory*, p. 212.

74 *a typical year*: The staff lists and salary figures are at EOLP. See also Heilbron and Seidel, p. 223.

75 *To obtain radio tubes*: Heilbron and Seidel, *Lawrence and His Laboratory*, p. 115.

75 *"This company"*: Hockenblamer to Leuschner, September 18, 1931, EOLP.

76 *Don Cooksey, who settled*: Kamen, *Radiant Science, Dark Politics*, pp. 141–42.

76 *"quite affluent"*: Molly Lawrence, address at the forty-fifth-anniversary dinner of the Lawrence Berkeley National Laboratory, October 1976, collection of the LBNL.

76 *if it fell silent*: Livingood recollections, HCP.

77 *"There appeared to be"*: Molly Lawrence, forty-fifth-anniversary address.

78 *"working day and night"*: Lawrence to C. R. Haupt, March 11, 1932, EOLP.

78 *"He'd come in"*: Livingood recollections, HCP.

78 *"a sort of laboratory slang"*: For an early use of the term "cyclotron" by Lawrence, see Lawrence to Tuve, September 23, 1933, EOLP: "we are calling the proton accelerator the 'cyclotron' now." The term was identified as "slang" in E. O. Lawrence, E. M. McMillan, and F. M. Thornton, "The Transmutation Functions for Some Cases of Deuteron-Induced Radioactivity," *Physical Review* 48 (September 15, 1935): pp. 493–99. The Nobel Prize presentation speech is at www.nobelprize.org/nobel_prizes/physics/laureates/1939/press.html [accessed March 17, 2013].

78 *"Everyone had to wait"*: Henderson recollections, HCP.

79 *One morning in 1934*: Livingood recollections, HCP.

79 *"a different color"*: Henderson recollections, HCP.

79 *"impress the fellows"*: Lawrence to Cooksey, December 29, 1934, EOLP.
80 *"all the horror stories"*: Molly Lawrence, forty-fifth-anniversary address.
80 *"We have been giving"*: Lawrence to Cockcroft, September 12, 1935, EOLP.
81 *"neither particularly"*: Tuve, "Memorandum Regarding Mr. Cowie's Eyes," June 18, 1947, EOLP.
81 *the Carnegie Institution*: Tuve to Kenneth Priestly, Rad Lab, June 20, 1947, EOLP. See also "Cyclotron Cataracts," *Time*, January 3, 1949.
81 *An amiable gentleman*: Childs, *American Genius*, pp. 252–53.
82 *But the tube failed*: Ibid., p. 192.
83 *"I have warned"*: Poillon to Knight, October 11, 1932, cited in Heilbron and Seidel, *Lawrence and His Laboratory*, p. 123. The first of several patents covering elements of the X-ray tube, No. 2,009,457, was issued to Sloan on July 30, 1935, and assigned to the Research Corp.
83 *"I am told"*: Lawrence to Poillon, August 18, 1932, EOLP.
83 *"had it running"*: Livingston, AIP.
84 *"As the youngest"*: Wilson, "My Fight Against Team Research," in Holton, *Twentieth-Century Sciences*, pp. 468ff.
85 *"I don't know"*: Henderson recollections, HCP.
85 *"Livingston looks tired"*: Hall to Lawrence, June 16, 1932, EOLP.
85 *"Ernest had enough"*: Henderson recollections, HCP
86 *"I'm running this"*: Brady recollections, HCP.
86 *"Certain things will occur"*: Loeb to Herbert Childs, October 4, 1960, HCP.
87 *Cornell had hired*: Details of the Cornell cyclotron and Livingston's role come from Courant, Ernest D., "Milton Stanley Livingston, A Biographical Memoir," National Academy of Sciences, 1997.
87 *"brought a certain idolization"*: Interview of M. Stanley Livingston by Charles Weiner and Neil Goldman, August 21, 1967, AIP.

Chapter Five: Oppie

90 *"Lawrence the experimentalist"*: Brady recollections, HCP.
91 *"had no theoretical physics"*: Oppenheimer recollections, HCP.
92 *"he intellectually looted"*: Rhodes, *Making of the Atomic Bomb*, p. 121.
92 *University of Leiden*: Bird and Sherwin, *American Prometheus*, p. 74.
92 *"I don't think"*: Smith and Weiner, *Robert Oppenheimer: Letters*, p. 121.
92 *"It won't be any trouble"*: Smith and Weiner, p. 149.
92 *"argued himself"*: Pais and Crease, *J. Robert Oppenheimer*, p. 29.
93 *"Oppie was extremely"*: Serber, interviewed by Martin Sherwin, 1/19/82, quoted in Bird and Sherwin, *American Prometheus*, p. 88.
93 *"His physics was good"*: Pais and Crease, *J. Robert Oppenheimer*, p. 25.
93 *"That's impossible"*: Alvarez, *Adventures*, pp. 75–76.

94 *"Pied Piper"*: Serber, *Peace & War*, p. 28.

94 *"We weren't supposed to"*: Bird and Sherwin, *American Prometheus*, p. 96.

94 *"I still visualize"*: Gerjuoy in Kelly, *Oppenheimer and the Manhattan Project*, p. 122.

94 *"Oppie, is it a* secret?*"*: Carl Anderson oral history, Caltech archives.

95 *"In those days"*: James Brady recollections, HCP.

95 *"I went to their seminar"*: Segrè, *Enrico Fermi, Physicist*, p. 134.

96 *"Oppie—highly cerebral"*: Kamen, *Radiant Science, Dark Politics*, p. 178.

96 *"Theorists tend to be"*: Edwin McMillan oral history by Charles Weiner, October 31, 1972, AIP.

97 *"So we went"*: Brady recollections, HCP.

97 *"Very often the things"*: Oppenheimer recollections, HCP.

98 *"very proper"*: Oppenheimer recollections, HCP.

98 *"We talked about"*: Ibid.

98 *"I feel pretty awful"*: Oppenheimer to Lawrence, October 12, 1931, EOLP.

98 *"Break it up"*: Smith and Weiner, *Robert Oppenheimer: Letters*, p. 147. The anecdote's source was Else Uhlenbeck, wife of the Dutch physicist George Uhlenbeck.

99 *"It was like you"*: Oppenheimer to Lawrence, undat. but presumably January 3, 1932, EOLP.

99 *"became a relatively prominent guy"*: Oppenheimer recollections, HCP.

100 *At Oppie's request*: Serber, *Peace & War*, p. 42.

100 *"One Jew"*: Ibid., p. 50.

100 *scribbled an announcement*: Alvarez, *Adventures*, p. 78.

102 *"has definitely established"*: Robert Oppenheimer to Frank Oppenheimer, October 7, 1933, in Smith and Weiner, *Robert Oppenheimer: Letters*. p. 165.

Chapter Six: The Deuton Affair

106 *Lewis conceived an electrolytic*: Interview of M. Stanley Livingston by Charles Weiner and Neil Goldman, August 21, 1967, AIP.

106 *"He liked to tell"*: McMillan, Edwin, "Early Days in the Lawrence Laboratory," speech at forty-fifth anniversary of Lawrence Berkeley Laboratory, October 30, 1976.

107 *"the world supply"*: Childs, *American Genius*, p. 197.

107 *"As soon as we used"*: Livingston, AIP.

107 *the yield jumped a hundredfold*: G. N. Lewis, M. S. Livingston, and E. O. Lawrence, "The Emission of Alpha-Particles from Various Targets Bombarded by Deutons of High Speed," *Physical Review* 44 (July 1, 1933): p. 56.

107 *"Ernest's love affair"*: Alvarez, *Adventures*, p. 52.

108 *"All of a sudden"*: Livingston, AIP.

108 *"at this rate"*: Berkeley Science service, May 20, 1933, quoted in Heilbron and Seidel, *Lawrence and His Laboratory*, p. 154.

108 *"I am almost"*: Heilbron and Seidel, *Lawrence and His Laboratory*, p. 155.

109 *"marvelous advancement"*: *New York Times*, May 21, 1933.

109 *"a new miracle worker"*: *New York Times*, June 20, 1933.

110 *"scouting party"*: *New York Times*, June 24, 1933.

110 *"It was much easier"*: "Complementarity in Chicago," *Time*, July 3, 1933.

110 *he acquired a tiny supply*: Oliphant, "Working with Rutherford," in Hendry, *Cambridge Physics*, p. 186.

110 *"Lawrence and his colleagues"*: Rutherford to Lewis, May 30, 1933, quoted in Heilbron and Seidel, *Lawrence and His Laboratory*, pp. 157–58.

111 *gold resisted*: On the Cavendish experience with contamination, see Boyce to Lawrence, January 23, 1933, EOLP.

111 *"sticking to the target"*: Oliphant, "Working with Rutherford."

112 *"Energy produced"*: Childs, *American Genius*, p. 205. See *Scientific American*, November 1933, for the fullest contemporary replication of the "moonshine" quote, which has appeared in many formulations, most of them paraphrases put into Rutherford's mouth.

112 *"purely a matter"*: *New York Herald Tribune*, September 12 1933.

113 *angrily crossing out*: Oliphant, "The Two Ernests-I," *Physics Today* 19, no. 9 (September 1966), pp. 35-49.

113 *Werner Heisenberg argued*: See Heilbron and Seidel, *Lawrence and His Laboratory*, pp. 164–66.

114 *Lawrence both made"*: Livingston OH, AIP.

114 *"He is just"*: Oliphant, "The Two Ernests-I." Interview of James Chadwick by Charles Weiner, April 17, 1969, AIP.

114 *"I gathered that"*: Pollard to Lawrence, December 6, 1933, EOLP.

114 *"He was a bit abrupt"*: Lawrence to Pollard, December 20, 1933, EOLP.

114 *"He's a brash"*: Oliphant, "The Two Ernests-I."

115 *"perhaps before long"*: Lawrence to Curtis R. Haupt, December 4, 1933, EOLP.

115 *"It seems to me"*: Lawrence to Livingston, January 26, 1934, EOLP.

115 *"It would seem now"*: Lawrence to Fowler, December 28, 1933, EOLP.

115 *"A series of measurements"*. G. N. Lewis, M. S. Livingston, M. C. Henderson, and E. O. Lawrence, "The Disintegration of Deutons by High Speed Protons and the Instability of the Deuton," *Physical Review* 45 (February 15, 1934): pp. 242–44.

115 *"This first definite"*: Lawrence to Poillon, December 15, 1933, cited in Heilbron and Seidel, *Lawrence and His Laboratory*, p. 167.

115 *His report*: Crane and Lauritsen, "On the Production of Neutrons from Lithium," appeared on page 783 of *Physical Review* 44 (November 1,1933); the Rad Lab's paper (Livingston, Henderson, and Lawrence, "Neutrons from Deutons and the Mass of the Neutron") appeared on pages 781–82.

116 *"After working up"*: See Tuve to Lawrence, April 17, 1934, EOLP.

116 *"very good justification"*: Cockcroft to Lawrence, February 28, 1934, EOLP.

116 *"Do you think"*: Oliphant to Lawrence, March 28, 1934, EOLP.

117 *"alternative and reasonable"*: G. N. Lewis, M. S. Livingston, M. C. Henderson, and E. O. Lawrence, "On the Hypothesis of the Instability of the Deuton," *Physical Review* 45 (April 1, 1934): p. 497.

117 *"I can not understand"*: The quotation is from Lawrence to Cockcroft, March 14, 1934, EOLP. Lawrence wrote to Tuve in the same vein on the same date.

117 *"for a long time"*: Fowler to Lawrence, March 14, 1934, EOLP.

117 *"one of judgment"*: Heilbron and Seidel, *Lawrence and His Laboratory*, p. 172.

118 *"In the face"*: Tuve to Lawrence, April 17, 1934, EOLP.

118 *at one point, Raymond Birge*: Childs, *American Genius*, p. 219.

118 *"are not contradictory"*: "The Berkeley Meeting of the American Association for the Advancement of Science," *Science* 80, no. 2064 (July 20, 1934): pp. 43-44.

119 *"erroneous and misleading"*: Tuve, "Nuclear-Physics Symposium: A Correction," *Science* 80, no. 2068 (August 17, 1934.): pp. 161–62.

119 *"I'm not going to have"*: Chadwick, AIP.

120 *"You would be surprised"*: Chadwick to Lawrence, December 29, 1935, EOLP. See also Andrew P. Brown, "Liverpool and Berkeley: The Chadwick-Lawrence Letters," *Physics Today* 49, no. 5 (May 1996): pp. 34-40.

121 *"wallowing in cash"*: Pollard to Cooksey, August 22, 1937, cited in Heilbron and Seidel, *Lawrence and His Laboratory*, p. 337.

121 *"This was a mistake"*: Livingston, AIP.

Chapter Seven: The Cyclotron Republic

123 *When they ceased*: Joliot, F., and I. Curie, "Artificial Production of a New Kind of Radio-Element," *Nature* 133, no. 3354 (February 10, 1934): pp. 201–202. See also W. Palmaer, presentation speech for the 1935 Nobel Prize in Chemistry, at www.nobel prize.org/nobel_prizes/chemistry/laureates/1935/press.html [accessed March 16, 2013].

124 *"I can still see"*: Goldsmith, *Frederic Joliot-Curie*, p. 57.

124 *"roaring into the lab"*: Livingston, AIP.

125 *"We have had"*: Lawrence to Boyce, February 27, 1934, EOLP.

125 *The unfortunate wiring*: Prominent among those who dismiss the miswiring as an explanation are Heilbron and Seidel, *Lawrence and His Laboratory*, p. 179.

125 *"would always be a matter"*: Lawrence to Weaver, February 21, 1940, EOLP.

126 *"I always felt"*: Henderson recollections, HCP.

126 *"We felt like"*: Davis, *Lawrence & Oppenheimer*, p. 60.

126 *"the field is getting"*: Kurie to Cooksey, March 4, 1934, EOLP.

127 *"in these nuclear"*: G. N. Lewis, M. C. Henderson, M. S. Livingston, and E. O. Lawrence, "Artificial Radioactivity Produced by Deuton Bombardment," *Physical Review* 45 (March 15, 1934): pp. 428-429.

127 *They provided precise*: Crane and Lauritsen, "Radioactivity from Carbon and Boron Oxide Bombarded with Deuterons and the Conversion of Positrons into Radiation," ibid.

127 *"To our surprise"*: Lawrence to Boyce, February 27, 1934, EOLP.

128 *In March, Enrico Fermi*: Enrico Fermi, "Radioactivity Induced by Neutron Bombardment," *Nature* 133, no. 3368 (May 19, 1934), p. 757.

128 *Martin Kamen*: Kamen, "The Birthplace of Big Science," *Bulletin of the Atomic Scientists*, November 1974.

129 *"We are finding"*: Lawrence to Beams, February 27, 1934, EOLP.

129 *"the most exciting"*: Alvarez, *Adventures*, p. 35.

130 *"enormous superiority"*: Heilbron and Seidel, *Lawrence and His Laboratory*, p. 187.

130 *Rasetti calculated*: Ibid.

130 *"radiation equal"*: Ibid., p. 190.

131 *"We are now"*: Lawrence to Kast, May 3, 1934, EOLP.

132 *"Doubtless radio-sodium"*: E. O. Lawrence, "Transmutations of Sodium by Deuterons," *Physical Review* 47 (Jan 1, 1935): pp. 17–27. The preliminary letter appeared in the *Physical Review* 46 (October 15, 1934): p. 746.

132 Fermi scoffed: Emilio Segrè, "Fifty Years Up and Down a Strenuous and Scenic Trail," *Annual Review of Nuclear Science* 31 (1981): pp. 1–18.

132 *"before the accompanying"*: Lawrence to Poillon, September 29, 1934, EOLP.

132 *They objected that*: For the objections and some responses, see USPO actions, March 9, 1935, and May 9, 1936; and Knight to Lawrence, April 17, 1937, all EOLP.

133 *"just what parts"*: Knight to Lawrence, April 8, 1935, EOLP.

133 *"the radio-sodium experiments"*: Lawrence to Knight, April 12, 1935, EOLP.

133 *"I know how repugnant"*: Poillon to Lawrence, Oct, 10, 1935. Poillon's mention of "powerful Katinka" is a slightly garbled reference to "the powerful Katrinka," a burly farm girl appearing in Fontaine Fox's popular newspaper comic strip "Toonerville Trolley."

134 *"Although prosecuting"*: Lawrence to Poillon, October 19, 1935, EOLP.

134 *"The more I think"*: Lawrence to Knight, April 26, 1939, EOLP.

135 *the patent office rejected*: Knight to Lawrence, April 25, 1939, EOLP.

135 *Cyclotron Republic*: See Segrè, *A Mind Always in Motion*, p. 136.

135 *"an abnormal competitive element"*: Karl T. Compton to M. C. Winternitz, November 24, 1941, EOLP. Winternitz was then a member of the government's wartime Committee on Medical Research, and Compton's letter was in the nature of an appeal for a federal contract.

136 *"seem to be rousing 'hits' "*: Lawrence to McMillan, May 11, 1935, EOLP.

136 *"It has been extraordinarily"*: Lawrence to McMillan, May 5, 1935, EOLP.

137 *The difficulty that even Lawrence:* Birge, Raymond T., *History of the Physics Department, University of California, Berkeley*, p. 418.

137 *"We were all supposed"*: Interview of L. Jackson Laslett by Charles Weiner, October 18, 1970, AIP.

137 *In Copenhagen*: Ibid.

138 *"all of whom know"*: Cooksey to "Dodie," March 30, 1938, cited in Heilbron and Seidel, *Lawrence and His Laboratory*, p. 234.

138 *"it can be used"*: W. Palmaer, presentation speech, December 10, 1935, www.nobelprize.org/nobel_prizes/chemistry/laureates/1935/press.html [accessed March 25, 2013].

Chapter Eight: John Lawrence's Mice

139 *writing each other*: This and other personal reminiscences are from "John H. Lawrence, M.D.: Nuclear Medicine Pioneer and Director of Donner Laboratory, University of California, Berkeley," interview by Sally Smith Hughes, 1979–1980, BANC. Henceforth John Lawrence OH.

140 *"Medical students were advised"*: Ibid.

141 *"got up and made"*: Birge OH, BANC.

141 *"the radiation is so"*: Lawrence to Poillon, October 4, 1933, EOLP.

142 *"told me facetiously"*: Lawrence to Poillon, September 30, 1936, EOLP.

142 *"No one ever"*: John Lawrence OH, BANC. The story of the dead mouse became one of the treasured legends of the early Rad Lab, though with slight variations. Alvarez, in *Adventures*, p. 63, times the irradiation at fifteen minutes, but he was not present and is reporting at second- or third-hand; Childs, whose source is unidentified, places it at three minutes (p. 228). The one-minute estimate is John Lawrence's and is accepted here because he was the designer of the experiment. There is common agreement, in any case, that the impact of the mouse's death was, as Edwin McMillan reports, "very dramatic." (Edwin M. McMillan, "History of the Cyclotron-II," *Physics Today* 12, no. 10 [October 1959], p. 24.)

142 *an "amusing" distraction*: Lawrence to Milton White, October 30, 1935, EOLP.

142 *"complete blood studies"*: John Lawrence to Ernest Lawrence, March 24, 1936, EOLP.

143 *"not unlike a billiard"*: John H. Lawrence and Ernest O. Lawrence, "The Biological Action of Neutron Rays," *Proceedings of the National Academy of Sciences of the United States of America* 22, no. 2 (February 15, 1936): pp. 124–33.

143 *"a development"*: Lawrence to Poillon, February 27, 1936, EOLP.

144 *"led to much coarse humor"*: Kamen, *Radiant Science, Dark Politics*, p. 67.

145 *intellectually stimulating*: For Oppenheimer at Harvard, see Bird and Sherwin, *American Prometheus*, pp. 29–31.

145 *"It's our duty"*: Childs, *American Genius*, p. 237.

146 *"was pure research"*: Birge OH, BANC.

146 *"To have a quarrel"*: A transcription of Birge's index cards is at EOLP.

147 *"Hell, he made me"*: Mary Blumer Lawrence oral history, interview by Suzanne Riess, 1984, Robert Gorden Sproul Oral History Project, BANC.

147 *Right away, President Sproul*: Lawrence to Poillon, February 26, 1936, EOLP.

148 *"Many of the leading"*: Lawrence to Sproul, February 20, 1936, EOLP.

149 *He ran his finger*: Lawrence described the meeting in his letter to Poillon, February 26.

150 *"I can report only"*: Sproul to Lawrence, February 28, 1936, EOLP.

150 *"It was a question"*: Mary Blumer Lawrence oral history, BANC.

151 *"It was like a secondhand"*: Neylan recollections, HCP. Neylan's memory appears to have betrayed him here; he placed his first meeting with Lawrence in 1928, shortly after he became a regent and Lawrence arrived on campus; but that is inconsistent with his recollection of the site of the encounter being the Rad Lab, which did not come into existence until several years later.

152 *"Neylan kind of considered"*: Mary Blumer Lawrence OH, BANC.

153 *"When can you begin"*: Alvarez, *Adventures*, p. 39.

153 *"that would signal"*: Ibid., p. 42.

154 *"epidemic of trouble"*: Lawrence to Tuve, September 12, 1935, EOLP.

154 *"our apparatus runs only"*: Lawrence to Poillon, October 16, 1935, EOLP.

155 *"one of the formerly"*: P. Gerald Kruger and G. K. Green, "The Construction and Operation of a Cyclotron to Produce One Million Volt Deuterons," *Physical Review*, 51 (April 30, 1937): p. 699.

155 *The Rad Lab called*: Lawrence and Cooksey, "On the Apparatus for the Multiple Acceleration of Light Ions to High Speeds," *Physical Review* 50 (December 15, 1936): pp. 1131–40.

156 *"signs of professionalism"*: Edwin M. McMillan, "History of the Cyclotron."

156 *By mid-1937*: Ibid. McMillan places the number of non-Berkeley cyclotrons around the world at 20 by the end of 1936. This is almost certainly an overestimate, resulting possibly from the imperfect memory of two decades later, when he gave the count. More authoritative lists of U.S. and foreign machines and their dates of development can be found in Heilbron and Seidel, *Lawrence and His Laboratory*, pp. 301, 310, and 321.

156 *"pushes a slide rule"*: William M. Brobeck oral history, interviews by Graham Hale, June 1975–January 1976, BANC.

157 *"was very pleased"*: Ibid.

158 *"no one objected"*: Brobeck, William M., "Early Days at the Radiation Laboratory," *IEEE Transactions in Nuclear Science* NS-28, no. 3 (June 1981): pp. 2004–2006.

158 *"amazed at how"*: Brobeck OH.

158 *preventive maintenance. . . . checklist*: Brobeck, "Suggested Maintenance Operations on 37" Cyclotron," July 29, 1938, EOLP. See also Heilbron and Seidel, *Lawrence and His Laboratory*, p. 241.

159 *"sitting open mouthed"*: Stanley Van Voorhis to Lawrence, February 5, 1939, EOLP.

159 *"The boys are all complaining"*: Cooksey to Barnes, April 19, 1937, EOLP.

159 *"two dozen physicists"*: Heilbron and Seidel, *Lawrence and His Laboratory*, p. 236.

159 "We hope very soon": McMillan to Lawrence, October 27, 1937, cited in Heilbron and Seidel, *Lawrence and His Laboratory*, p. 238.

160 *"too painful"*: Kamen, *Radiant Science, Dark Politics*, p. 70.

160 *"a mania for gadgets"*: Quoted in Heilbron and Seidel, *Lawrence and His Laboratory*, p. 240.

160 *"as a radioactivity"*: Alvarez, *Adventures*, pp. 55–56.

161 *"knack for ingenious"*: Kamen, *Radiant Science, Dark Politics*, p. 308.

161 *"We are trying"*: Lawrence to Henderson, January 24, 1937, EOLP.

161 *"a full-time"*: Kamen, *Radiant Science, Dark Politics*, p. 80.

162 *"She was the first"*: Birge OH, p. 165, BANC.

162 *Carl reported*: Childs, *American Genius*, p. 278.

162 *"I'd stand by"*: John Lawrence OH, p. 51, BANC.

163 *("[T]he patients will")*: Lawrence to Frank Exner, September 22, 1938, EOLP.

163 *"I could see"*: John H. Lawrence OH.

164 *So it was*: For an influential review of the Berkeley treatments, see Sheline et. al., "Effects of Fast Neutrons on Human Skin," *American Journal of Roentgenology* 111, no. 1 (January 1971): pp. 31–41.

Chapter Nine: Laureate

165 *the Radiation Laboratory was leading*: See Fig. 8.4 in Heilbron and Seidel, *Lawrence and His Laboratory*, p. 388, for the comparative role in isotope discovery by Berkeley, the Cavendish, and five other leading nuclear laboratories.

165 *This was an extraordinary*: Ibid., p. 387, for Birge's estimate.

166 *"the infallibility"*: Alvarez, *Adventures*, p. 53.

166 *"Bethe's Bible"*: Bethe, et. al., "Nuclear Physics," appeared in *Reviews of Modern Physics* 8, no. 2 (April 1936): pp. 82–229; vol. 9, no. 2 (April 1937): pp. 69–244; and vol. 9, no. 3 (July 1937): pp. 245–390.

166 *"practically unobservable"*: Hans Bethe and R. F. Bacher, "Nuclear Physics: Stationary States of Nuclei" (Bethe Bible-I).

167 *"The cyclotron evidently"*: Segrè to Lawrence, February 7, 1937, EOLP.

167 *"I would beg you"*: Segrè to Lawrence, June 13, 1937, EOLP.

167 *"Of course all of us"*: Lawrence to Buffum, April 5, 1937, EOLP.

168 *"Ernest Rutherford was"*: *Time*, November 1, 1937.

168 *"blew his top"*: Childs, *American Genius*, p. 261.

169 *"I realize it is none"*: Alan Wells to Lawrence, February 7, 1940, EOLP.

169 *In a letter*: Bethe and Rose, "The Maximum Energy Obtainable from the Cyclotron," *Physical Review* 52 (December 14, 1937): p. 1254.

170 *"I am awfully glad"*: Lawrence to DuBridge, December 4, 1937, EOLP.

170 *Robert R. Wilson had been*: R. R. Wilson, "Magnetic and Electrostatic Focusing in the Cyclotron," *Physical Review* 53 (March 1, 1938): pp. 408–20. See also Lawrence to DuBridge, December 4, 1937, and Heilbron and Seidel, *Lawrence and His Laboratory*, p. 468.

171 *"we considered"*: Bethe to McMillan, quoted in Heilbron and Seidel, *Lawrence and His Laboratory*, p. 470.

171 *"Although the principle"*: J. D. Cockcroft, "The Cyclotron and Its Applications," *Journal of Scientific Instruments* 16, no. 2 (February 1939): pp. 2–34.

171 *"The real limitation"*: Lawrence to Oliphant, August 2, 1938, EOLP.

171 *"an elegant theory"*: Kamen, *Radiant Science, Dark Politics*, p. 76.

172 *"fantastic number"*: Edwin McMillan and Martin Kamen, "Neutron-Induced Radioactivity of the Noble Metals," *Physical Review* 15 (August 15, 1937): pp. 375–77.

173 *"Why, he just"*: Cooksey to Lawrence, April 29, 1938, EOLP.

173 *"there should be"*: Lawrence to A. L. Hughes, May 26, 1938, quoted in Heilbron and Seidel, *Lawrence and His Laboratory*, p. 311.

173 *"whether or not"*: Evans to Cooksey, June 1, 1938, EOLP.

174 *Ernest called MIT's bluff*: Lawrence to Evans, June 2, 1938, cited in Heilbron and Seidel, *Lawrence and His Laboratory*, p. 235.

174 *"Don't let this"*: Livingston to Cooksey, July 28, 1938, EOLP.

174 *"For medical purposes"*: Lawrence to A. L. Hughes, July 5, 1939, quoted in Heilbron and Seidel, p. 283.

174 *"because we can get"*: Ibid., p. 284.

174 *"its neutrons would reach"*: Arthur Snell, cited in Heilbron and Seidel, *Lawrence and His Laboratory*, p. 283.

175 *the magnet alone*: See Lawrence to Ludvig Hektoen, November 13, 1937, EOLP.

175 *"immediate and urgent"*: Ibid.

175 *"It is not necessary"*: Compton to Lawrence, November 29, 1937, EOLP.

175 *Ernest drew up*: The list is in Heilbron and Seidel, *Lawrence and His Laboratory*, p. 268, n. 162.

176 *"special concern"*: Weaver, "The Program in the Natural Sciences," Trustees Confidential Report, March 1950, RF.

177 *Weaver mistakenly described*: Weaver Diary, January 25, 1937, RF.

178 *"who had obvious"*: John Lawrence OH, BANC.

178 *"congress of cripples"*: Quoted by Weaver, memo to Raymond B. Fosdick, "Cyclotron Project—Professor E. O. Lawrence, University of California," November 23, 1937, RF.

178 *"an unexpected emergency"*: Ibid.

178 *"as a biological"*: Lawrence to Weaver, November 10, 1937, RF.

178 *"There is a spirit"*: Sproul to Weaver, November 26, 1937, RF.

179 *"subversive to the spirit"*: Kast to Poillon, October 13, 1936, quoted in Heilbron and Seidel, *Lawrence and His Laboratory*, p. 217.

179 *"in such a way"*: Hanson, Frank B., memorandum of visit to the University of California, Berkeley, April 13–23, 1938, RF.

179 *In January 1939*: The chronology of construction is from Lawrence, "The First Ten Years of Cyclotrons, 1930–1939, inclusive," EOLP.

179 *"no hand-me-down"*: Heilbron and Seidel, *Lawrence and His Laboratory*, p. 284.

180 *"We are convinced"*: E. O. Lawrence, et al., "Initial Performance of the 60-Inch Cyclotron of the William H. Crocker Radiation Laboratory, University of California," *Physical Review* 56 (July 1, 1939): p. 124.

181 *"Lawrence apparatus"*: Josephson, *Physics and Politics in Revolutionary Russia*, p. 181.

181 *but news reporters*: Childs, *American Genius*, p. 287.

182 *"The new cyclotron"*: Kamen to Hahn, November 21, 1939, EOLP.

182 *"kept me in a steady"*: Martin Kamen, "Early History of Carbon-14," *Science* 140, no. 3567 (May 10, 1963): pp. 584–90.

182 *they draped these*: Kamen, *Radiant Science, Dark Politics*, p. 79.

183 *"almost driven"*: Lawrence to Kruger, October 1, 1935, EOLP; "amazing smoothness" is from Lawrence to Kruger, March 14, 1940, quoted in Heilbron and Seidel, *Lawrence and His Laboratory*, p. 297.

183 *"no less than prodigious"*: Lawrence to Foster, March 26, 1940, EOLP.

183 *Hydrogen's naturally occurring*: Luis W. Alvarez and Robert Cornog, "Helium and Hydrogen of Mass 3," *Physical Review* 56 (Sept. 15, 1939): p. 613.

184 *The young chemist sprinted*: Kamen, *Radiant Science, Dark Politics*, p. 127ff.

185 *Aebersold's contribution*: Aebersold file, EOLP.

186 *"Your career is showing"*: Childs, *American Genius*, p. 296.

186 *"without comparison": Nobel Prize in Physics 1939—Presentation Speech.*

186 *"Well, what has Lawrence done?"*: Bohr to Thomson, quoted in Heilbron and Seidel, *Lawrence and His Laboratory*, p. 491.

186 *"It is extremely"*: Oliphant to Lawrence, November 20, 1939, ibid., p. 492.

187 *Kamen was alone*: Kamen, *Radiant Science, Dark Politics*, p. 130.

188 *What they did not*: See Samuel Ruben and Martin D. Kamen, "Radioactive Carbon of Long Half-Life," *Physical Review* 57 (March 15, 1940): p. 549.

189 *"immediate practical significance"*: Lawrence, Nobel banquet speech, Feb. 29, 1940, at http://www.nobelprize.org/nobel_prizes/physics/laureates/1939/lawrence-speech.html [accessed June 28, 2014].

Chapter Ten: Mr. Loomis

191 *"peculiar accomplishment"*: "Amateur of the Sciences," *Fortune*, March 1946, uncorrected typescript, EOLP.

192 *"the last of the great"*: Alvarez, "Alfred Lee Loomis," biographical memoir, National Academy of Sciences, 1980.

193 *"his homes contained"*: Ibid.

193 *"gadgeteering"*: Ibid.

193 *Winthrop & Stimson*: Jennet Conant, *Tuxedo Park*, p. 28.

193 *Aberdeen Chronograph*: See U.S. Patent 1,376,890, "Chronograph."

194 *"Loomis had ninety"*: "Amateur of the Sciences," *Fortune*.

194 *"He thought [TVA] would"*: Jennet Conant, *Tuxedo Park*, p. 89.
194 *$50 million*: "Amateur of the Sciences," *Fortune*.
194 *"Without so much"*: Jennet Conant, *Tuxedo Park*, p. 87.
195 *"He suggested"*: Seabrook, *Dr. Wood*, p. 214.
195 *"Queer things"*: "Amateur of the Sciences," *Fortune*.
195 *"You damned American"*: Loomis recollections, HCP.
196 *a conference in January*: The papers were all published in the *Journal of the Franklin Institute*, April 1928; see also Seabrook, *Dr. Wood*, p. 221.
196 *"anonymous friend"*: Alvarez, "Alfred Lee Loomis."
196 *"Every famous scientist"*: Loomis recollections, HCP.
197 *"gaps in time"*: Ibid.
197 *"had all the earmarks"*: Alvarez, "Alfred Lee Loomis."
197 *"He was just a"*: Loomis recollections, HCP.
197 *"He probably didn't"*: Birge OH, BANC.
198 "in any way": Loomis to Sproul, November 28, 1940, EOLP.
198 *"subject to your discretion"*: R. W. Kettler (university controller) to Lawrence, March 17, 1955, EOLP.
198 *Every day*: Birge oral history, BANC.
198 *"to learn about"*: Alvarez, *Adventures*, p. 80.
199 *"I hope your"*: Chadwick to Lawrence, April 16, 1938, EOLP.
200 *"in some quarters"*: Lawrence to Weaver, September 16, 1939, cited in Heilbron and Seidel, *Lawrence and His Laboratory*, p. 472.
201 *"going ahead splendidly"*: Lawrence to Loomis, December 27, 1939, EOLP.
201 *"If it looks"*: Childs, *American Genius*, p. 298.
201 *"not an entirely"*: Weaver to Sproul, January 23, 1940, EOLP.
202 *"Dr. Weaver has come"*: Lawrence to Loomis, January 14, 1940, EOLP.
202 *"there seems every"*: Weaver to Sproul, January 23, 1940, EOLP.
203 *"I may be sort of panicky"*: Transcript, telephone conversation, Professor E. O. Lawrence and Dr. Warren Weaver, January 29, 1940, EOLP.
204 *"link the names"*: Morris to Ford, Nov. 14, 1939, EOLP. Although historians often identify him as "David," Morris's given first name was "Dave."
205 *"one of the most interesting"*: Compton, Karl, to Weaver, January 29, 1940, EOLP.
205 *"a considered statement"*: The quotation is from the version Weaver sent to Karl Compton, January 25, 1940, EOLP.
206 *"Is someone going"*: Weaver to Lawrence, February 13, 1940, EOLP.
206 *This query bore*: See Heilbron and Seidel, *Lawrence and His Laboratory*, p. 472.
206 *He disposed of*: Lawrence to Weaver, February 21, 1940, EOLP.
208 *"We were agreed"*: Bush, *Pieces of the Action*, p. 33.
209 *"You can't get a group"*: Jennet Conant, *Tuxedo Park*, p. 149.
209 *"Our trustees voted"*: Heilbron and Seidel, *Lawrence and His Laboratory*, p. 482.
210 *"walking on air"*: Ibid.

210 *"Great and small"*: Ibid.

210 *"After spending some time"*: Alvarez, "Alfred Lee Loomis."

Chapter Eleven: "Ernest, Are You Ready?"

213 *"It was a cool"*: Compton, *Atomic Quest*, p. 7.

214 *"It may be"*: Heilbron and Seidel, *Lawrence and His Laboratory*, p. 444.

214 *"within a week"*: Weiner, *Exploring the History of Nuclear Physics*, pp. 90–91.

214 *"In how many ways"*: Oppenheimer to Fowler, January 28, 1939, Smith and Weiner, *Robert Oppenheimer: Letters*, pp. 205–6.

214 *"I think it really"*: Oppenheimer to George Uhlenbeck, February 5, 1939, ibid., p. 209.

214 *"You know what"*: Rhodes, *Making of the Atomic Bomb*, p. 281.

215 *he had kept two suitcases*: "Leo Szilard: His version of the facts," Part II, *Bulletin of the Atomic Scientists*, March 1979.

215 *"We both wanted"*: Weart, Spencer R., "Scientists with a Secret," *Physics Today*, February 1976.

216 *"quickie experiment"*: Alvarez, *Adventures*, pp. 76–77.

217 *"Chances for reaction"*: Rhodes, *Making of the Atomic Bomb*, p. 292.

218 *"what you are after"*: Hewlett and Anderson, *New World*, p. 17. See also Rhodes, *Making of the Atomic Bomb*, p. 314.

219 *"we have a goat"*: Rhodes, *Making of the Atomic Bomb*, p. 315.

219 *"All right"*: Hewlett and Anderson, *New World*, p. 20.

219 *"swimming in syrup"*: Kevles, *The Physicists*, p. 324.

219 *"I had assumed"*: "Leo Szilard: His version of the facts," Part III, *Bulletin of the Atomic Scientists*, April 1979.

220 *the* Athenia *had been sunk*: Childs, *American Genius*, p. 293; see also Aebersold recollections, HCP.

220 *"regarding the war situation"*: Lawrence to Loomis, May 20, 1940.

221 *"hard pressed"*: Heilbron and Seidel, p. 451.

221 *In June 1940*: Rhodes, *Making of the Atomic Bomb*, p. 351. The McMillan-Abelson letter is Edwin McMillan and Philip Hauge Abelson, "Radioactive Element 93," *Physical Review* 57 (June 15, 1940): p. 1185.

222 *"Just come over"*: Interview of Otto Frisch by Charles Weiner, May 3, 1967, AIP.

223 *They were "electrified"*: Oliphant, "The Beginning: Chadwick and the Neutron," *Bulletin of the Atomic Scientists*, December 1982.

223 *"All wheels"*: Bush, *Pieces of the Action*, p. 36.

223 *"an end run"*: Ibid., p. 32.

223 *For Fermi*: Hewlett and Anderson, *New World*, p. 27.

224 *"a sort of"*: Childs, *American Genius*, p. 306.

224 *after hosting*: Alvarez, *Adventures*, p. 84.

224 *one of the "vaudeville"*: Interview of Lee DuBridge by Thomas D. Cornell, March 6, 1987, AIP.

224 *"I can't tell you"*: Buderi, *Invention That Changed the World*, p. 46.

224 *"If Lawrence was interested"*: Guerlac, *Radar in World War II*, p. 260.

225 *"We just got"*: DuBridge interview by Thomas D. Cornell, AIP.

225 *"It was essentially"*: Interview of Edwin McMillan by Charles Weiner, June 2, 1972, AIP.

225 *"light a fire"*: Childs, *American Genius*, p. 311.

226 *"slow, conservative"*: Rhodes, *Making of the Atomic Bomb*, p. 361.

226 *"very vindictive"*: Interview of Conant by John C. Landers, 1974, quoted in Hershberg, *James B. Conant*, p. 147.

226 *"I told him flatly"*: Ibid.

227 *"over my head"*: Ibid.

227 *"You did a very fine"*: Bush to Lawrence, July 14, 1941, text in HCP.

227 *"Two facts"*: Compton, *Atomic Quest*, p. 46.

228 *The report discussed*: Hewlett and Anderson, *New World*, pp. 37–38.

228 *"on the whole"*: Compton, *Atomic Quest*, p. 47.

228 *"certainly no clear-cut"*: Ibid.

228 *"an extremely important"*: Compton, pp. 49–50.

229 *But he was stuck*: Childs, *American Genius*, p. 315.

229 *The MAUD Committee's*: The MAUD Report can be found in Cantelon, Hewlett, and Williams, eds., *The American Atom:* pp. 16–20.

230 *"discreet enquiries"*: Oliphant, "The Beginning: Chadwick and the Neutron."

230 *"I'll even fly"*: Childs, *American Genius*, p. 315. Childs attributes this wire to Lawrence; but since Oliphant was in Washington and Lawrence in Berkeley, it would seem that Oliphant instigated the meeting. (See Oliphant, "The Beginning.")

230 *"in your hands"*: Childs, ibid., p. 317.

232 *"involuntary conference"*: The quote is from Conant's personal "history" of the atomic bomb project, prepared in 1943 and cited by Rhodes, *Making of the Atomic Bomb*, p. 376.

232 *"Much of the difficulty"*: Rhodes, ibid.

233 *"throw themselves"*: Compton, *Atomic Quest*, p. 53.

233 *"Oppenheimer has"*: Smith and Weiner, *Robert Oppenheimer: Letters*, p. 223.

233 *"leftwandering"*: This may have been Oppenheimer's description rather than Lawrence's; see Oppenheimer recollections, HCP, which is the original source for the term.

233 *"I embarrassedly asked"*: Kamen, *Radiant Science, Dark Politics*, p. 185.

233 *"Federal Union of Democracies"*: Urey to Lawrence, August 10, 1940; Lawrence to Urey, August 20, 1940. Both in EOLP.

234 *"blew a gasket"*: Bird and Sherwin, *American Prometheus*, p. 175.

234 *"there will be no"*: Oppenheimer to Lawrence, November 12, 1941, in Smith and Wiener, *Robert Oppenheimer: Letters*, p. 220.

235 *Compton's report*: Hewlett and Anderson, *New World*, pp. 47–48.

235 *"V.B. OK"*: Ibid., p. 49.

235 *"The matter would . . ."*: Quoted in Rhodes, *Making of the Atomic Bomb*, p. 387.

235 *"whether atomic bombs"*: Compton, *Atomic Quest*, p. 63.

236 *"Before, we'd been jogging"*: Seaborg, *Adventures in the Atomic Age*, pp. 81–82.

Chapter Twelve: The Racetrack

238 *Driven by frustration*: Hewlett and Anderson, *New World*, pp. 56–57.

238 *"certain experimentation"*: Hewlett and Anderson, *New World*, p. 51.

238 *$340,000*: Ibid., p. 53.

240 *"In all the years"*: Alvarez, *Adventures*, p. 113.

240 *"You'll never get"*: Compton, *Atomic Quest*, p. 81.

240 *"difficult . . . to suggest"*: Heilbron and Seidel, *Lawrence and His Laboratory*, p. 510.

241 *"mind-boggling"*: Kamen, *Radiant Science, Dark Politics*, p. 141.

241 *"Podbielniak fractional distillation"*: Ibid.

241 *"Plutonium is so unusual"*: Seaborg, "Recollections and Reminiscences at the 25th Anniversary of the First Weighing of Plutonium," delivered September 10, 1967, U.S. Atomic Energy Commission.

244 *"whether at the laboratory"*: Seaborg, *Nuclear Milestones*, p. 7.

244 *"war project"*: Seaborg, *Journals*, p. 13 (November 28, 1940).

245 *"Things look good"*: Seaborg to McMillan, January 20, 1941, cited in Heilbron and Seidel, p. 461.

245 *He staked*: G. T. Seaborg, E. M. McMillan, J. W. Kennedy, and A. C. Wahl, "Radioactive Element 94 from Deuterons on Uranium," *Physical Review* 69 (April 1, 1946): p. 366.

245 *"We felt like shouting"*: Seaborg, *Adventures in the Atomic Age*, p. 71.

246 *There they separated*: Ibid., p. 75.

247 *"rather hasty"*: Ibid., p. 81.

247 *"Seaborg tells me"*: Compton, *Atomic Quest*, p. 71.

247 *"It wasn't a cozy"*: Kamen to McMillan, December 25, 1940, cited in Heilbron and Seidel, p. 520.

248 *"I need somebody"*: Edward W. Strong OH, conducted by Harriet Nathan, 1988, BANC.

248 *"stimulating" and "refreshing"*: Bush to Murphree, February 23, 1942, quoted in Hewlett and Anderson, *New World*, p. 60.

249 *"All I knew"*: *Oakland Tribune*, January 9, 1986.

249 *"We had more trouble"*: Transcript, *In the Matter of J. Robert Oppenheimer*, Personnel Security Board, April 12, 1954, p. 272 (henceforth IMJRO).

250 *"Ernest Lawrence yelled"*: Ibid., p. 268.

250 *"The crushing responsibilities"*: Kamen, *Radiant Science, Dark Politics*, p. 181.

250 *"He is Jewish"*: Lawrence to Allen, September 22, 1937, EOLP.

250 *He was aware*: The saga of Kamen's security case is from Kamen, *Radiant Science, Dark Politics*, pp. 164–67.

251 *"EOL thinks I told"*: Seidel, "The national laboratories of the Atomic Energy Commission in the early Cold War," *Historical Studies in the Physical and Biological Sciences* 32, no. 1 (2001): pp. 145-162.

251 *"felt no doubt"*: The quote is from Compton's deposition in Kamen's 1954 libel case against the *Chicago Tribune*. See Kamen, *Radiant Science, Dark Politics*, p. 316, n. 17.

251 *Arthur Compton, who faced*: Seaborg, *Adventures in the Atomic Age*, pp. 94–95.

252 *"It's none of your business"*: Molly Lawrence recollections, HCP.

252 *by mid-January*: Ibid., p. 57.

253 *He telephoned*: Ibid., p. 59.

253 *Vannevar Bush balked*: Weaver recollections, HCP.

254 *"for expediting the construction"*: Fosdick, *Story of the Rockefeller Foundation*, p. 174.

254 *"In the whole"*: Ibid., p. 175.

255 *"We all knew"*: Parkins, "The Uranium Bomb or the Calutron, and the Spare-Change Problem," *Physics Today*, May 2005.

255 *"the possession of"*: Conant to Bush, May 14, 1942, cited in Hershberg, *James B. Conant*, p. 160.

256 *Earlier that spring*: Seaborg, *Journals*, p. 158.

256 *With his customary*: Hewlett and Anderson, *New World*, p. 70.

258 *"picking a horse"*: Childs, *American Genius*, p. 336.

258 *There would be*: Hewlett and Anderson, *New World*, pp. 142–43.

Chapter Thirteen: Oak Ridge

259 *"a group of men"*: Ibid., p. 80.

260 *"I was hoping"*: Leslie R. Groves, "The Atom General Answers his Critics," *Saturday Evening Post*, June 19, 1948.

260 *"Having seen General Groves"*: Groves, *Now It Can Be Told*, p. 20.

260 *He arrived at the Rad Lab*: See Rhodes, *Making of the Atomic Bomb*, p. 489.

261 *On November 5*: Smyth Report, p. 195.

261 *"Groves walked in"*: Serber, *Los Alamos Primer*, p. xxxii.

261 *"designed and fabricated"*: Groves, *Now It Can Be Told*, p. 60.

262 *"setting up an organization"*: Nichols, *Road to Trinity*, p. 73.

262 *"I have known"*: Lawrence [to Groves], January 15, 1943, EOLP.

263 *"invisible materials"*: Hewlett and Anderson, *New World*, p. 90.

263 *"Unfortunately I cannot"*: Seaborg, *Adventures in the Atomic Age*, p. 89.

264 *Fermi's most recent prototype*: Hewlett and Anderson, *New World*, p. 48.

265 *"I don't see"*: Seaborg, *Adventures in the Atomic Age*, p. 100.

265 *The mass and purity*: See Smyth Report, p. 196.

266 *Vacuum pumps*: Hewlett and Anderson, *New World*, p. 143.

267 *Ernest now found himself*: For Groves's drive during construction, see Hewlett and Anderson, *New World*, p. 147–48.

268 *"certain special precautions"*: Groves to Lawrence, July 29, 1943, HCP.

268 *Groves calculated*: Groves, *Now It Can Be Told*, p. 97.

268 *During a two-week*: Ibid., p. 104.

269 *"When you see the magnitude"*: Hewlett and Anderson, *New World*, p. 154–55.

269 *"so many PhDs"*: Ibid., p. 148.

269 *"Would you like"*: Childs, *American Genius*, p. 344.

269 *"unfinished movie set"*: Seaborg, *Adventures in the Atomic Age*, p. 110.

270 *"He felt it"*: Childs, *American Genius*, p. 342.

270 *"essentially illiterate"*: Kamen, *Radiant Science, Dark Politics*, p. 157.

270 *"watching meters"*: Gladys Owens, quoted in "The Calutron Girls" at smithday1.net /angeltowns/or/go.htm [accessed December 9, 2012].

271 *"no receiving dock"*: Parkins, "The Uranium Bomb, the Calutron, and the Space-Charge Problem."

271 *Colonel Nichols, exasperated*: Nichols, *Road to Trinity*, p. 131.

271 the *eleven thousand Tennessee Eastman employees*: Ibid., p. 129.

272 *71 different types*: Smyth Report, p. 201.

272 *"The problem of where"*: Nichols, *Road to Trinity*, p. 92.

273 *Some material always*: Smyth Report, p. 202.

273 *construction was well along*: Hewlett and Anderson, *New World*, p. 159.

274 *"I had never seen"*: Alvarez recollections, HCP.

275 *the Alpha II racetracks*: Hewlett and Anderson, *New World*, p. 164.

275 *"The primary fact"*: Lawrence to Conant, May 31, 1944, cited in Hewlett and Anderson, *New World*, p. 166.

277 *"you'll be interested"*: Compton, *Atomic Quest*, p. 144.

277 *"Of course"*: Seaborg, *Adventures in the Atomic Age*, p. 103.

Chapter Fourteen: The Road to Trinity

280 *For four dispiriting*: Hewlett and Anderson, *New World*, p. 342.

280 *"not using the bomb"*: Alice Kimball Smith, "Behind the Decision to Use the Atomic Bomb: Chicago 1944–1945," *Bulletin of the Atomic Scientists*, October 1958.

282 *He proposed several*: Hewlett and Anderson, *New World*, p. 323.

283 *The proposal landed*: Ibid., pp. 326–29.

283 *The weapon in development*: Ibid., p. 329.

283 *"know in our hearts"*: Franck memo of April 21, 1945, quoted by Smith, "Behind the Decision."

284 *Groves had begun training*: Hewlett and Anderson, *New World*, p. 334.

284 *"it was considered"*: Henry L. Stimson, "The Decision to Use the Atomic Bomb," *Harper's*, February 1947.

284 *"something akin"*: Hewlett and Anderson, *New World*, p. 339.

285 *"Vannevar Bush and Jim Conant"*: Stimson, "The Decision."

285 *"the world in its present state"*: The nearly complete memorandum is published in Stimson, ibid.

286 *Conant later described*: James B. Conant, *My Several Lives*, p. 302.

286 *Japan "was an overpowered"*: Compton, *Atomic Quest*, p. 226.

287 *"moral isolation"*: Ibid., p. 235.

287 *"This demonstration might"*: The Bush-Conant memo can be found in Sherwin, *World Destroyed*, appendix F, p. 286–88. Also see Hewlett and Anderson, *New World*, p. 329.

287 *Compton estimated*: Hewlett and Anderson, *New World*, p. 356. Byrnes's reaction is at Byrnes, *All in One Lifetime*, p. 283.

288 *"vigorous program"*: Hewlett and Anderson, *New World*, p. 356.

288 *"it seemed to be"*: Compton, *Atomic Quest*, p. 238.

288 *"It was discussed"*: Lawrence to Karl K. Darrow, August 17, 1945, EOLP.

288 *"An atomic bomb"*: Compton, *Atomic Quest*, p. 239.

289 *"We didn't think"*: Transcript, IMJRO, p. 34.

289 *"the number of people"*: Lawrence to Darrow, August 17, 1945.

290 *"We don't think"*: Rhodes, *Making of the Atomic Bomb*, p. 642.

290 *"could not be considered"*: Stimson, "The Decision."

290 *"as to whether"*: Compton, *Atomic Quest*, p. 239. Alice Kimball Smith, *A Peril and a Hope*, p. 48, reports Oppenheimer's recollection that the panel did not receive the assignment until it had already convened in Los Alamos, where he recalled it was delivered verbally by Compton.

290 *"as soon as possible"*: Stimson, "The Decision."

290 *Bard, presently dissented*: Smith, "Behind the Decision."

291 *"We thought of"*: Compton, *Atomic Quest*, p. 240.

291 *Ryokichi Sagane and Tameichi Yasaki*: Heilbron and Seidel, pp. 317–18.

291 *The memo suggested*: "Recommendations on the Immediate Use of Nuclear Weapons," June 16, 1945, copy in EOLP.

292 *"It was hard to behave"*: Rhodes, *Making of the Atomic Bomb*, p. 657.

293 *"Anytime after the fifteenth"*: Oppenheimer to Lawrence, July 5, 1945, EOLP.

293 *Groves arrived*: Herken, *Brotherhood of the Bomb*, p. 136.

293 *Richard Feynman*: Rhodes, *Making of the Atomic Bomb*, p. 668.

294 *I decided the best*: "Thoughts by E. O. Lawrence," July 16, 1945, appended to Groves, Memorandum for the Secretary of War, July 18, 1945. The author wishes to thank Elaine B. McConnell of the special collections and archives department, U.S. Military Academy Library at West Point, for her assistance in locating this document.

295 *" 'Now I am become' "*: See Bird and Sherwin, *American Prometheus*, p. 309. It has sometimes been pointed out that this is not the only possible translation of the text: it can also be rendered as: "Now I am become *time*, the destroyer of worlds"—perhaps a subtler and more sinister thought than Oppenheimer's version.

295 *"Operated on this morning"*: Hewlett and Anderson, *New World*, p. 383.

296 *"I casually mentioned"*: Truman, *1945*, p. 416.

296 *"That was carrying"*: Rhodes, *Making of the Atomic Bomb*, p. 690.

296 *"suitable warning"*: Compton, *Atomic Quest*, p. 242.

297 *"Are not the men"*: Ibid.

297 *"Suddenly a bright flash"*: Alvarez, *Adventures*, p. 7.

298 *"From three of your former"*: Ibid., p. 145. The letter, in facsimile, is reproduced in Compton, *Atomic Quest*, p. 258.

298 *"I am sure"*: Lawrence to Akeley, August 16, 1945, EOLP, quoted in Bernstein, Barton J., "Four Physicists and the Bomb," *Historical Studies in the Physical and Biological Sciences* 18, no. 2 (1988): pp. 231–63.

299 *"I hope you will"*: Darrow to Lawrence, August 9, 1945, EOLP.

299 *"I am inclined"*: Lawrence to Darrow, August 17, 1945, EOLP.

301 *as finally prepared*: Oppenheimer to the Secretary of War, August 17, 1945, in Smith and Weiner, *Robert Oppenheimer: Letters*, pp. 293–94.

301 *"For the time being"*: Hewlett and Anderson, *New World*, p. 417.

301 *"it was a bad time"*: Oppenheimer to Lawrence, August 30, 1945, EOLP.

Chapter Fifteen: The Postwar Bonanza

303 *"tunic of Superman"*: *Life*, August 20, 1945, quoted in Kevles, *The Physicists*, p. 334.

303 *"In the past"*: Franck et al., Preamble, "Report of the Committee on Political and Social Problems" (The Franck Report), June 11, 1945.

303 *"Mr. President, I feel"*: Bird and Sherwin, *American Prometheus*, p. 332.

304 *"to protect him from"*: Neylan Recollections, HCP.

304 *mid-1944*: Robert Seidel, "Accelerating Science: The Postwar Transformation of the Lawrence Radiation Laboratory," *Historical Studies in the Physical Sciences* 13, no. 2 (1983), pp. 375–400.

305 *"Our stock was"*: Brobeck OH, BANC.

305 *"lying in bed"*: McMillan, AIP.

306 *"no reference"*: V. Veksler, "Concerning Some New Methods of Acceleration of Relativistic Particles," *Physical Review* 69 (February 28, 1946): p. 244.

306 *McMillan acknowledged*: Edwin M. McMillan, "The Origin of the Synchrotron," *Physical Review* 69 (May 1, 1946): pp. 534–35.

306 *Known as SCR-268s*: Alvarez, *Adventures*, p. 154. See also "Program for the Radiation Laboratory," April 1, 1946, appendix C, EOLP.

307 *Lawrence pried*: Seidel, "Accelerating Science."

307 *At its inception*: Birge OH, BANC.

307 *The Rad Lab budget*: The estimates are from "Program for the Radiation Laboratory," April 1, 1946.

307 *He also knew*: Seidel, "Accelerating Science."

308 *"if it hadn't been"*: Fosdick to Weaver, Inter-office correspondence, September 20, 1945, RF.

308 *"brought our civilization"*: The Rockefeller Foundation Annual Report, 1945.

309 *"We bet a hundred million"*: Childs, *American Genius*, p. 374.

310 *"very drastic and successful"*: The speaker was Bernard Peters, quoted in Seidel, Robert, "A Home for Big Science: The Atomic Energy Commission's Laboratory System," *Historical Studies in the Physical and Biological Sciences* 16, no. 1 (1986): pp. 135–75.

310 *Instead, army engineers*: Yoshio Nishina, "A Japanese Scientist Describes the Destruction of His Cyclotrons," *Bulletin of the Atomic Scientists*, June 1947.

311 *"an act of utter stupidity"*: *New York Times*, December 6, 1945.

311 *"the present time"*: Lawrence to Groves, January 20, 1946, quoted in Seidel, "Accelerating Science."

311 *"I hope very much"*: Childs, *American Genius*, p. 369.

311 *Groves granted the Rad Lab*: Hewlett and Anderson, *New World*, p. 628.

311 *"We ran it"*: Greenberg, *Politics of Pure Science*, p. 132.

311 *the launch of*: See spending figures in *Science and Public Policy, vol. 1: A Program for the Nation*, August 27, 1947, pp. 10 and 12. The document is known as the "Steelman Report" after John R. Steelman, chairman of the President's Scientific Research Board.

312 *"for every dollar"*: Philip Morrison, "The Laboratory Demobilizes," *Bulletin of the Atomic Scientists*, November 1946.

312 *"The Berkeley laboratory"*: Groves, *Now It Can Be Told*, p. 377.

313 *"first totalitarian bill"*: *New York Times*, October 31, 1945.

313 *"I must confess"*: Anderson to Higinbotham, October 11, 1945, quoted in Hewlett and Anderson, *New World*, p. 432.

314 *"He was kind of boyish"*: Neylan recollections, HCP.

315 *Federation of American Scientists*: Smith, *A Peril and a Hope*, p. 383.

315 *"My own feeling"*: Childs, *American Genius*, p. 378.

316 *"free run"*: Seidel, "The national laboratories."

316 *Seaborg had accepted*: Seaborg, *Journals*, p. 747 (August 28, 1945).

316 *$170,000 from the Manhattan District*: See Weaver to Fosdick, Inter-Office Correspondence, February 11, 1948, RF.

317 *The machine first*: Childs, *American Genius*, p. 386.

317 *"a kind of smoked"*: Lilienthal, *Atomic Energy Years 1945–1950*, pp. 107–8 (journal entry of November 18, 1946).

318 *"from irrationality to idiocy"*: Smith, *The Colonel*. p. 462.

318 *"pass no judgment"*: Seidel, "Accelerating Science."

319 *"Somebody asked"*: Brobeck OH, BANC.

319 *"good, big dinners"*: Hewlett and Duncan, *Atomic Shield*, p. 109.

319 *"Members of the committee"*: Cooksey to Loomis, August 24, 1947, quoted in Seidel, "Accelerating Science."

320 *"He regarded me"*: Oppenheimer recollections, HCP.

320 *"It may seem odd"*: Oppenheimer to Lawrence, August 30, 1945, EOLP.

320 *"Good, I can clip"*: Oppenheimer recollections, HCP.

321 *Brobeck had designed*: See Brobeck OH, BANC.

322 *It would harm science*: Seidel, "Accelerating Science."

322 *"Take something"*: Rigden, *Rabi: Scientist and Citizen*, p. 186.

323 *"What we submitted"*: Brobeck OH, BANC.

323 *"discouragement of the Berkeley group"*: quoted in Seidel, "Accelerating Science."

324 *Lawrence was summoned*: Alvarez, "Ernest Orlando Lawrence: A Biographical Memoir," *National Academy of Sciences*, 1970, p. 279.

324 *Brookhaven settled*: Seidel, "Accelerating Science."

Chapter Sixteen: Oaths and Loyalties

325 *"E. O. Lawrence personally"*: Panofsky, *Panofsky on Physics*, p. 39.

325 *"real cash"*: Hewlett and Duncan, *Nuclear Navy*, p. 49.

325 *"a subject which was"*: IMJRO, p. 776.

326 *Lawrence's concern*: For Neylan's opinion of communists, see "John Francis Neylan: Politics, Law, and the University of California," oral history interview by Corinne L. Gilb and Walton E. Bean, 1961, BANC.

326 *The star witness*: Barrett, *Tenney Committee*, p. 33.

327 *The chairman*: Stephanie Young, "Something Resembling Justice: John Francis Neylan and the AEC Personnel Security Hearings at Berkeley, 1948–49," in Carson and Hollinger, *Reappraising Oppenheimer*, p. 225.

327 *The Neylan board's first*: Herken, *Brotherhood of the Bomb*, p. 191.

327 *"he was the smart-alec"*: Neylan recollections, HCP.

327 *"if some kindly"*: Young, p. 236.

328 *"'I hope you'll give him'"*: Ibid.

328 *"If somebody sympathetic"*: Young, p. 240.

329 *"I had found"*: Serber, *Peace & War*, p. 165.

329 *"strengthened and reaffirmed"*: Young, p. 243.

329 *Frank had joined*: Cole, *Something Incredibly Wonderful Happens*, pp. 47–48.

330 *"I warned him"*: IMJRO, p. 117.

330 *"Why do you fool around"*: Frank Oppenheimer recollections, HCP. See also Childs, *American Genius*, p. 354.

330 *"Now look what you've done"*: Frank Oppenheimer recollections, HCP. See also Childs, p. 405.

330 *"one of the most useful"*: Lawrence to J. W. Buchta, October 16, 1946, EOLP.

331 *"meets with your approval"*: Williams to Lawrence, October 26, 1948, EOLP.

331 *Dear Lawrence*: Frank Oppenheimer to Lawrence, October 26, 1948, EOLP. Bird and Sherwin, *American Prometheus*, pp. 403–4, dated this letter "circa 1949"; their text suggests late 1949, after Frank Oppenheimer had publicly confessed to having been a member of the Communist Party. According to their source citation, they viewed an undated copy in the Oppenheimer Papers at Berkeley; the dated copy cited here, showing it was written prior to that public confession, can be found in EOLP, also at Berkeley.

331 *But he knew well*: Bird and Sherwin, *American Prometheus*, p. 402.

331 *The reporter's source*: Hewlett and Duncan, *Atomic Shield*, p. 13.

332 *That hurt*: For the McMillan-Oppenheimer friendship, see McMillan, AIP.

332 *"When we ran into"*: Robert Oppenheimer recollections, HCP.

332 *"any party or organization"*: Text from Gardner, *California Oath Controversy*, p. 25.

333 *"I was convinced"*: Neylan OH, BANC.

333 *he seized the reins*: For Neylan's role, see Gardner, *California Oath Controversy*, pp. 113–16.

334 *"byzantine quibbles"*: Segrè, *A Mind Always in Motion*, p. 235.

334 *"get the hell out"*: Ibid. See also Alvarez recollections, HCP.

334 *"my secret weapon"*: Alvarez, *Adventures*, p. 139.

335 *"Don't do anything"*: Panofsky, *Panofsky on Physics*, p. 43.

335 *"Now, listen"*: John David Jackson, "Panofsky Agonistes: The 1950 Loyalty Oath at Berkeley," *Physics Today* 62, no. 1 (January 2009): pp. 1–7.

335 *"lectured by Alvarez"*: Steinberger, *Learning About Particles*, p. 39.

336 *"One can only speculate"*: Gardner, *California Oath Controversy*, p. 248.

336 *"transient lunacies"*: Segrè, *A Mind Always in Motion*, pp. 235–36.

336 *"While I thought"*: Seaborg, *Adventures in the Atomic Age*, pp. 143–44.

337 *"Ernest felt emotionally"*: Alvarez recollections, HCP.

337 *Discovering that Neylan*: Herken, *Brotherhood of the Bomb*, p. 221.

337 *"There was one exception"*: Teller to Maria Goppert Meyer, undat., quoted by Herken, ibid.

338 *"Outstanding people left"*: Symposium, "The University Loyalty Oath," October 7, 1999.

Chapter Seventeen: The Shadow of the Super

339 *"We have to do"*: IMJRO, p. 659.

340 *"had essentially not"*: Ibid., p. 775.

341 *"even went so far"*: Rhodes, *Dark Sun*, p. 385.

341 *"A whopping big [program]"*: Lilienthal, *Atomic Energy Years*, p. 577 (journal entry of October 10, 1949).

342 *"shocked about his behavior"*: IMJRO, pp. 777–78.

342 *double the size*: The estimate comes from Herbert F. York, Livermore's first director. See York, *The Advisors*, p. 11.

342 *"A very great change"*: Oppenheimer to Conant, October 21, 1949. Text from IMJRO, pp. 242–43.

343 *"Rabi was worried"*: Alvarez, *Adventures*, p. 171.

343 *"They were extremely"*: IMJRO, pp. 460–61.

344 *"if it can be done"*: Rhodes, *Dark Sun*, p. 387.

344 *"felt he could count"*: IMJRO, p. 782.

344 *"the greatest misgivings"*: Hans Bethe, "Comments on the History of the H-Bomb," *Los Alamos Science*, Fall 1982.

345 *"We both had to"*: IMJRO, p. 329.

345 *"I am going"*: Ibid., p. 781.

345 *"There seemed to be a lack"*: Ibid., p. 783.

346 *"Keep your shirts"*: Ibid.

346 *"The East was evidently"*: Serber, *Peace & War*, p. 169.

346 *"my friends and any number"*: Alvarez, *Adventures*, p. 172.

346 *"eyes light up"*: Lilienthal, *Atomic Energy Years*, p. 581 (journal entry of October 29, 1949).

347 *"quite strongly negative"*: IMJRO, p. 247.

347 *"The main reason"*: Alvarez, *Adventures*, p. 172.

347 *"Pretty foggy thinking"*: IMJRO, p. 784.

347 *"No member of"*: The text of the report can be found in York, *The Advisors*, pp. 150–59.

348 *It has been suggested*: See, for example, Rhodes, *Dark Sun*, pp. 402–3.

348 *"rather awful"*: Lilienthal, *Atomic Energy Years*, p. 582 (journal entry of October 30, 1949).

349 *"a desperate need"*: Joseph Alsop and Stewart Alsop, "We Accuse," *Harper's*, October 1954.

349 *"an absolute dither"*: Jeremy Bernstein, "Physicist-II," *New Yorker*, October 20, 1975.

350 *"I listened as carefully"*: Beisner, *Dean Acheson*, p. 232.

350 *On November 18*: Hewlett and Duncan, p. 394.

350 *The White House meeting*: Lilienthal, *Atomic Energy Years*, p. 632 (journal entry of January 31, 1950).

350 *"What the hell"*: Rhodes, *Dark Sun*, p. 407.

350 *"say 'No' to a steamroller"*: Lilienthal, Atomic Energy Years, p. 633 (journal entry of January 31, 1950).

351 *Not even when*: Pfau, *No Sacrifice Too Great*, p. 123.

351 *"I never forgave"*: Bernstein, "Physicist."

Chapter Eighteen: Livermore

354 *"absolutely no experience"*: Serber, *Peace & War*, p. 169.

354 *"I went back"*: Alvarez, *Adventures*, p. 172.

354 *The nation's entire*: Hewlett and Duncan, pp. 147, 173–74.

355 *"If you have"*: Lawrence testimony, Joint Atomic Energy Committee, April 11, 1951, EOLP.

355 *obtaining the end product*: York, *The Advisors*, p. 123.

355 *"Ernest wants to see"*: Gow recollections, HCP.

356 *"The first priority"*: Brobeck OH, BANC.

357 *"occupied most"*: Alvarez, *Adventures*, p. 173.

358 *"spectacular stalagmites"*: Panofsky, *Panofsky on Physics*, p. 41.

358 *"Night after night"*: Alvarez, *Adventures*, p. 175.

358 *"I am not waiting"*: Lawrence testimony, JAEC, April 11, 1951.

358 *The key*: Hewlett and Duncan, p. 551.

359 *"MTA breaks the bottleneck"*: Lawrence to Smyth, April 16, 1952, EOLP.

359 *"I don't understand"*: Smyth to Lawrence, April 3, 1952, EOLP.

360 *"the wisdom of continuing"*: Powell to Gordon Dean (AEC chairman), July 11, 1952, EOLP.

360 *"The prototype MTA"*: Alvarez, *Adventures*, p. 176.

361 *"If the Russians"*: Rhodes, *Dark Sun*, p. 390.

362 *"the first thermonuclear"*: York, *The Advisors*, p. 127.

362 *"It's a boy"*: Hewlett and Duncan, p. 542.

363 *"dominated by irresistible"*: Segrè, *A Mind Always in Motion*, p. 238.

363 *"wouldn't work with him"*: Norris E. Bradbury OH, BANC.

363 *"like waving a red flag"*: Rogers M. Anders, *Forging the Atomic Shield*, p. 164.

364 *"the somewhat ironic"*: Hewlett and Duncan, p. 569.

364 *At a physics department*: York, *Making Weapons, Talking Peace*, p. 62.

365 *"all those other physicists"*: York, ibid., p. 53.

365 *Lawrence drove Edward Teller*: Hewlett and Duncan, p. 582.

366 *"We debated this"*: *IMJRO*, p. 311.

366 *"because of Teller"*: Ibid., p. 314.

366 *"You get a bunch"*: York, *Making Weapons, Talking Peace*, p. 62.

367 *"development and experimentation"*: York, *The Advisors*, p. 132.

367 *"well-lubricated"*: York, *Making Weapons*, p. 68.

368 *"obvious special status"*: York, *The Advisors*, p. 132.

368 *"he simply instructed"*: York, *Making Weapons*, p. 66.

368 *"It certainly was"*: Childs, *American Genius*, p. 445.

369 *"some folks would"*: Peter Moulthrop in Sybil Francis, "Warhead Politics: Livermore and the Competitive System of Nuclear Weapon Design," unpublished Ph.D. dissertation, 1996, p. 64.

369 *"simply coming"*: York, *The Advisors*, p. 134.

370 *"built a laboratory"*: Statement by Norris E. Bradbury, September 24, 1954, Los Alamos National Laboratory archives.

370 *Livermore's work*: Edward Teller, "The Work of Many People," *Science* 121, no. 3139 (February 25, 1955): pp. 267-275.

370 *At the first explosion*: Francis, "Warhead Politics."

372 *"complete revolution"*: Hewlett and Holl, *Atoms for Peace and War*, p. 180.

372 *"It was not surprising"*: York, *The Advisors*, p. 135.

372 *"believed in some quarters"*: Francis, "Warhead Politics"

372 *"had [not] been an effective"*: Ibid.

373 *Livermore's first-year budget*: York, *The Advisors*, p. 135.

373 *"coming of age"*: Francis, "Warhead Politics."

Chapter Nineteen: The Oppenheimer Affair

376 *"We do not operate well"*: The text of the speech was published as J. Robert Oppenheimer, "Atomic Weapons and American Policy," *Foreign Affairs* 31, no. 4 (July 1953): pp. 525-235.

376 *"The campaign is dangerous"*: Pfau, *No Sacrifice Too Great*, p. 145.

377 *"blank wall"*: Hewlett and Holl, p. 69.

377 *"set in motion"*: Bird and Sherwin, *American Prometheus*, p. 487.

378 *"should never again"*: Hewlett and Holl, *Atoms for Peace and War*, p. 87.

379 *"slippery sonofabitch"*: Bird and Sherwin, *American Prometheus*, p. 542.

379 *"I personally would feel"*: *IMJRO*, p. 710

379 *"we have an A-bomb"*: Ibid., p. 468.

380 *" 'But that is treason' "*: Ibid., p. 130.

381 *As late as Friday*: Herken, *Brotherhood of the Bomb*, p. 290.

381 *"barely civil"*: Ibid., p. 291.

381 *"would seem to reflect"*: Alvarez recollections, HCP.

382 he summoned his fellow: Childs, *American Genius*, p. 473.

382 *"I had never seen"*: Alvarez, *Adventures*, p. 180.

382 *"admiration and respect"*: Ibid.

383 *"Every time I have found"*: *IMJRO*, p. 802.

383 *"one of two men"*: Ibid., p. 805.

383 *"I think the reason"*: Ibid., p. 787.

384 *"because he expressed"*: Ibid., p. 567.

384 *"I am quite sure"*: Ibid., pp. 910–11.

384 *"I remember driving"*: Ibid., p. 969.

385 *"I think there was"*: Oppenheimer recollections, HCP.

385 *"a man so conceited"*: Neylan recollections, HCP.

386 *"How does this"*: Brady recollections, HCP.

Chapter Twenty: The Return of Small Science

387 *"The work of"*: "Statement by the Atomic Energy Commission," June 29, 1954, p. 6, copy in EOLP.

388 *"It was not likely"*: Hewlett and Holl, *Atoms for Peace and War*, p. 112.

389 *"not prepared"*: For the libel suit and Lawrence's role, see Kamen, *Radiant Science, Dark Politics*, pp. 278–89.

390 *"I knew that"*: Neylan recollections, HCP.

391 *"an old wreck"*: Childs, *American Genius*, p. 377.

391 *"The remodeling was"*: Margaret (Lawrence) Casady to Herbert Childs, April 15, 1963, HCP.

391 *The very idea*: Alvarez, *Adventures*, p. 165.

391 *"Ed and I"*: Ibid., p. 166.

393 *"a place where"*: Molly Lawrence recollections, HCP.

393 *"sense of urgency"*: Don Gow recollections, HCP.

395 *On one occasion*: Childs, *American Genius*, p. 450.

395 *Dr. John Sherrick*: Casady to Childs.

396 *Soon after taking office*: Hewlett and Holl, *Atoms for Peace and War*, pp. 261–62.

Chapter Twenty-one: The "Clean Bomb"

399 *Ernest's colitis recurred*: Childs, *American Genius*, p. 501.

399 *"We're still in business!"*: Herken, p. 301.

399 *"Exactly what is"*: Francis, "Warhead Politics."

399 *fallout was detected*: Hewlett and Holl, p. 290.

400 *"far below the levels"*: Divine, *Blowing on the Wind*, p. 12.

400 *"Lewis, I wouldn't"*: Hagerty, *Diary of James C. Hagerty*, p. 36.

400 *"working philosophy"*: York, *Making Weapons*, p. 75.

401 *"That certainly expresses"*: Notes of Stassen-Eisenhower conference, March 22, 1955, *Foreign Relations of the United States, 1955–1957*, vol. 20, p. 61. [Henceforth *FRUS.*]

401 *"experienced men"*: Ibid., p. 60.

401 *of the Lawrence panel's twelve*: Herken, p. 305.

402 *"Edward was always"*: York, *Making Weapons*, p. 82.

402 *"All that the man"*: Appleby, *Eisenhower and Arms Control, 1953–1961*, p. 148.

402 *His estimate came*: "Memorandum of Discussion at the 271st Meeting of the National Security Council, December 22, 1955," *FRUS, 1955–1957, Vol. 20*, pp. 250–55.

403 *The proposal won him*: Hewlett and Holl, p. 334.

403 *"mandatory to the defense"*: Dunning, Gordon M., "Effects of Nuclear Weapons Testing," *The Scientific Monthly* 81, no. 6 (December 1955): pp. 265–70.

403 *"It is possible"*: See Libby, Willard F., "Radioactive Strontium Fallout," *Proceedings of the National Academy of Sciences* 42, no. 6 (June 1956): pp. 365–90.

404 *"God in His almighty"*: Hewlett and Holl, p. 337.

404 *"not to make a bigger"*: Eisenhower press conference, April 25, 1956, at www.presidency.ucsb.edu/ws/index.php?pid=10787 [accessed August 14, 2013].

405 *"maximum effect"*: Divine, *Blowing on the Wind*, p. 82.

405 *"Thus the current"*: *Bulletin of the Atomic Scientists*, September 1956, p. 263.

405 *"The superbomb can be"*: Lapp, Ralph E., "The 'Humanitarian' H-Bomb," *Bulletin of the Atomic Scientists*, September 1956.

405 *The study concluded*: "The Biological Effects of Atomic Radiation," National Academy of Sciences, 1956.

406 *"The concept of a safe rate"*: Ibid., p. 16.

406 *"uncontrollable forces"*: Divine, *Blowing on the Wind*, p. 104.

406 *"no sure method": New York Times*, November 6, 1956.

407 *"half of our"*: Francis, "Warhead Politics," p. 113.

408 *"a catastrophe"*: Divine, *Blowing on the Wind*, p. 122.

409 *The conversation inevitably*: Hewlett and Holl, *Atoms for Peace and War*, p. 398.

409 *"the gleam in the scientists' "*: Ibid., p. 399.

409 *"If we know"*: *FRUS 1955–1957*, vol. 20, p. 641.

410 *"to produce nuclear"*: *New York Times*, June 25, 1957.

410 *"The irony of this"*: Lilienthal, *Road to Change*, p. 204 (journal entry of June 25, 1957).

411 *"Madison Avenue–type"*: Ibid., p. 239 (journal entry of May 5, 1958).

411 *"an unequivocal yes"*: Eisenhower press conference, June 26, 1957, at www.presidency.ucsb.edu/ws/index.php?pid=10822 [accessed August 16, 2013].

412 *"very lucky breaks"*: Starbird to Strauss, July 11, 1957, cited in Hansen, *Swords of Armageddon*, vol. 4, p. 293.

412 *"What an 'absolutely clean' "*: *Newsweek*, July 8, 1957, cited in Divine, *Blowing on the Wind*, p. 151.

412 *"Everything has worked"*: Hansen, *Swords of Armageddon*, vol. 4, p. 287.

412 *"You may detect"*: Ibid.

413 *"I mark off"*: Pfau, *No Sacrifice Too Great*, p. 204.

Chapter Twenty-two: Element 103

415 *"a silly bauble"*: Divine, *Blowing on the Wind*, p. 170.

416 *"a matter of self interest"*: Diary entry by the president, October 29, 1957, *FRUS, 1955–1957*, vol. 20, pp. 754–55. See also Herken, *Brotherhood of the Bomb*, p. 316.

417 *"my 'wizard' "*: Eisenhower, *White House Years*, p. 224.

417 *The last hurrah*: Minutes of the NSC meeting of January 6, 1958, are at *FRUS, 1958–1960*, vol. III, pp. 533–45.

419 *"weigh very carefully"*: Memorandum of Conference with President Eisenhower, January 22, 1958, *FRUS, 1958–1960*, vol. 3, p. 553.

419 *at a White House meeting*: Memorandum of Conference with President Eisenhower, March 24, 1958, *FRUS, 1958–1960*, vol. 3, pp. 567–72.

419 *"an extremely difficult position"*: Ibid.

420 *"Mankind insists"*: Divine, *Blowing on the Wind*, p. 202.

420 *"There is far more"*: Hansen, *Swords of Armageddon*, vol. 4, p. 340.

421 *"The blunt fact"*: Bradbury to Starbird, January 8, 1958, in Hansen, *Swords of Armageddon*, vol. 4, pp. 335–37.

421 *"fit as a fiddle"*: Childs, *American Genius*, p. 518.

422 *"would be greatly"*: Report by the President's Science Advisory Committee, April 11, 1958, *FRUS, 1958–1960*, vol. 3, p. 598.

422 *"had never been"*: Memorandum of Conference with President Eisenhower, April 17, 1958, *FRUS, 1958–1960*, vol. 3, p. 604.

423 *"We helped start"*: Childs, *American Genius*, p. 523.

423 *"practically lived together"*: Bacher recollections, HCP.

424 *"a purely technical scientific"*: Divine, *Blowing on the Wind*, p. 216.

424 *It did not help*: Ibid., p. 217.

424 *"We were going"*: Bacher recollections, HCP.

424 *a "shaggy man"*: Hewlett and Holl, *Atoms in Peace and War*, p. 540.

426 *"I wish I'd taken"*: Molly Lawrence recollections, HCP.

426 *"Molly, I'm ready"*: Ibid.

426 *"A man of Ernest's"*: Cooksey to Lauriston Marshall, September 25, 1958, Lawrence Berkeley Laboratory archives.

427 *"the one touch"*: Lilienthal, *Road to Change*, p. 307 (journal entry of January 31, 1959).

428 *"satisfactory progress"*: Divine, *Blowing on the Wind*, p. 229. See also Hewlett and Holl, p. 546.

428 *Since Hiroshima*: Divine, *Blowing on the Wind*, p. 238.

429 *"the only person"*: Seaborg, *Adventures in the Atomic Age*, p. 144.

Epilogue: The Twilight of Big Science?

431 *"Manlike creatures"*: Clark Kerr, "Tribute to Professor Ernest O. Lawrence," August 30, 1958, copy in HCP.

432 *"a mighty symbol"*: "President's Review," The Rockefeller Foundation Annual Report 1940.

432 *"He will always"*: Alvarez, Luis W., "Ernest Orlando Lawrence, 1901–1958, A Biographical Memoir," National Academy of Sciences, 1970.

432 *"where physicists"*: Adams, John B., address, Proceedings of the Celebration of the 50th Anniversary of the Lawrence Berkeley National Laboratory, October 1981, LBNL.

432 *"tended to shun"*: Galison, Peter, "The Many Faces of Big Science," in Galison and Hevly, *Big Science*.

435 *"We simply do not"*: Panofsky in Galison and Hevly, *Big Science*, p. 7.

436 *"A 20-year honeymoon"*: Abelson, "National Science Policy," *Science*, January 28, 1966.

436 *"basic science was worth"*: Price, Don K., "Federal money and University Research," *Science*, January 21, 1966.

436 *"The President's Science Advisor"*: Sanders, Ralph, "The Autumn of Power: The Scientist in the Political Establishment," *Bulletin of the Atomic Scientists*, October 1966.

436 *federal government spending*: Ibid.

437 *Mansfield Amendment*: See Kevles, *The Physicists*, p. 414.

437 *"a generation of people"*: Klaw, Spencer, "Letter from MIT," *Harper's*, May 1972

438 *"or heart disease, or stroke"*: Fischer, John, "The Editor's Easy Chair," *Harpers*, September 1966.

438 *"The great ideas"*: Ibid.

439 *"the loss will not"*: Sheldon L. Glashow, and Leon M. Lederman, "The SSC: A Machine for the Nineties," *Physics Today*, March 1985. See also Kevles, *The Physicists*, p. xix.

439 *"expensive irrelevance"*: *New York Times*, July 16, 1967, cited in Kevles, *The Physicists*, p. *xi*.

439 *"He said that he"*: Weinberg, Steven, "The Crisis of Big Science," *New York Review of Books*, May 10, 2012.

440 *In the first decades*: *Science Resources Statistics Info Brief*, National Science Foundation, January 2010.

441 *"I heard that"*: *East Bay Express*, March 23, 1984.

Index

Photography Credits

Courtesy of Robert Lawrence: 1

Lawrence Berkeley National Laboratory: 2–4, 6–29, 33–36

US Patent and Trademark Office: 5

Los Alamos National Laboratory: 30–31

Arthur Holly Compton, *Atomic Quest*: 32